Two-Dimensional Materials in Nanophotonics

Two-Dimensional Materials in Nanophotonics

Developments, Devices, and Applications

edited by

Yuerui Lu

JENNY STANFORD
PUBLISHING

Published by

Jenny Stanford Publishing Pte. Ltd.
Level 34, Centennial Tower
3 Temasek Avenue
Singapore 039190

Email: editorial@jennystanford.com
Web: www.jennystanford.com

British Library Cataloguing-in-Publication Data
A catalogue record for this book is available from the British Library.

Two-Dimensional Materials in Nanophotonics:
Developments, Devices, and Applications

ISBN 978-981-4800-22-8 (Hardcover)
ISBN 978-0-429-42877-7 (eBook)

Contents

Preface

Approaching the nano domain and searching for quantum phenomena has always been a emerging topic in science since the discovery of the atom, the fabrication of graphene, the synthesis of further 2D materials and the possibility of nanofabrication techniques. As materials and devices are scaled down into the atomic region, a plethora of scientific discoveries have been discovered with more discoveries emerging all the time; such examples include, strong many body interactions, enhanced light–matter coupling, direct band-gap semiconductors, superior mechanical properties and many more. I edited this book intended for a variety of readers from scientists and engineers working in this field to the undergraduate student who is starting their endeavour towards nano devices, 2D materials, strong light matter coupling and nanophotonics. I was always interested in applied physics and the potential for manipulating new materials for new scientific applications as this has always been the way technology increases, from the transistor in the late 1940's until today where concepts of quantum computing are now a real possibility. My inspiration in science has always led me towards this path of research and working in the fields of 2D materials, optics and nano devices, and this book should serve readers with useful information and insight into the underlying physics as well as the applications of these materials and devices in the future. It is my sincere hope that the work in this book can help scientists in their research endeavours in addition to inspiring young students and researchers in always pursuing high quality impact research for practical applications. I have always known that inspiration and motivation are key for driving new science and the persistence to continue even after many experimental failures. I will always encourage young scientists to purse what they are passionate about and to never give up on their dreams.

This book starts with the background physics needed to understand 2D nanophotonic devices before it digs deeper into the realm of lenses and gratings, optical pulses and the integration of these systems into more complex devices. This book came about from my own thoughts and passions towards 2D nanophotonic research and my aim is that this book will direct and inspire future work in this field and attract many new researchers towards aiming for 2D nanophotonic devices. I believe this book will serve as an exemplary reference for the potential of 2D nanophotonic devices for future devices and applications ranging from biomedical measurements, optoelectronic devices and many more. While this book starts with fundamental physics pertaining to nanophotonics, it ends with the ability of nano plasmonic structures and their manipulation and on chip devices which will be essential for future devices involving nanophotonic applications. I particularly wanted to arrange this book in a simple but effective manner to guide individuals into the realm of 2D materials and 2D nanophotonic devices. In preparing this manuscript, I thought about the structure for a long time to give it the most effective way to communicate the technical work, relate theory and experiments, inspire and motivate individuals. I believe the structure and the chapters written in this book are detailed for explaining each topic and highly commend the authors of all sections in this book for their detailed expertise, time and commitment.

I have been doing research in these fields since my undergraduate degree, doctorate and have steadily pursued in this field since that time. My passion for nanotechnology has grown over the years as technology has consistently upgraded and more avenues and possibilities emerge daily. Nanophotonic device and applications will play a significant part in the future of technology and will enable improved solar cell efficiency and nanotechnology applications for space. This growing field makes always pursue new avenues of research, fabricating devices and finding new applications as well as improving efficiency in other technologies. I really had strong motivation towards nanophotonic devices due to their high applicability in solar cells which is something the world is looking for steadily. I hope this book will also inspire many students and other scientists in exploring new devices and applications to allow the field to grow and bring about many new applications. It is also my hope that this book may be used a classroom reference in addition in the future as nanotechnology, 2D materials and photonics is rapidly increasing around the world. Moreover, it is my intention that this book can be available in labs around the world so that all may benefit from it as I believe science is a strong motivator for the individual and for society.

During preparation of this book, I would like to thank my family, friends and students for their support and welcomed feedback. I would particularly like to thank the author of the chapters for the high quality research and technical expertise in addition to their outstanding professionalism. Moreover, I highly thank the publishing group for their tireless efforts in helping to arrange the book, copyediting and their feedback and advice. Without the help of all these possible the quality and presentation in this book isn't possible, so you have my sincerest thanks. Whilst many hours went into conceiving this book, writing it and preparing it, there were many difficulties to overcome, however, thanks to my family for their outstanding support in being patient with me and always giving me welcomed support. I am also grateful to all of my own supervisors throughout university for developing me into the scientist I am today, without their teaching this work would also not be possible. I am also highly grateful to my colleagues at the Australian National University for their feedback and always having me as a great friend. To my collaborators around the world, your commitment to science and high quality work has always been outstanding and it has been my greatest pleasure to be working with so many distinguished researchers. Everyone will always be my friend and I am sure we will continue to work for many years to come. Without the help of all these people, this book is not possible and they always have my sincerest gratitude. To the readers of this book, I will always say that science is about exploring the unknown and generating ideas is key so never give up and always pursue your goals and dreams.

Chapter 1

An Introduction of 2D Materials for Nanophotonic Devices

Jiong Yang[a,b] **and Yuerui Lu**[a,c]
[a]*College of Chemistry and Environmental Engineering, Shenzhen University, Shenzhen 518060, China*
[b]*Key Laboratory of Optoelectronic Devices and Systems of Ministry of Education and Guangdong Province, College of Optoelectronic Engineering, Shenzhen University, Shenzhen 518060, China*
[c]*Research School of Engineering, College of Engineering and Computer Science, Australian National University, Canberra ACT 2601, Australia*
yuerui.lu@anu.edu.au; yrlu@szu.edu.cn

In 1704, Sir Isaac Newton demonstrated with a prism that white light could be resolved into its component colours and would form a rainbow, and then he refracted the resolved spectrum back together with a second prism, proving that prisms would not colour light [1]. Now we know that white light is composed of lights of different frequencies, and they will be refracted to different angles when interacting with a prism. Light–matter interactions can be classified into inelastic and elastic interactions [2]. Inelastic interactions involve energy transfer between photons and electrons or phonons. Photon–electron and photon–phonon interactions will result in photo-luminescence (PL) phenomenon and the Raman effect,

Two-Dimensional Materials in Nanophotonics: Developments, Devices, and Applications
Edited by Yuerui Lu

ISBN 978-981-4800-22-8 (Hardcover), 978-0-429-42877-7 (eBook)
www.jennystanford.com

respectively. By using PL and Raman spectroscopy to characterise 2D semiconductors, their intrinsic properties can be well understood, including the optical bandgap [3], carrier recombination mechanism [4], atomic bonds and thermal properties [5, 6]. In contrast, elastic interactions do not involve energy transfer and are responsible for controlling light propagation. Optical components, from conventional cavities [7], waveguides [8], lenses [9] and gratings [10] to recent optical meta-materials [11] and photonic crystals [12], all rely on strong elastic light–matter interactions to achieve sophisticated control of light propagation.

Two-dimensional materials are a family of crystalline materials consisting of a single layer or a few layers of atoms, including zero-bandgap graphene [13, 14], middle-bandgap transition metal dichalcogenide (TMD) and phosphorene [15, 16] and wide-bandgap hexagonal boron nitride (hBN) [17]. They own unique properties originated from their large surface-to-volume ratio and reduced dielectric screening [18–25]. And plenty of research has been conducted to investigate their thermal, electrical, optical and optoelectronic properties [15, 26–30]. Among those distinct characteristics, optical properties of 2D semiconductors are especially intriguing and much efforts have been drawn to this field, including Raman and PL spectroscopic analysis [5, 6, 29, 31], exciton (a quasi-particle consisting of one electron and one hole), trion (charged exciton) and higher-order complex study [32–35], refractive index measurement [36–39] and investigation of light–matter interactions [40–46].

Light–matter interactions of 2D materials can be further modified by directly processing them into specific shapes [45, 47–49] or by coupling them with artificial nanostructures [50–52], and thus nanophotonic devices based on 2D materials can be fabricated. By using laser micro-/nanomanufacturing and the photo-chemical reaction technique, Lu et al. fabricated thin-layer TMDs and black phosphorus (BP) colour-sensitive structures [53, 54]. And by using gallium ion beam to mill thin-layer MoS_2 into certain shapes, Yang et al. successfully fabricated atomically thin optical lenses and gratings to modify light propagation [49]. Besides, Li et al. have successfully observed pulse generation from graphene and BP [55–57]. In addition to the aforementioned nanophotonic devices based on 2D materials, researchers have coupled 2D materials with plasmonic, photonic and other on-chip devices to demonstrate other applications, including second harmonic generation (SHG) enhancement and light-emitting devices [58–60].

In this book, research groups around the world working on light–matter interactions of 2D semiconductors will conclude their recent research progress, including optical property characterisation of 2D semiconductors, and ultra-fast lasers, non-linear optical devices, photonic crystals, plasmonic cavities and micro-lens and gratings fabricated from 2D semiconductors and their performance characterisation.

1.1 Two-Dimensional Semiconductors

In this section, we will briefly introduce recent research advances of several typical 2D semiconductors, including graphene, transition metal dichalcogenides (TMDs) and phosphorene. Graphene has been most widely studied as a 2D zero-bandgap semiconductor since its first demonstration by Novoselov and Geim et al. [61]. Later, TMD semiconductors rise up due to their intrinsic bandgap and the direct-bandgap nature when thinning down to a single layer [15, 62]. And recently, phosphorene has attracted much attention because of its highly anisotropic nature and direct bandgap in the infrared range [63]. In this section, we will briefly introduce the physical, electrical and optical properties of those 2D semiconductors.

1.1.1 Graphene

Graphene has been most widely studied in the family of 2D materials because of its high mobility, optical transparency, flexibility, robustness and environmental stability, making it a promising candidate for fundamental research and electrical, optical and optoelectronic applications [64]. Graphene is a flat single layer (or a few layers) of carbon atoms that are tightly packed into a 2D hexagonal honeycomb lattice, and its schematic plot is shown in Fig. 1.1. Graphene is well known for its large intrinsic strength predicted to exceed that of any other materials [65], huge thermal conductivity of ~5000 W/mK at room temperature [27] and gate-tunable high electric conductivity [14].

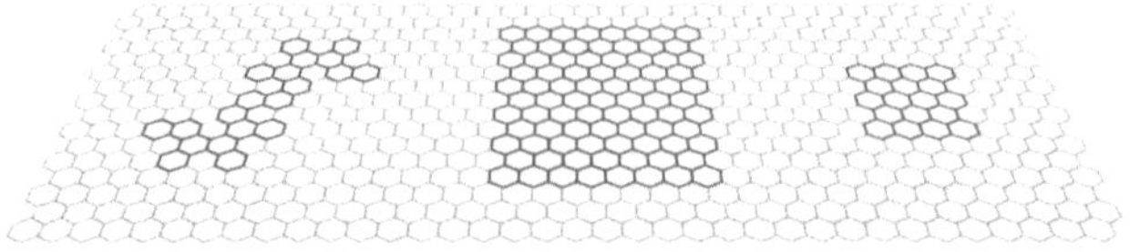

Figure 1.1 Schematic plot of graphene. Reprinted with permission from Springer Customer Service Centre GmbH: Springer Nature, *Nature Materials*, Ref. [13], Copyright (2007).

Lee et al. measured the elastic properties and intrinsic strength of monolayer (1L) graphene by nanoindentation in atomic force microscopy (AFM) [65]. The measured second- and third-order elastic stiffness is 340 Nm^{-1} and –690 Nm^{-1}, respectively. The measured effective Young's modulus is about 1 TPa, and the measured effective pre-stress and pre-tension can reach ~0.7 GPa and 0.4 Nm^{-1}. All these measured values from pristine graphene indicate that carbon fibre reinforcements can be used in advanced composites. Novoselov et al. firstly demonstrated the massless Dirac fermions in graphene and measured the electric field effect in graphene [14]. Electric conductivity of graphene increases with the impact of either positive or negative gate voltage and reaches its minimum when the gate voltage is zero. For different devices made from 1L graphene with different carrier mobility ranging from 4000 to over 12,000 $cm^2V^{-1}s^{-1}$, they all show a minimum conductivity at neutral point and no metal–insulator transition can be observed, indicating that conductivity might be quantised. What limits graphene in application in conventional semiconductor industry is its zero-bandgap nature. And Zhang et al. successfully demonstrated that bandgap can be opened to a value of ~250 meV in bilayer (2L) graphene at strong electric gating [66]. Raman spectroscopy has been used to probe the intrinsic properties of graphene [26, 67]. D, G and 2D (or G') phonon modes can be observed in Raman spectra at ~1350 cm^{-1}, ~1580 cm^{-1} and ~2700 cm^{-1} [67]. Berciaud et al. showed the G-mode frequency of freestanding graphene exhibited blue shift and the G-mode width is larger than that of graphene on a SiO_2 substrate, indicating that interaction with substrate can vary intrinsic properties of graphene. And by analysing Raman spectra of graphene with different layer numbers, Ferrari et al. captures its electronic structure evolution [26].

1.1.2 Transition Metal Dichalcogenides

Due to the absence of a bandgap in pristine graphene, its unique optical and electronic properties cannot be fully exploited, making it necessary to investigate other 2D semiconductors. Single-layer TMD semiconductors own a bandgap of ~2 eV in the visible range and are very promising for optoelectronic applications. TMD semiconductors can be represented as MX_2 (M represents transition metal elements, such as Mo and W, and X represents chalcogenide elements, such as S, Se and Te). Transition metal and chalcogenide atoms are covalently bonded in a single layer and different layers are stacked together by weak van der Waals interaction, as illustrated in Fig. 1.2.

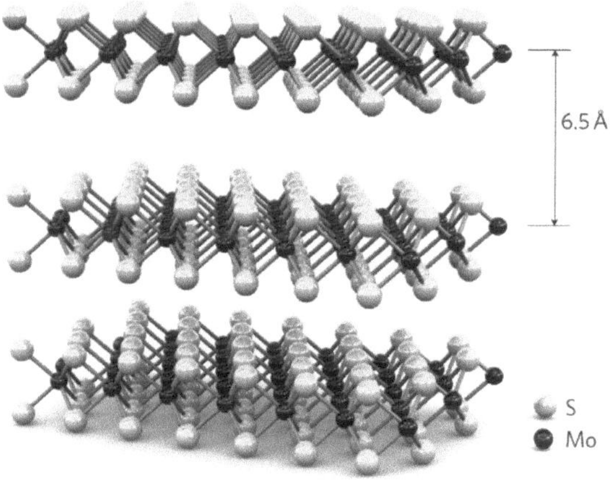

Figure 1.2 Schematic plot of MoS_2, a typical member of TMD semiconductors. Reprinted with permission from Springer Customer Service Centre GmbH: Springer Nature, *Nature Nanotechnology*, Ref. [15], Copyright (2011).

Right after the demonstration of 1L graphene in ambient conditions, Novoselov et al. reported other freestanding 2D atomic crystals prepared by micromechanical cleavage, including hBN, niobium diselenide ($NbSe_2$) and molybdenum disulphide (MoS_2), opening up the research of 2D TMD semiconductors [68]. In the year of 2010, Castellanos-Gomez et al. optically identified atomically thin TMD crystals, including $NbSe_2$ and MoS_2 [36]. In the same year, Mak et al. reported the indirect-to-direct transition when the layer number of MoS_2 is thinned to a single layer, pointing out that 1L MoS_2 is a direct-bandgap semiconductor by theoretically calculating its band diagram and by comparing its absorption and photo-luminescence (PL) spectra [62]. Later, the in-plane stiffness of 1L MoS_2 is measured by AFM to be ~200 Nm^{-1}, corresponding to an effective Young's modulus of ~300 GPa, comparable to that of steel [69]. And in the year after, Radisavljevic et al. successfully demonstrated, for the first time, a 1L MoS_2 field-effect transistor (FET) [15]. The measured carrier mobility is over 200 $cm^2V^{-1}s^{-1}$ and the room-temperature current on/off ratio of 1L MoS_2 transistor exceeds 10^8. The measured carrier mobility of 1L MoS_2 is comparable to the mobility achieved in thin silicon films [70] and graphene nanoribbons [71], and the on/off ratio is much larger than those made from zero-bandgap graphene [72, 73]. Ever since then, the study on 2D TMD semiconductors has boomed. Raman and PL spectroscopies revealed their layer-dependent electronic band structures and different lattice vibration modes [6, 31, 74–76]. Dynamics of exciton, trion and other higher-order complexes in 2D TMDs were systematically studied by means of low-temperature and circular-polarised PL spectroscopy together with absorption spectroscopy [33–35, 77–79]. Exciton, trion and biexciton binding energies of 1L TMDs were measured to be ~0.5 eV, ~30 meV and ~60 meV, respectively [80]. All the aforementioned characterisation methods, including Raman, PL and absorption spectroscopy, rely on inelastic interactions between the light source and 2D TMD semiconductors. As we will discuss later, the elastic light–matter interactions in 2D TMDs are also extraordinarily strong and can benefit the miniaturisation of future optical devices.

1.1.3 Phosphorene

Comparing to 2D TMDs, phosphorene is a relatively new member in the family of 2D semiconductors, and it has shown unique optical and optoelectronic properties due to its highly anisotropic nature originated from its puckering structure [63, 81]. The schematic plot of phosphorene is illustrated in Fig. 1.3, showing its unique puckering structure. Similarly, phosphorus atoms are bonded together by covalent bonds to form a continuous single layer of phosphorene, and different layers are stacked by van der Waals force, making it convenient to separate 1L phosphorene by mechanical exfoliation.

With its unique puckering structure, 1L phosphorene can sustain up to 30% tensile strain and this strain effectively flattens the pucker of phosphorene rather than extending the covalent bonds of phosphorus atom [82]. Besides, the Young's modulus of 1L phosphorene can reach up to 18 GPa and 8 GPa in the zigzag and armchair directions, respectively [82]. Owning a direct bandgap from the visible to the near-infrared range in 1L to few-layer phosphorene, thin-layer phosphorene can bridge the zero-bandgap graphene and comparatively large-bandgap TMDs. Its direct bandgap varies with layer number, from ~2 eV for 1L phosphorene gradually down to ~0.3 eV for bulk black phosphorus (BP) [83, 84]. With the direct bandgap in the visible to near-infrared range, phosphorene can serve as the foundation material for optical and optoelectronic applications in the infrared range. Since the first demonstration of phosphorene FET by Li et al., showing a high carrier mobility up to ~1000 cm2V–1s–1 and a drain current modulation of ~105 [16], intense interest has been drawn into the study of phosphorene. Isolation and preparation of thin-layer phosphorene [85, 86], FETs with improved performance and stability based on few-layer phosphorene [87–89], anisotropic and tightly bound excitons [90, 91] and various photo-voltaic and photo-detection applications [92, 93] have been reported. However, layer-dependent PL characterisation of thin-layer phosphorene, exciton and trion dynamics of 1L phosphorene and the study of localised excitons in 1L phosphorene are still underexplored.

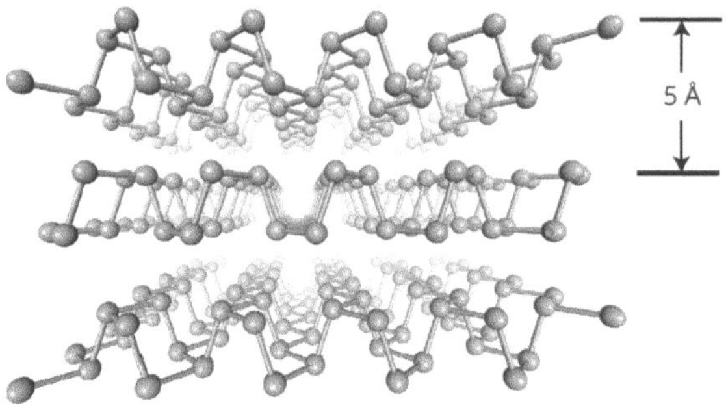

Figure 1.3 Schematic plot of phosphorene. Reprinted with permission from Springer Customer Service Centre GmbH: Springer Nature, *Nature Nanotechnology*, Ref. [16], Copyright (2014).

1.2 Light–Matter Interaction

In the previous section, we have introduced that optical properties of 2D semiconductors can be investigated by conducting Raman and PL measurements. During these measurements, laser excitation source will inelastically interact with 2D semiconductors, and we can detect their internal properties by analysing the energy difference of laser after interaction or the photon emission from 2D semiconductors. So far, the study of light–matter interaction of 2D semiconductors has mainly focused on inelastic ones. In this section, we will briefly introduce researches on inelastic light–matter interactions of 2D semiconductors, and we will also summarise reports on elastic light–matter interactions.

1.2.1 Inelastic Light–Matter Interactions of 2D Semiconductors

Due to the atomically thin nature of 2D semiconductors, their interactions with light are inherently weak, which is the key challenge in practical device applications. Several different approaches have been studied to enhance the inelastic light–matter interactions in 2D semiconductors, including light management around 2D semiconductors with plasmonic nanostructures [50–52, 94, 95], optical confinement with micro-cavities [40, 96], complementary advantages of different 2D semiconductors with heterostructures [95, 97], etc.

Light management around 2D semiconductors with plasmonic nanostructures is one of most frequently used methods to enhance their inelastic light–matter interactions. Figure 1.4 illustrates the schematic plot of plasmonic nanostructures around 2D semiconductor molybdenum diselenide ($MoSe_2$). Similar plasmonic nanostructures are applied to other 2D semiconductors. Buntun et al. used Ag nanodisc arrays to enhance light emission from 1L MoS_2 of up to 12 times [50]. Kern et al. used Au nanoantenna to enhance the light–matter interaction in atomically thin tungsten disulphide (WS_2) [94]. Chen et al. reported both enhanced and quenched PL emission of 2D $MoSe_2$ with Au nanoantennas [52]. In both Kern and Chen's works, nanoantennas are in the shape of a rectangle and the manipulation of light emission from 2D semiconductors is dependent on the polarisation of incident light. Besides, the space between 2D semiconductors and nanoantennas is another critical parameter in influencing light–matter interactions, that is, a compromise of quenching effect from metal and plasmonic enhancement from nanostructures exists.

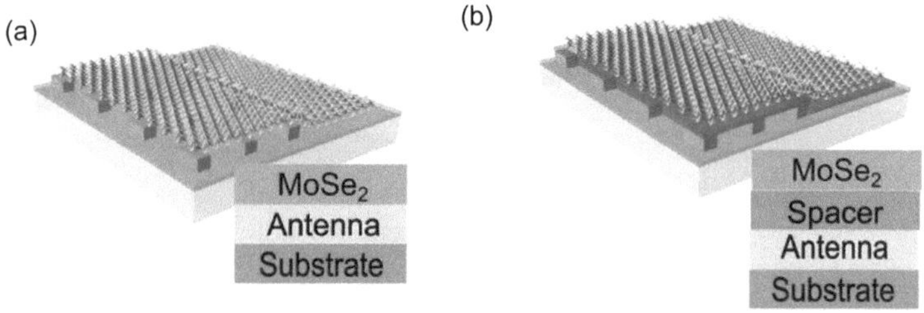

Figure 1.4 Schematic plot of plasmonic nanostructures around 2D semiconductors. The condition (a) without and (b) with oxide spacers. Adapted from Ref. [52]. https://creativecommons.org/licenses/by/4.0/.

Optical confinement with nanocavities is another commonly used method to enhance inelastic light–matter interactions of 2D semiconductors. Plenty of relevant reports have been published [40, 96, 98–100]. Figure 1.5 lists three different kinds of nanocavities used for optical confinement of 2D semiconductors. In 2012, Engel et al. reported the control of light–matter interaction of a graphene transistor with micro-cavity, which can act as a spectrally selective light detector and emitter with significantly enhanced sensitivity [96]. Liu et al. used distributed Bragg reflector (DBR) to confine excitons from 2D semiconductor MoS_2 and couple it with cavity photons to form polaritons, with its schematic plot showing in Fig. 1.5a [40]. Ye et al. and Wu et al. successfully used nanocavities to enhance the spontaneous emission from 2D semiconductors, and excitonic lasers were demonstrated with 1L WS_2 and tungsten diselenide (WSe_2), respectively [98, 99]. With the cavity structure shown in Fig. 1.5b, electric field polarised in the plane of 2D semiconductor WS_2 will couple with in-plane dipoles of excitons and generate coherent light radiation. Similarly, the photonic crystal cavity structure shown in Fig. 1.5c with 1L 2D semiconductor WSe_2 as the gain medium can generate lasers with ultra-low thresholds.

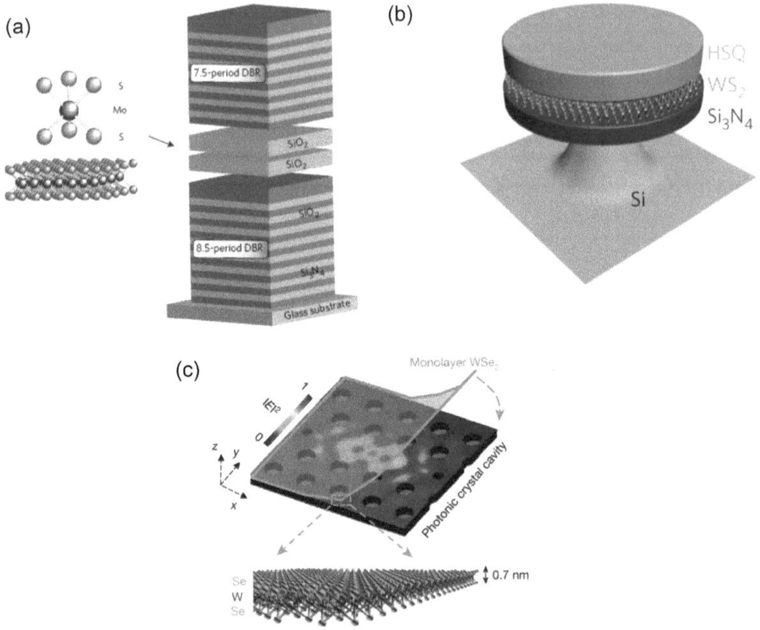

Figure 1.5 Schematic plots of three types of nanocavities used to enhance inelastic light–matter interactions of 2D semiconductors. (a) Schematic plot of DBR used to couple 2D semiconductor excitons and cavity photons. Reprinted with permission from Springer Customer Service Centre GmbH: Springer Nature, *Nature Photonics*, Ref. [40], Copyright (2014). (b) Schematic plot of the nanocavity to generate coherent light radiation from 2D semiconductor. Reprinted with permission from Springer Customer Service Centre GmbH: Springer Nature, *Nature Photonics*, Ref. [98], Copyright (2015). (c) Schematic plot of the photonic crystal cavity to generate lasers with 2D semiconductors. Reprinted with permission from Springer Customer Service Centre GmbH: Springer Nature, *Nature*, Ref. [99], Copyright (2015).

Except for using deliberately fabricated nanostructures, one can also simply use the combination of different 2D semiconductors to form a heterostructure and enhance the inelastic light–matter interactions of 2D semiconductors, considering the fact that plenty of 2D semiconductors have been investigated on their optical properties. Wang et al. used a graphene/silicon heterostructure to fabricate high-responsivity photo-detectors and Britnell et al. demonstrated flexible photo-voltaic devices with a sandwiched graphene-WS_2-graphene heterostructure, where graphene serves as a good transparent electrode due to its high conductivity [97, 103].

Different TMD semiconductors can also form vertical and in-plane heterostructures, as illustrated in Fig. 1.6a,b, and their inelastic light–matter interactions will be significantly enhanced through charge transfer within the heterostructure. Ultra-fast transfer of holes in the valence band of MoS_2 to the valence band of WS_2 has been observed in the vertical MoS_2-WS_2 heterostructure. PL spectra from different spots of the in-plane WS_2-MoS_2 heterostructure 1L exhibited different peak energies and peak widths, demonstrating the manipulation of inelastic light–matter interaction [102].

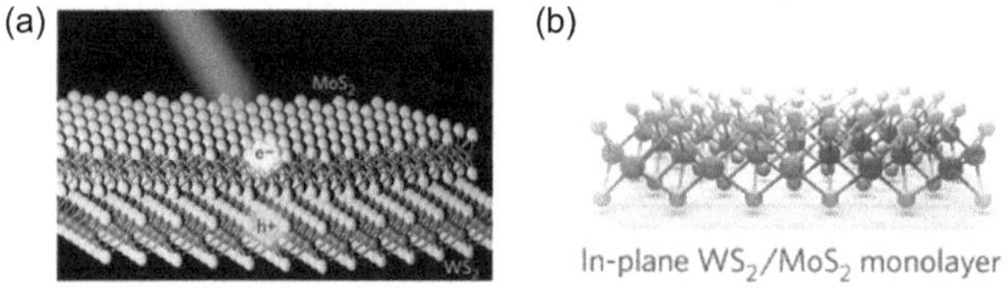

Figure 1.6 Schematic plots of different types of heterostructures of 2D semiconductors. (a) Schematic plot of a vertical MoS_2-WS_2 heterostructure. (b) Schematic plot of an in-plane WS_2-MoS_2 heterostructure. Reprinted with permission from Springer Customer Service Centre GmbH: Springer Nature, *Nature Materials*, Ref. [102], Copyright (2014).

1.2.2 Elastic Light–Matter Interactions of 2D Semiconductors

Light can be reflected, refracted and diffracted after elastically interacting with matter, and no energy transfer will occur in this process. Conventional optical components, such as lenses, prisms and gratings, rely on the elastic light–matter interaction to work properly. Miniaturisation of conventional optical components has always been a crucial and pressing task to fabricate more compacted devices and systems. With the rapid research progress of 2D semiconductors, it is natural to investigate their elastic light–matter interactions for optical device miniaturisation. However, elastic light–matter interactions are normally weak in 2D semiconductors because their structures are atomically thin. To enhance the light–matter interaction of 2D semiconductors remains a big issue. There are mainly two approaches to enhance elastic light–matter interactions in miniaturised devices. One is to use specifically designed structures, and the other is to use advanced materials.

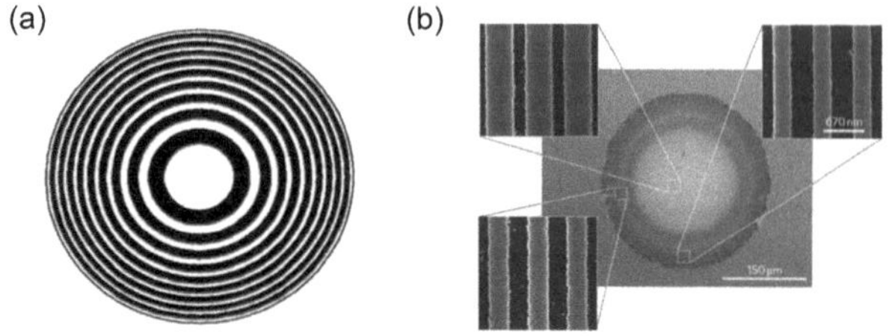

Figure 1.7 Different structures to enhance elastic light–matter interaction in miniaturised devices. (a) Schematic plot of the photo-mask showing Fresnel zone structure. Adapted from Ref. [105] with permission of AIP Publishing. (b) Optical microscope image of a fabricated SWG mirror. Reprinted with permission from Springer Customer Service Centre GmbH: Springer Nature, *Nature Photonics*, Ref. [106], Copyright (2010).

Fresnel lens is a type of compact lens with certain structures and it can be constructed with less mass and volume comparing to conventional lenses. Figure 1.7a illustrates the schematic plot of a Fresnel lens. It realises device miniaturisation, while maintaining equivalent light–matter interaction by following Fresnel's condition:

$$(k-1)\lambda \leq \varphi(u, v) \leq k\lambda;\ k = 1, 2, \ldots, m, \tag{1.1}$$

where λ is the wavelength [104]. A Fresnel lens can also be made completely flat with zone plate patterns. Figure 1.7b displays the photo-mask showing a Fresnel zone structure. The transparent odd zones and opaque even zone patterns will be transferred to certain materials after photo-lithography and subsequent fabrication process, resulting in a Fresnel lens. Ren et al. proposed such a tunable flat Fresnel lens in which the radius of the nth zone r_n is given by

$$r_n^2 = nr_1^2, \tag{1.2}$$

where n is the zone number [105]. A similar design was proposed by Fattal et al. to be flat dielectric sub-wavelength gratings (SWGs) and it owns focusing abilities [106]. Figure 1.7c shows the optical microscopy image of fabricated spherical SWG mirror and scanning electron microscopy (SEM) images of various locations. Besides, plasmonic meta-surfaces have been demonstrated to control the light propagation [107, 108]. Tight lateral confinement of light can be achieved with, but not limited to, the aforementioned structures, and thus elastic light–matter interactions can be enhanced in the micron or even the nanometre scale.

To use advanced materials is another approach to realise enhanced light–matter interaction in miniaturised devices. Considerable effort has been devoted to this issue, but success has been only achieved in the mid- to far-infrared where the plasmonic resonance in graphene can enhance the elastic optical response [109–111]. During conducting my research project, other works of strong light–matter interactions in ultra-thin lenses based on different materials have come into notice. For example, Kim et al. reported self-assembled nanoscale spherical lenses based on organic materials calix[4] hydroquinone (CHQ) with near-field focusing and magnification capabilities [112]. Gu et al. have also reported ultra-thin lenses with 3D sub-wavelength focusing based on graphene oxide [47]. We have concluded different ultra-thin lenses and generated Table 1.1 consisting of lens parameters and performances.

Table 1.1 Comparison between different ultra-thin lenses

Lens	Thickness	NA	Focusing efficiency
Nanoscale CHQ lens [112]	<800 nm	–	–
Flat a-Si lens [106]	~450 nm	~0.01	80%–90%[a]
Ultra-thin graphene oxide lens [47]	~200 nm	–	>32%
Meta-surface lens [107]	60 nm	0.015	1%
Graphene lens [45]	1.8	–	2.9%
WSe_2 lens [48]	0.7	–	31%
Tunable Fresnel lens [105]	15 μm	~0.01	~35%

[a] This value is the total reflection.

But all in all, elastic light–matter interactions of 2D semiconductors are still under-explored. In this dissertation, we will show that 2D TMD semiconductors will highly enhance the elastic light–matter interaction in the nanometre scale. And by utilising such effect, miniaturised optical lenses and gratings will be fabricated.

1.3 Summary

In this chapter, we have introduced typical 2D semiconductors by summarising their physical and optical properties, including zero-bandgap graphene, mid-bandgap TMDs and phosphorene. We have also concluded both inelastic and elastic light–matter interactions of 2D semiconductors. On the basis of these inelastic and elastic light–matter interactions in 2D materials, nanophotonic devices can be fabricated, which will pave the way for device miniaturisation and other on-chip devices.

References

1. Newton, I. (1704). *Opticks: or A Treatise of the Reflexions, Refractions, Inflexions and Colours of Light* (Smith and Walford, London).
2. Weiner, J. and Ho, P.-T. (2008). *Light-Matter Interaction, Fundamentals and Applications*, Vol. 1 (John Wiley & Sons, Hoboken, NJ).
3. Conley, H. J., et al. (2013). Bandgap engineering of strained monolayer and bilayer MoS_2, *Nano Lett.*, **13**(8), pp. 3626–3630.
4. Kozawa, D., et al. (2014). Photocarrier relaxation pathway in two-dimensional semiconducting transition metal dichalcogenides, *Nat. Commun.*, **5**, p. 4543.
5. Sahoo, S., et al. (2013). Temperature-dependent Raman studies and thermal conductivity of few-layer MoS_2, *J. Phys. Chem. C*, **117**(17), pp. 9042–9047.
6. Li, H., et al. (2012). From bulk to monolayer MoS_2: evolution of Raman scattering, *Adv. Funct. Mater.*, **22**(7), pp. 1385–1390.
7. Favero, I. and Karrai, K. (2009). Optomechanics of deformable optical cavities, *Nat. Photonics*, **3**(4), pp. 201–205.
8. Adams, M. J. (1981). *An Introduction to Optical Waveguides*, Vol. 14 (Wiley New York).
9. Ziegler, J. (1970). Fabrication or correction of optical lenses, *IBM Tech. Disclosure Bull.*, **12**(10), pp. 1573–1575.
10. Stroke, G. W. (1967). Diffraction gratings, in *Optische Instrumente/Optical Instruments* (Springer, New York), pp. 426–754.
11. Cai, W. and Shalaev, V. M. (2010). *Optical Metamaterials*, Vol. 10 (Springer).
12. Joannopoulos, J. D., et al. (2011). *Photonic Crystals: Molding the Flow of Light* (Princeton University Press).
13. Geim, A. K. and Novoselov, K. S. (2007). The rise of graphene, *Nat. Mater.*, **6**(3), pp. 183–191.

14. Novoselov, K. S., et al. (2005). Two-dimensional gas of massless Dirac fermions in graphene. *Nature*, **438**(7065), pp. 197–200.

15. Radisavljevic, B., et al. (2011). Single-layer MoS_2 transistors, *Nat. Nanotechnol.*, **6**(3), pp. 147–150.

16. Li, L., et al. (2014). Black phosphorus field-effect transistors, *Nat. Nanotechnol.*, **9**(5), pp. 372–377.

17. Lee, C., et al. (2010). Frictional characteristics of atomically thin sheets, *Science*, **328**(5974), pp. 76–80.

18. Chernikov, A., et al. (2014). Exciton binding energy and nonhydrogenic Rydberg series in monolayer WS_2, *Phys. Rev. Lett.*, **113**(7), p. 076802.

19. Cudazzo, P., Tokatly, I. V. and Rubio, A. (2011). Dielectric screening in two-dimensional insulators: implications for excitonic and impurity states in graphane, *Phys. Rev. B*, **84**(8), p. 085406.

20. Prins, F., Goodman, A. J. and Tisdale, W. A. (2014). Reduced dielectric screening and enhanced energy transfer in single- and few-layer MoS_2, *Nano Lett.*, **14**(11), pp. 6087–6091.

21. Hüser, F., Olsen, T. and Thygesen, K. S. (2013). How dielectric screening in two-dimensional crystals affects the convergence of excited-state calculations: monolayer MoS_2, *Phys. Rev. B*, **88**(24), p. 245309.

22. Zou, X., et al. (2016). Dielectric engineering of a boron nitride/hafnium oxide heterostructure for high-performance 2D field effect transistors, *Adv. Mater.*, **28**(10), pp. 2062–2069.

23. O'Brien, M., et al. (2014). Plasma assisted synthesis of WS_2 for gas sensing applications, *Chem. Phys. Lett.*, **615**, pp. 6–10.

24. Cho, B., et al. (2015). Chemical sensing of 2D graphene/MoS_2 heterostructure device, *ACS Appl. Mater. Interfaces*, **7**(30), pp. 16775–16780.

25. Tan, C. and Zhang, H. (2015). Two-dimensional transition metal dichalcogenide nanosheet-based composites, *Chem. Soc. Rev.*, **44**(9), pp. 2713–2731.

26. Ferrari, A. C., et al. (2006). Raman spectrum of graphene and graphene layers, *Phys. Rev. Lett.*, **97**(18), p. 187401.

27. Balandin, A. A., et al. (2008). Superior thermal conductivity of single-layer graphene, *Nano Lett.*, **8**(3), pp. 902–907.

28. Castro Neto, A. H., et al. (2009). The electronic properties of graphene, *Rev. Mod. Phys.*, **81**(1), pp. 109–162.

29. Splendiani, A., et al. (2010). Emerging photoluminescence in monolayer MoS_2, *Nano Lett.*, **10**(4), pp. 1271–1275.

30. Lopez-Sanchez, O., et al. (2013). Ultrasensitive photodetectors based on monolayer MoS_2, *Nat. Nanotechnol.*, **8**(7), pp. 497–501.

31. Eda, G., et al. (2011). Photoluminescence from chemically exfoliated MoS_2, *Nano Lett.*, **11**(12), pp. 5111–5116.

32. Shi, H., et al. (2012). Exciton dynamics in suspended monolayer and few-layer MoS_2 2D crystals, *ACS Nano*, **7**(2), pp. 1072–1080.

33. Mak, K. F., et al. (2013). Tightly bound trions in monolayer MoS_2, *Nat. Mater.*, **12**(3), pp. 207–211.

34. Ross, J. S., et al. (2013). Electrical control of neutral and charged excitons in a monolayer semiconductor, *Nat. Commun.*, **4**, p. 1474.

35. You, Y., et al. (2015). Observation of biexcitons in monolayer WSe_2, *Nat. Phys.*, **11**(6), pp. 477–481.

36. Castellanos-Gomez, A., Agraït, N. and Rubio-Bollinger, G. (2010). Optical identification of atomically thin dichalcogenide crystals, *Appl. Phys. Lett.*, **96**(21), p. 213116.

37. Santos, E. J. G. and Kaxiras, E. (2013). Electrically driven tuning of the dielectric constant in MoS_2 layers, *ACS Nano*, **7**(12), pp. 10741–10746.

38. Zhang, H., et al. (2015). Measuring the refractive index of highly crystalline monolayer MoS_2 with high confidence, *Sci. Rep.*, **5**, p. 8440.

39. Venkatachalam, D. K., et al. (2011). Rapid, substrate-independent thickness determination of large area graphene layers, *Appl. Phys. Lett.*, **99**(23), p. 234106.

40. Liu, X., et al. (2015). Strong light–matter coupling in two-dimensional atomic crystals, *Nat. Photonics*, **9**(1), pp. 30–34.

41. Niu, L., et al. (2015). Controlled synthesis of organic/inorganic van der Waals solid for tunable light–matter interactions, *Adv. Mater.*, **27**(47), pp. 7800–7808.

42. Peng, B., Ang, P. K. and Loh, K. P. (2015). Two-dimensional dichalcogenides for light-harvesting applications, *Nano Today*, **10**(2), pp. 128–137.

43. Lu, J., et al. (2016). Interactions between lasers and two-dimensional transition metal dichalcogenides, *Chem. Soc. Rev.*, **45**(9), pp. 2494–2515.

44. Lu, J., et al. (2016). Light–matter interactions in phosphorene, *Acc. Chem. Res.*, **49**(9), pp. 1806–1815.

45. Kong, X.-T., et al. (2015). Graphene-based ultrathin flat lenses, *ACS Photonics*, **2**(2), pp. 200–207.

46. Wang, P., et al. (2016). Graphene-based superlens for subwavelength optical imaging by graphene plasmon resonances, *Plasmonics*, **11**(2), pp. 515–522.

47. Zheng, X., et al. (2015). Highly efficient and ultra-broadband graphene oxide ultrathin lenses with three-dimensional subwavelength focusing, *Nat. Commun.*, **6**, p. 8433.

48. Lin, H., et al. (2016). High performance atomically thin flat lenses, arXiv preprint arXiv:1611.06457.

49. Yang, J., et al. (2016). Atomically thin optical lenses and gratings, *Light Sci. Appl.*, **5**, p. e16046.

50. Butun, S., Tongay, S. and Aydin, K. (2015). Enhanced light emission from large-area monolayer MoS_2 using plasmonic nanodisc arrays, *Nano Lett.*, **15**(4), pp. 2700–2704.

51. Liu, W., et al. (2016). Strong exciton–plasmon coupling in MoS_2 coupled with plasmonic lattice, *Nano Lett.*, **16**(2), pp. 1262–1269.

52. Chen, H., et al. (2016). Manipulation of photoluminescence of two-dimensional $MoSe_2$ by gold nanoantennas, *Sci. Rep.*, **6**, p. 22296.

53. Lu, J., et al. (2014). Improved photoelectrical properties of MoS_2 films after laser micromachining, *ACS Nano*, **8**(6), pp. 6334–6343.

54. Lu, J., et al. (2015). Bandgap engineering of phosphorene by laser oxidation toward functional 2D materials, *ACS Nano*, **9**(10), pp. 10411–10421.

55. Li, D., et al. (2015). Polarization and thickness dependent absorption properties of black phosphorus: new saturable absorber for ultrafast pulse generation, *Sci. Rep.*, **5**, p. 15899.

56. Diao, L., et al. (2017). Graphene actively Q-switched lasers, *2D Mater.*, **4**(2), p. 025095.

57. Li, D., et al. (2016). Black phosphorus polycarbonate polymer composite for pulsed fibre lasers, *Appl. Mater. Today*, **4**, pp. 17–23.

58. Wang, Z., et al. (2016). Giant photoluminescence enhancement in tungsten-diselenide–gold plasmonic hybrid structures, *Nat. Commun.*, **7**, p. 11283.

59. Chen, H., et al. (2017). Enhanced second-harmonic generation from two-dimensional $MoSe_2$ on a silicon waveguide, *Light Sci. Appl.*, **6**(10), p. e17060.

60. Liu, Y., et al. (2016). Van der Waals heterostructures and devices, *Nat. Rev. Mater.*, **1**(9), p. 16042.

61. Novoselov, K. S., et al. (2004). Electric field effect in atomically thin carbon films, *Science*, **306**(5696), pp. 666–669.

62. Mak, K. F., et al. (2010). Atomically thin MoS_2: a new direct-gap semiconductor, *Phys. Rev. Lett.*, **105**(13), p. 136805.

63. Qiao, J., et al. (2014). High-mobility transport anisotropy and linear dichroism in few-layer black phosphorus, *Nat. Commun.*, **5**, p. 4475.

64. Bonaccorso, F., et al. (2010). Graphene photonics and optoelectronics, *Nat. Photonics*, **4**(9), pp. 611–622.

65. Lee, C., et al. (2008). Measurement of the elastic properties and intrinsic strength of monolayer graphene, *Science*, **321**(5887), pp. 385–388.

66. Zhang, Y., et al. (2009). Direct observation of a widely tunable bandgap in bilayer graphene, *Nature*, **459**(7248), pp. 820–823.

67. Berciaud, S., et al. (2009). Probing the intrinsic properties of exfoliated graphene: Raman spectroscopy of free-standing monolayers, *Nano Lett.*, **9**(1), pp. 346–352.

68. Novoselov, K. S., et al. (2005). Two-dimensional atomic crystals, *Proc. Natl. Acad. Sci. U. S. A.*, **102**(30), pp. 10451–10453.

69. Bertolazzi, S., Brivio, J. and Kis, A. (2011). Stretching and breaking of ultrathin MoS_2, *ACS Nano*, **5**(12), pp. 9703–9709.

70. Gomez, L., Aberg, I. and Hoyt, J. (2007). Electron transport in strained-silicon directly on insulator ultrathin-body n-MOSFETs with body thickness ranging from 2 to 25 nm, *IEEE Electron Device Lett.*, **28**(4), pp. 285–287.

71. Li, X., et al. (2008). Chemically derived, ultrasmooth graphene nanoribbon semiconductors, *Science*, **319**(5867), pp. 1229–1232.

72. Xia, F., et al. (2010). Graphene field-effect transistors with high on/off current ratio and large transport band gap at room temperature, *Nano Lett.*, **10**(2), pp. 715–718.

73. Britnell, L., et al. (2012). Field-effect tunneling transistor based on vertical graphene heterostructures, *Science*, **335**(6071), pp. 947–950.

74. Lee, C., et al. (2010). Anomalous lattice vibrations of single- and few-layer MoS_2, *ACS Nano*, **4**(5), pp. 2695–2700.

75. Zhao, W., et al. (2013). Evolution of electronic structure in atomically thin sheets of WS_2 and WSe_2, *ACS Nano*, **7**(1), pp. 791–797.

76. Tongay, S., et al. (2012). Thermally driven crossover from indirect toward direct bandgap in 2D semiconductors: $MoSe_2$ versus MoS_2, *Nano Lett.*, **12**(11), pp. 5576–5580.

77. Jones, A. M., et al. (2013). Optical generation of excitonic valley coherence in monolayer WSe_2, *Nat. Nanotechnol.*, **8**(9), pp. 634–638.

78. Plechinger, G., et al. (2015). Identification of excitons, trions and biexcitons in single-layer WS_2, *Phys. Status Solidi RRL*, **9**(8), pp. 457–461.

79. Shang, J., et al. (2015). Observation of excitonic fine structure in a 2D transition-metal dichalcogenide semiconductor, *ACS Nano*, **9**(1), pp. 647–655.

80. Zhang, D. K., Kidd, D. W. and Varga, K. (2015). Excited biexcitons in transition metal dichalcogenides, *Nano Lett.*, **15**(10), pp. 7002–7005.

81. Xia, F., Wang, H. and Jia, Y. (2014). Rediscovering black phosphorus as an anisotropic layered material for optoelectronics and electronics, *Nat. Commun.*, **5**, p. 4458.

82. Wei, Q. and Peng, X. (2014). Superior mechanical flexibility of phosphorene and few-layer black phosphorus, *Appl. Phys. Lett.*, **104**(25), p. 251915.

83. Li, L., et al. (2017). Direct observation of the layer-dependent electronic structure in phosphorene, *Nat. Nanotechnol.*, **12**(1), pp. 21–25.

84. Liang, L., et al. (2014). Electronic bandgap and edge reconstruction in phosphorene materials, *Nano Lett.*, **14**(11), pp. 6400–6406.

85. Andres, C.-G., et al. (2014). Isolation and characterization of few-layer black phosphorus, *2D Mater.*, **1**(2), p. 025001.

86. Yasaei, P., et al. (2015). High-quality black phosphorus atomic layers by liquid-phase exfoliation, *Adv. Mater.*, **27**(11), pp. 1887–1892.

87. Avsar, A., et al. (2015). Air-stable transport in graphene-contacted, fully encapsulated ultrathin black phosphorus-based field-effect transistors, *ACS Nano*, **9**(4), pp. 4138–4145.

88. Na, J., et al. (2014). Few-layer black phosphorus field-effect transistors with reduced current fluctuation, *ACS Nano*, **8**(11), pp. 11753–11762.

89. Buscema, M., et al. (2014). Fast and broadband photoresponse of few-layer black phosphorus field-effect transistors, *Nano Lett.*, **14**(6), pp. 3347–3352.

90. Wang, X., et al. (2015). Highly anisotropic and robust excitons in monolayer black phosphorus, *Nat. Nanotechnol.*, **10**(6), pp. 517–521.

91. Tran, V., et al. (2014). Layer-controlled band gap and anisotropic excitons in few-layer black phosphorus, *Phys. Rev. B*, **89**(23), p. 235319.

92. Youngblood, N., et al. (2015). Waveguide-integrated black phosphorus photodetector with high responsivity and low dark current, *Nat. Photonics*, **9**, pp. 247–252.
93. Buscema, M., et al. (2014). Photovoltaic effect in few-layer black phosphorus PN junctions defined by local electrostatic gating, *Nat. Commun.*, **5**, p. 4651.
94. Kern, J., et al. (2015). Nanoantenna-enhanced light–matter interaction in atomically thin WS_2, *ACS Photonics*, **2**(9), pp. 1260–1265.
95. Eda, G. and Maier, S. A. (2013). Two-dimensional crystals: managing light for optoelectronics, *ACS Nano*, **7**(7), pp. 5660–5665.
96. Engel, M., et al. (2012). Light–matter interaction in a microcavity-controlled graphene transistor, *Nat. Commun.*, **3**, p. 906.
97. Britnell, L., et al. (2013). Strong light-matter interactions in heterostructures of atomically thin films, *Science*, **340**(6138), pp. 1311–1314.
98. Ye, Y., et al. (2015). Monolayer excitonic laser, *Nat. Photonics*, **9**(11), pp. 733–737.
99. Wu, S., et al. (2015). Monolayer semiconductor nanocavity lasers with ultralow thresholds, *Nature*, **520**(7545), pp. 69–72.
100. Schwarz, S., et al. (2014). Two-dimensional metal–chalcogenide films in tunable optical microcavities, *Nano Lett.*, **14**(12), pp. 7003–7008.
101. Hong, X., et al. (2014). Ultrafast charge transfer in atomically thin MoS_2/WS_2 heterostructures, *Nat. Nanotechnol.*, **9**(9), pp. 682–686.
102. Gong, Y., et al. (2014). Vertical and in-plane heterostructures from WS_2/MoS_2 monolayers, *Nat. Mater.*, **13**(12), pp. 1135–1142.
103. Wang, X., et al. (2013). High-responsivity graphene/silicon-heterostructure waveguide photodetectors, *Nat. Photonics*, **7**(11), pp. 888–891.
104. Miyamoto, K. (1961). The phase Fresnel lens, *J. Opt. Soc. Am.*, **51**(1), pp. 17–20.
105. Ren, H., Fan, Y.-H. and Wu, S.-T. (2003). Tunable Fresnel lens using nanoscale polymer-dispersed liquid crystals, *Appl. Phys. Lett.*, **83**(8), pp. 1515–1517.
106. Fattal, D., et al. (2010). Flat dielectric grating reflectors with focusing abilities, *Nat. Photonics*, **4**(7), pp. 466–470.
107. Aieta, F., et al. (2012). Aberration-free ultrathin flat lenses and axicons at telecom wavelengths based on plasmonic metasurfaces, *Nano Lett.*, **12**(9), pp. 4932–4936.

108. Yu, N. and Capasso, F. (2014). Flat optics with designer metasurfaces, *Nat. Mater.*, **13**(2), pp. 139–150.

109. Lee, S. H., et al. (2012). Switching terahertz waves with gate-controlled active graphene metamaterials, *Nat. Mater.*, **11**(11), pp. 936–941.

110. Ju, L., et al. (2011). Graphene plasmonics for tunable terahertz metamaterials, *Nat. Nanotechnol.*, **6**(10), pp. 630–634.

111. Vakil, A. and Engheta, N. (2011). Transformation optics using graphene, *Science*, **332**(6035), pp. 1291–1294.

112. Lee, J. Y., et al. (2009). Near-field focusing and magnification through self-assembled nanoscale spherical lenses, *Nature*, **460**(7254), pp. 498–501.

Chapter 2

Light–Matter Interactions in 2D Materials

Junpeng Lu,[a] Zhenliang Hu[b] and Hongwei Liu[c]
[a]*Department of Physics, Southeast University, Nanjing 211189, China*
[b]*Department of Physics, National University of Singapore, 2 Science Drive 3, Singapore 117542*
[c]*Jiangsu Key Lab on Opto-Electronic Technology, School of Physics and Technology, Nanjing Normal University, Nanjing 210023, China*
fantasyvivi@gmail.com, phylhw@njnu.edu.cn

In this chapter, we present the light–matter interactions in 2D materials of graphene, transition metal dichalcogenides (TMDs) and black phosphorus (BP). The basic influence factor of the interactions is the electronic band structure of the materials, which can induce a variety of fundamental optical properties. In addition, the vibrational property and carrier dynamics in the 2D materials would contribute to the interactions. In turn, they could also be affected by the interactions. Moreover, we also discuss the destructive interactions in 2D materials triggered by light with high power.

Two-Dimensional Materials in Nanophotonics: Developments, Devices, and Applications
Edited by Yuerui Lu

ISBN 978-981-4800-22-8 (Hardcover), 978-0-429-42877-7 (eBook)
www.jennystanford.com

2.1 Introduction

Since the beginning of 2004, the invention of graphene has facilitated 2D materials to be promising candidates for versatile applications. As a forerunner of the 2D atomic material family, graphene has opened up a great upsurge in research interests of exploring of a large amount of future applications since the reduced dimensionality reinvest its novel properties in electronics, opto-electronics and mechanics [1, 2]. However, the gapless feature of pristine graphene lack the capacity to provide the on/off switching control for a transistor in the logic circuit. To replenish the shortage of graphene, layered transition metal dichalcogenides (TMDs) were revitalised and attracted giant attentions in recent research as their various chemical compositions and unique properties [3]. In this chapter, we focus on the group VI aspects of TMDs, including MoS_2, $MoSe_2$, WS_2 and WSe_2, which attract most of the attention at the current stage due to their semiconducting behaviour and the sizable energy bandgaps covering the spectral range from visible to near infrared (IR) [4]. Nevertheless, the relatively lower mobility of TMDs monolayers restricts the applications of them in electronic devices. Since 2014, the re-introduction of black phosphorus (BP) has been attracting increasing interest of scientists from physics, chemistry, material science and engineering [5–13]. It has been demonstrated to present both high carrier mobility and a high on/off ratio [5].

Given the thickness-dependent band structures, 2D materials are considered as promising candidates for novel applications in nanophotonics and optoelectronics. The performance of opto-electronic devices is significantly affected by the factors of light absorbency, photo-carrier excitation, relaxation, separation, free-carrier diffusion, trapping/detrapping, recombination, etc. [14–16]. All these processes would be directly enabled, incidentally facilitated or indirectly affected by the light–matter interactions within the working materials. Therefore, a deep insight into the light–matter interactions in 2D materials could render directive significance to optimise the optoelectronic devices based on 2D materials. In this chapter, we present the light–matter interactions in 2D materials. The discussion is expanded from fundamental theoretical calculations to experimental observations.

2.2 Electronic Band Structure

The optical fundamentals of a material are determined by the electronic band structure. One of the key features of 2D materials is that the electronic band structures strongly depend on the number of layers. The bandgaps of 2D materials cover a wide range from the zero bandgap of graphene to ~2.5 eV of TMDs. The wide tunable range of the bandgaps facilitates the spectral photo-response of 2D materials crossing from visible to far IR (Fig. 2.1).

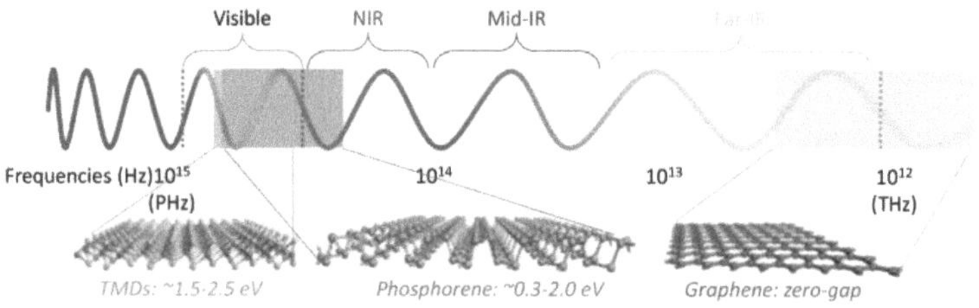

Figure 2.1 Spectral response range of 2D materials. Reproduced from Ref. [17] by permission of The Royal Society of Chemistry.

2.2.1 Electronic Band Structure of Graphene

Graphene is a single layer of carbon atoms packed in a regular honeycomb pattern. The carbon atom has six electrons and four exterior electrons take part in the formation of covalent chemical bonds with other atoms. In graphene, the 2s orbital is hybridised with other two 2p orbitals, forming three sp^2 hybridisation orbitals lying symmetrically in the graphene plane, which is perpendicular to the left one 2p orbital. The three sp^2 orbitals are totally the same, thus giving rise to the hexagonal structure of graphene with the in-plane σ bonds. Each σ bond between any two adjacent carbon atoms is formed by the overlapping of sp^2 orbitals, while the overlapping of remaining 2p orbitals results in the delocalised π bonds and π^* bonds, which form the highest occupied valence band and the lowest unoccupied conduction band. Since the binding energy of in-plane σ bonds (615 kJ/mol) [18, 19] is much stronger than the π bonds, it's easy to mechanically exfoliate the graphene layer with the intact in-plane hexagonal structure [20].

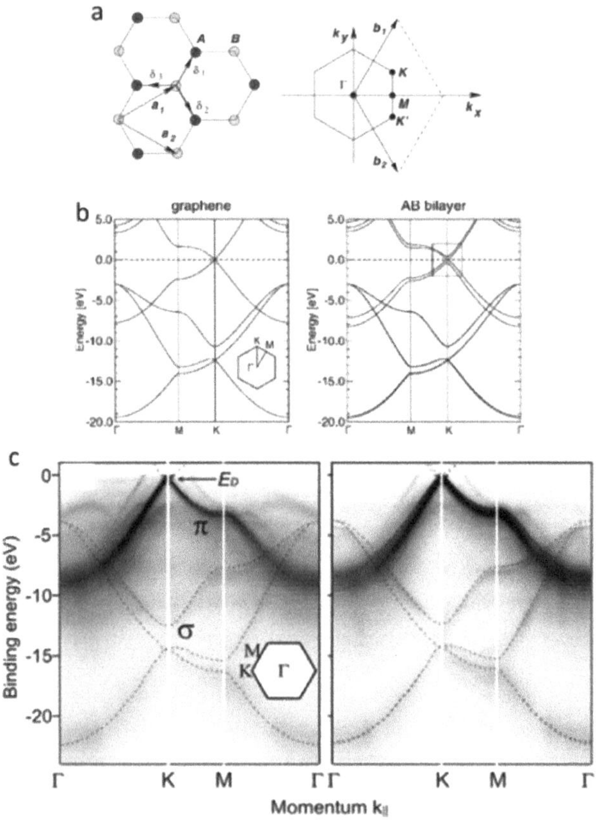

Figure 2.2 (a) Honeycomb lattice structure of graphene and its reciprocal space with high-symmetry points. $\mathbf{a_1}$ and $\mathbf{a_2}$ are the lattice unit vectors, and $\mathbf{b_1}$ and $\mathbf{b_2}$ are the reciprocal lattice unit vectors. Reprinted (figure) with permission from Ref. [2]. Copyright (2009) by the American Physical Society. (b) Electronic band structures of (left panel) single-layer and (right panel) bilayer (AB stacking) graphene through density functional theory simulation. Inset: Brillouin zone of graphene and some high-symmetry points. Reprinted (figure) with permission from Ref. [24]. Copyright (2006) by the American Physical Society. (c) Photoemission images of (left panel) single-layer and (right panel) bilayer graphene through ARPES. The dashed lines are the scale electronic band structures in (b). Inset: Brillouin zone of graphene and some high-symmetry points. Reprinted (figure) with permission from Ref. [25]. Copyright (2007) by the American Physical Society.

The hexagonal structure of graphene is shown in Fig. 2.2a. The structure is made up of a triangular lattice with a basis of two non-equivalent carbon atoms (A and B) per unit cell. The carbon–carbon distances between any two adjacent carbon atoms are 1.42 Å [21]. The reciprocal lattice is also shown in Fig. 2.2a with high-symmetry points Γ, M, K, K′ indicated. Each reciprocal lattice point corresponds to a lattice plane in real space, and the reciprocal lattice can also be used to represent the electronic structures along the high-symmetry directions within the first Brillouin zone. By using the tight-binding model for electrons in graphene, the band structure close to one of the Dirac points can be obtained as $E_{\pm}(q) \approx \pm v_F |q| + O[(q/K)^2]$, where $q = K - k$ with $|q| << |K|$ [22], the plus and the minus sign represent the upper and the lower band, respectively, and υ_F is the Fermi velocity around $\upsilon_F \approx 1 \times 10^8$ m/s which is independent to the energy or the momentum [22]. At the exact Dirac point, the conduction band and the valence band touch each other, leading to the zero bandgap and zero electronic density of states [1]. Also, the linear energy dispersion leads to the massless Dirac quasiparticles in graphene, which contribute to a high intrinsic carrier mobility [23].

The full electronic band structure along the high-symmetry directions can be further investigated through the first-principle techniques [24]. The band structures of single-layer graphene and bilayer graphene with the AB (Bernal) arrangement are reproduced in Fig. 2.2b. Obviously, for the single graphene layer, the band structure shows the linear energy dispersion of π bond and π^* bond at the Fermi level close to the Dirac point, and three σ bonds are identified by other three occupied valence bands. When two single graphene layers stack to form a bilayer in AB arrangement, because of the inter-layer coupling, the π bond and π^* bond are split and the linear dispersion is lost, while the σ bonds remain with much less influence. Experimental identification of band structures are realised using angle-resolved photo-emission spectroscopy (ARPES), which is a direct experimental approach to observe the electron distribution in the reciprocal space [25]. The ARPES images for the single-layer and bilayer graphene epitaxially grown on 6H-SiC substrates are shown in Fig. 2.2c, with the scaled band structures in Fig. 2.2b plotted with dashed lines. Three σ bonds and the π bond with the splitting in the bilayer can be identified [25]. Because of the inequivalent on-site Coulomb potential in each layer of the bilayer on the substrate, the bandgap occurs between the π bond and π^* bond at the Dirac point, which is not seen in the simulation results (Fig. 2.2b) [26]. In fact, the energy bandgap can be controlled by adjusting the carrier concentration in each layer of bilayer graphene [26].

2.2.2 Electronic Band Structure of TMDs

In the TMD monolayer, each metal atom is fixed in the centre of a trigonal prismatic cage and bounded to six chalcogen atoms via primarily covalent intra-layer bonding (Fig. 2.3a) [27]. This structure of monolayer TMDs facilitates the breaking of the inversion symmetry since there are no matched positions of the chalcogen atoms when the metal atom works as an inversion centre (Fig. 2.3b) [28]. Similar to graphene, one of the key features of group VI TMDs is that the electronic band structures exhibits strong dependence on the number of layers [29]. The multi-layer or bulk materials are indirect-bandgap semiconductors, while the bandgap of the monolayer counterparts converts to a direct bandgap at the 2D limit [29–31]. In multi-layers, the indirect bandgap is determined by the conduction band minimum (CBM) at the Λ point (Λ_c, midpoint between K and Γ points) and the valence band maximum (VBM) coinciding at the Γ point (Γ_v). When the thickness of the material is thinned down to a monolayer, the CBM and VBM shift to and finally coincide at the K point, producing a direct bandgap [32]. The electronic states at the Γ point are dominated by the chalcogen P_z orbitals, which are sensitive to the layer number since their charge density extends to the outer surface and is strongly affected by inter-layer coupling. Whilst the electronic states at the K point are mainly determined by the d electrons of the metal atoms. These states are not sensitive to inter-layer actions since the metal atoms are confined at the middle of the X-M-X sandwich [30, 33]. Therefore, the evolution of the bandgap transition is due to the combined contributions of the quantum confinement effect and inter-layer orbital interactions. Figure 2.3c shows the calculated electronic band structure of MoS_2 with monolayer and multi-layers using density functional theory (DFT) calculations [34]. The bandgap gradually changes from 1.2 eV to 1.9 eV, with the thickness thinning down from bulk to monolayer. Simultaneously, the type of the bandgap makes a crossover from indirect to direct. The band structure of MoS_2 from bulk to few layers to monolayer was experimentally observed via micro-ARPES (Fig. 2.3d) [35] The experimental results are well consistent with the theoretical calculations.

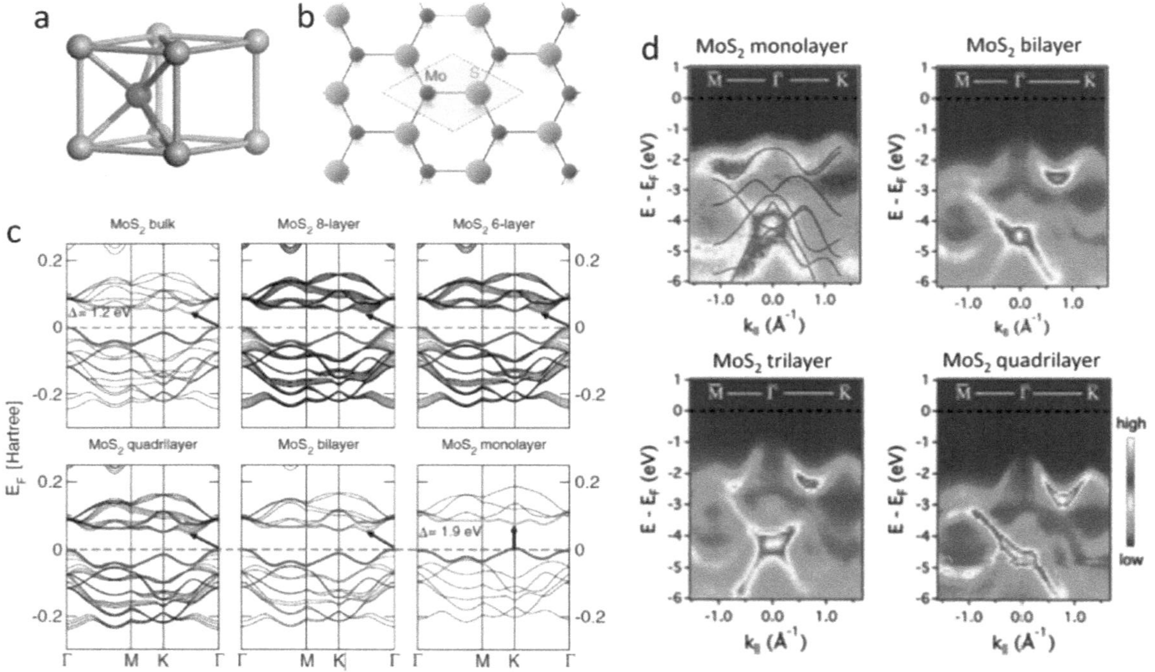

Figure 2.3 Schematic of (a) unit cell and (b) top view of lattice structure of TMD monolayer. Reprinted with permission from Springer Customer Service Centre GmbH: Springer Nature, *Nature Communications*, Ref. [27], Copyright (2012). (c) Calculated thickness dependence of band structure of MoS_2. Reprinted (figure) with permission from Ref. [34]. Copyright (2011) by the American Physical Society. (d) ARPES band map of MoS_2 with different thickness. Reprinted (figure) with permission from Ref. [35]. Copyright (2013) by the American Physical Society.

2.2.3 Electronic Band Structure of BP

BP has a puckered orthorhombic lattice structure obeying the D_{2h}^{18} space group symmetry. The inter-layer spacing between two single layers is 0.53 nm (Fig. 2.4a). The unit cell contains four atoms with two symmetry-inequivalent P–P bonds (Fig. 2.4a,b). Each phosphorus atom connects to three neighboring atoms. The interesting optical properties of 2D semiconductors arise from the electronic structure and dispersion of the electronic states at the 2D limit. The calculated electronic band structures of phosphorene with different thickness are shown in Fig. 2.4c. Different from the indirect- to direct-bandgap transition in TMDs from bulk to monolayer, the bandgap of phosphorene remains direct for any number of layers. However, the CBM and VBM coincide at the Z point in bulk BP, but they shift to the Γ point when the thickness is thinned down to few layers and a monolayer. Remarkably, the valence band top of monolayer phosphorene is very flat. Possibly the VBM might even be slightly off the Γ point [36]. But monolayer phosphorene is still approximately considered as direct bandgap since the separation between the VBM and the Γ point is less than 10 meV [37]. The theoretical evolution of the bandgap according to the layer numbers is shown in Fig. 2.4d. The estimated bandgap of the phosphorene varies from around 2.0 eV in the monolayer to around 0.3 eV in bulk, consistent with the experimental results [11, 38].

2.3 Fundamental Optical Property

2.3.1 Fundamental Optical Property of Grapheme

Under optical microscopy, the optical image contrast of graphene on the substrate can be used to determine the layer number and is the result of different reflection between graphene layer and substrate. The contrast shows high reliability and efficiency in number determination before other tools like atomic force microscopy (AFM) and Raman spectroscopy. The contrast spectra is obtained by the formula $C(\lambda) = (R_0(\lambda) - R(\lambda))/R_0(\lambda)$, where $R_0(\lambda)$ and $R(\lambda)$ are the reflection from the substrate and the graphene layer, respectively [41]. Figure 2.5a shows the optical image of exfoliated graphene with one, two, three and four layers on the Si/SiO_2 substrate.

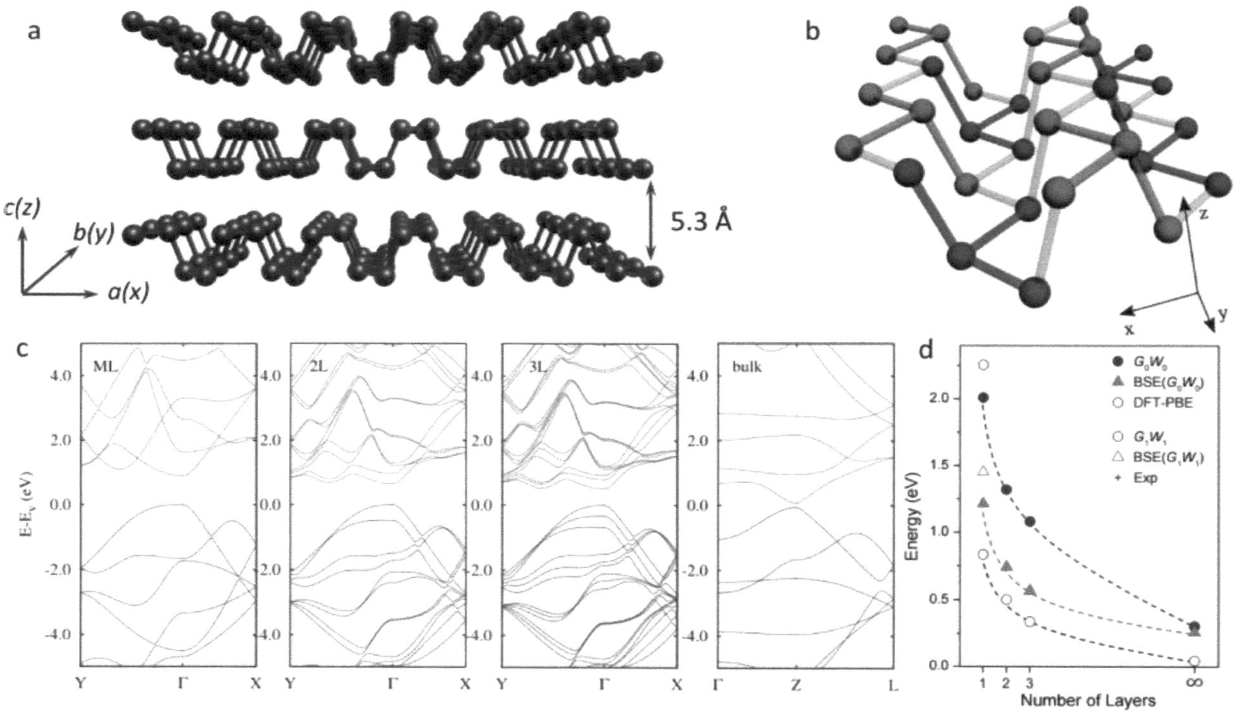

Figure 2.4 Crystal structure of (a) few-layer and (b) monolayer phosphorene. (c) Calculated thickness dependence of electronic band structures of phosphorene. (d) Evolution of the electronic bandgap as a function of the layer number, calculated by different methods. (a) Reproduced from Ref. [39] with permission from John Wiley and Sons. (b) Reprinted (figure) with permission from Ref. [8]. Copyright (2014) by the American Physical Society. (c) Reproduced from ref [40]. (d) Reproduced from ref [37].

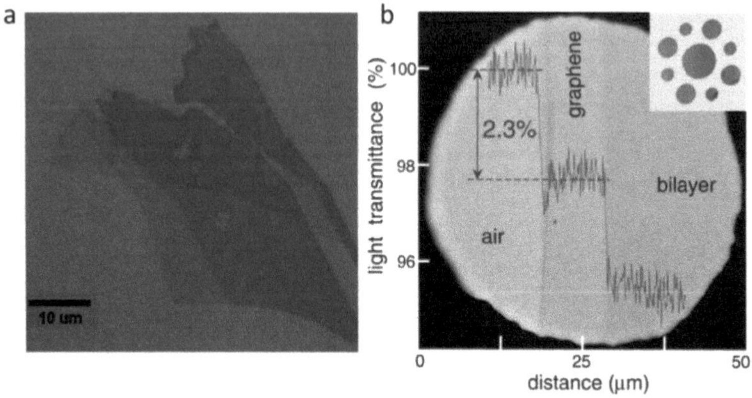

Figure 2.5 (a) Optical image of graphene with one, two, three, and four layers on the SiO_2/Si substrate. Reprinted with permission from Ref. [41]. Copyright (2007) American Chemical Society. (b) Photograph and transmittance of single- and bilayer graphene. Inset: Sample design for the experiment in Ref. [42]. From Ref. [42]. Reprinted with permission from AAAS.

The high-frequency (dynamic) conductivity G for Dirac fermions in graphene has been argued to be a universal constant $e^2/4\hbar$ [42]. This yields to observable quantities, such as the optical transmittance and reflectance given by $T = (1 + 2\pi G/c)^{-2} = (1 + \pi\alpha/2)^{-2}$ and $R = \pi^2\alpha^2/4$ for normal incident light. Then graphene's opacity can be derived as $(1 - T) \approx \pi\alpha \approx 2.3\%$ [42]. The transmittance, reflectance, opacity and other optical quantities are all universal constants that origin from the 2D nature and the zero-bandgap electronic band structure of graphene. The transmittance constant has been experimentally demonstrated for graphene in white light and over the visible range, as shown in Fig. 2.5b. The opacity for graphene is practically independent of the wavelength and is found to increase linearly with the number of layers since each layer can contribute to that despite the little perturbation from the inter-layer interaction [42]. For the graphene synthesised via chemical vapour deposition (CVD), the optical transmittance decreases slightly because of the contaminations produced during the growth and transfer process [43].

2.3.2 Fundamental Optical Property of TMDs

The conduction and valence band edges of TMDs at the K points are determined by the d_{z^2} and dxy / $d_{x^2-y^2}$ orbitals of the metal atom. The distinct d orbits induce obvious Zeeman-like spin–orbit coupling. DFT calculations indicate that the coupling facilitates significant splitting of the valence band of the TMD monolayers at the K points [44]. The splitting of the valence band facilitates two inter-band optical transition paths for the carriers. These two transitions could be monitored by the absorption spectra (Fig. 2.6a), as defined and corresponding to the A and B excitonic transitions. Besides the information of the spin–orbit splitting, the absorption spectra also exhibit strong absorption peaks at higher-energy ranges in TMDs monolayer (peak C in Fig. 2.6a). However, this peak is not observed in the PL spectra. As indicated by the calculated band structures (Fig. 2.6b), the giant absorption is caused by the resonance with the band nesting in the band structures. The band nesting is the region in the band structure where the valence and conduction bands are parallel [44]. The existence of band nesting results in the excitation dependence of the relaxation paths of the photo-carriers. Since the photo-excited electron and holes possess the same but opposite velocities at the band-nesting region, the electron–hole pairs prefer to separate in that space. Therefore, the radiative recombination (photo-luminescence [PL]) is suppressed for absorption in the resonance range with band nesting.

The thickness dependence of the electronic structure facilitates TMDs' interesting optical properties. The near-band-edge emissions can be detected via PL spectroscopy [29, 30, 45]. The respective PL spectra of MoS_2 with different layer numbers are shown in Fig. 2.6c. The emission peak shift (feature I in Fig. 2.6c) clearly demonstrates the broadness of the bandgap with thinning of the thickness, whilst the quantum yield (PL intensity) provides direct evidence to the transition of the type of the bandgap. Therefore, PL spectra are rendered as an effective tool to monitor the evolution of the band structure transition of TMDs. The much higher (several orders in magnitude) quantum yield in monolayer TMDs indicates the direct-bandgap behaviour since the involvements of defect or phonon scattering which restrict the radiative recombination are suppressed in direct-bandgap materials [28]. This feature makes TMDs monolayers promising building blocks for various photonics and optoelectronics applications.

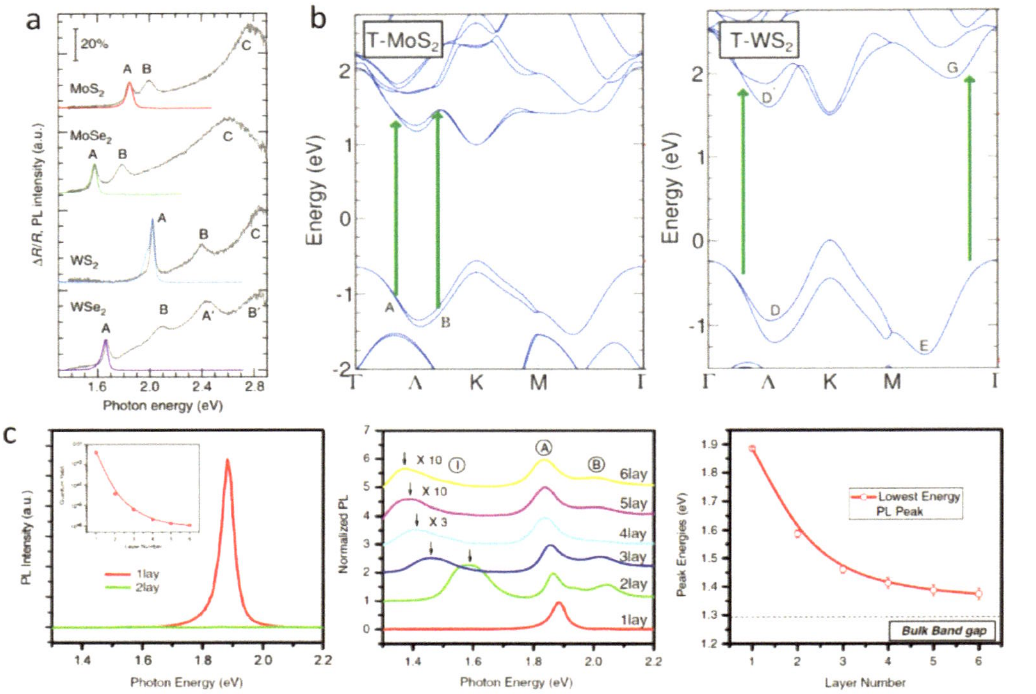

Figure 2.6 (a) Absorption (grey curves) and PL (coloured curves) spectra of TMD monolayers. Reprinted with permission from Springer Customer Service Centre GmbH: Springer Nature, *Nature Communications*, Ref. [46], Copyright (2014). (b) Calculated band structure of MoS_2 and WS_2 monolayers. Reprinted (figure) with permission from Ref. [44]. Copyright (2013) by the American Physical Society. (c) From left to right: PL spectra of MoS_2 monolayer and bilayer (inset shows the quantum yield as a function of layer numbers), PL spectra of MoS_2 with a thickness of 1–6 layers and bandgap energy as a function of layer numbers. Reprinted (figure) with permission from Ref. [29]. Copyright (2010) by the American Physical Society.

2.3.3 Fundamental Optical Property of BP

Arising from the puckered orthorhombic lattice structure with the D_{2h} symmetry, phosphorene exhibits exciting and novel property of in-plane anisotropy. It is induced by the difference of the effective mass of carriers in BP where the effective mass of carriers along the zigzag direction (y) is ~10 times heavier than that along the armchair direction (x) [13, 47]. This particularity can give rise to strong in-plane anisotropic optical, electrical and vibrational properties [9, 11, 48]. At the optical aspect, the anisotropy is determined by the special selection rules of the carrier transition. The occurrence of the optical transition relates to the momentum operator, $\hat{p}_{x/y}$ [36]. In the phosphorene monolayer, the $\hat{p}_x$ between the VBM and the CBM is finite. This allows the occurrence of electronic transition, while the transition along the y polarisation is forbidden by symmetry [49]. The anisotropy of the optical property is described by the theoretical calculation of the angle-dependent absorption coefficient, $A(\alpha)$ (Fig. 2.7a). Evidently, $A(\alpha)$ of phosphorene strongly depends on the polarisation angle. The calculation result agrees well with the experimentally measured extinction spectra which include the information of both light absorption and reflection of a phosphorene sample on a SiO_2 substrate. Since the anisotropy of the optical property is very significant, it can be used as an indicator to determine the crystallographic axes.

The evolution of the electronic band structure with thickness and the correspondingly induced optical property in phosphorene can be monitored by collection of the excitonic emissions using PL spectroscopy. The normalised PL spectra of phosphorene with varying thickness (from a monolayer to five-layer numbers) are shown in Fig. 2.7b. The significant blue shift of the peak position clearly indicates the broadening of the bandgap with decreasing thickness. The broadness of the gap size arises from the synergy of the quantum confinement effect and the interactions between the inter-layers [37, 50]. The optical gap, E_{opt}, can be estimated from the PL peak position. Accordingly, the E_{opt} values of phosphorene with thickness from a monolayer to five layers are ~1.75, ~1.29, ~0.97, ~0.84 and ~0.80 eV, respectively. These experimental values can be described well by a power law formula, $E_{opt} = 1.486/N^{0.686} + 0.295$, where N is the layer number (Fig. 2.7c). The variation trend

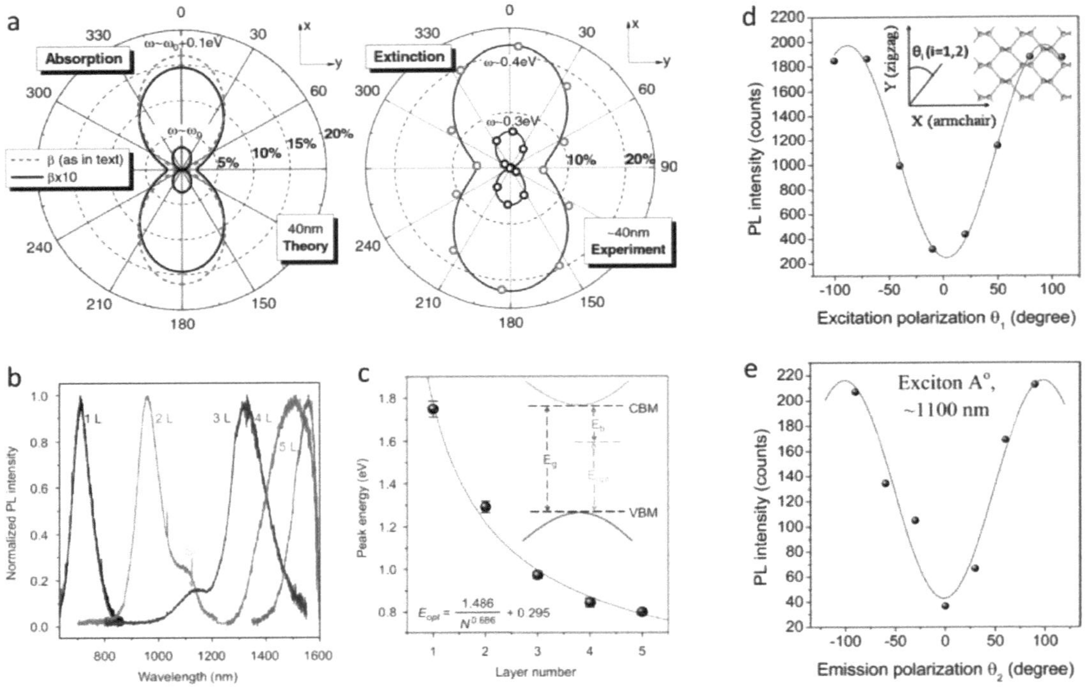

Figure 2.7 (Left) Predicted angle-dependent absorption coefficient $A(\alpha)$, where α is the light polarisation angle. (Right) Experimentally measured angle-dependent extinction spectra. Reprinted (figure) with permission from Ref. [40]. Copyright (2014) by the American Physical Society. (b) Normalised PL spectra of phosphorene with varying thickness. (c) E_{opt} as a function of layer number. Inset: Schematic illustration of the electronic bandgap and exciton binding energy. Reprinted with permission from Springer Customer Service Centre GmbH: Springer Nature, *Light: Science & Applications*, Ref. [51], Copyright (2015). Intensity of PL emission as a function of (d) excitation polarisation and (e) emission polarisation. Reprinted with permission from Ref. [52]. Copyright (2016) American Chemical Society.

of the optical gap as a function of layer numbers is consistent with the theoretical prediction [9, 37]. However, the obtained values of the optical gaps are smaller than the electronic bandgaps predicted by theoretical calculations [37] or measured by experiments [38]. This is caused by the exciton binding energy in phosphorene. The exciton binding energy (E_b) can be calculated from the difference between the electronic bandgap (E_g) and the optical gap (E_{opt}), as illustrated by the schematic in the inset of Fig. 2.7c. By employing the experimentally measured E_g of ~2.05 eV for phosphorene monolayer, the E_b of phosphorene monolayer can be estimated to be ~0.3 eV, which is well consistent with the theoretical prediction. The large value of E_b indicates the high stability of excitons in phosphorene. It arises from the strong confinement of electrons and holes and the reduced screening of the Coulomb interaction. Consequently, the optical properties of phosphorene are dominated by excitons. Because of anisotropic absorption property of phosphorene, the excitons are predicted to be confined in a quasi-1D space. The quasi-1D property can be validated by detection of completely linearly polarised exciton emissions. This linearly polarised emissions are observed by angle-resolved PL spectroscopy. The PL intensity is observed to show significant dependence on the polarisation of the excitation light (θ_1) (Fig. 2.7d). If the polarisation of excitation light is fixed, the emissions from excitons are observed to be completely linearly polarised along the armchair direction (Fig. 2.7e).

2.4 Vibrational Property

2.4.1 Vibrational Property of Graphene

In the specific case of graphene, Raman spectra observed for different numbers of graphene layers provide us with an effective approach to determine the layer number of graphene when stacked in the AB (Bernal) arrangement [53].

The most prominent features of graphene and graphite are the G peak at around 1580 cm^{-1} and the G' band at around 2700 cm^{-1} (Fig. 2.8a). For graphene with defects, such as the graphene edge, the Raman signal shows a different feature (Fig. 2.8b), with another intense D peak at about 1350 cm^{-1}. The G peak is linked to the doubly degenerate phonon mode (E_{2g} symmetry) at the Brillouin zone centre [54], while the D peak and G' peak origin from a second-order double resonance process between nonequivalent K points in the first Brillouin zone, contributing to the absence of a D peak in the well-ordered graphene and graphite [55]. As the G' peak is almost twice the D peak in peak position, people also refer to the G' peak as 2D. And D' is used to denote another week defect-caused feature appearing at about 1620 cm^{-1}.

In fact, in the Raman signal of graphene and graphite, only the G peak is from the natural first-order Raman scattering process [56], which is clearly depicted in Fig. 2.8c. A photo-excited electron–hole pair recombine after the phonon scattering, emitting a photon with energy loss or gain compared to the excitation. On the contrary, in the double-resonance process shown in the centre and right panels of Fig. 2.8c, the phonon wave vector **q** is linked to the electronic band structure. The electron induced by the laser excitation is scattered by a phonon or the defect with an exchange momentum **q**. The electron then goes through another electron–phonon scattering with exchange momentum –**q**, and eventually the electron and hole recombine with emission of a photon [57]. For the G' peak, phonons are involved in two inelastic scatterings, so the energy shift is nearly twice of the phonon energy and results in the G' Raman frequency twice the scattering phonon frequency q, which is determined by the double-resonance condition [56]. In the case of the D peak, only one scattering is completed by the phonon; another one is accomplished by the elastic electron defect scattering; therefore the D peak is absent in the well-ordered graphene and graphite and the resulting D peak frequency is almost half of the frequency of the G' peak [58]. As for the D' peak, the double-resonance process connects two points around one K point, leading to almost the same peak frequency as the G peak.

When the graphene layer number changes, the peak positions, shapes and relative intensities of G and G' peaks have a great change [53]. As a result, this property is used to help determine the layer number in a non-destructive and effective way. Figure 2.8d shows the Raman spectra of graphene with different layer numbers. When comparing the bilayer Raman spectra with that of a single layer, the G' peak is much broader and its position shifts up. Actually, the G' peak of bilayer graphene has four components coming from four

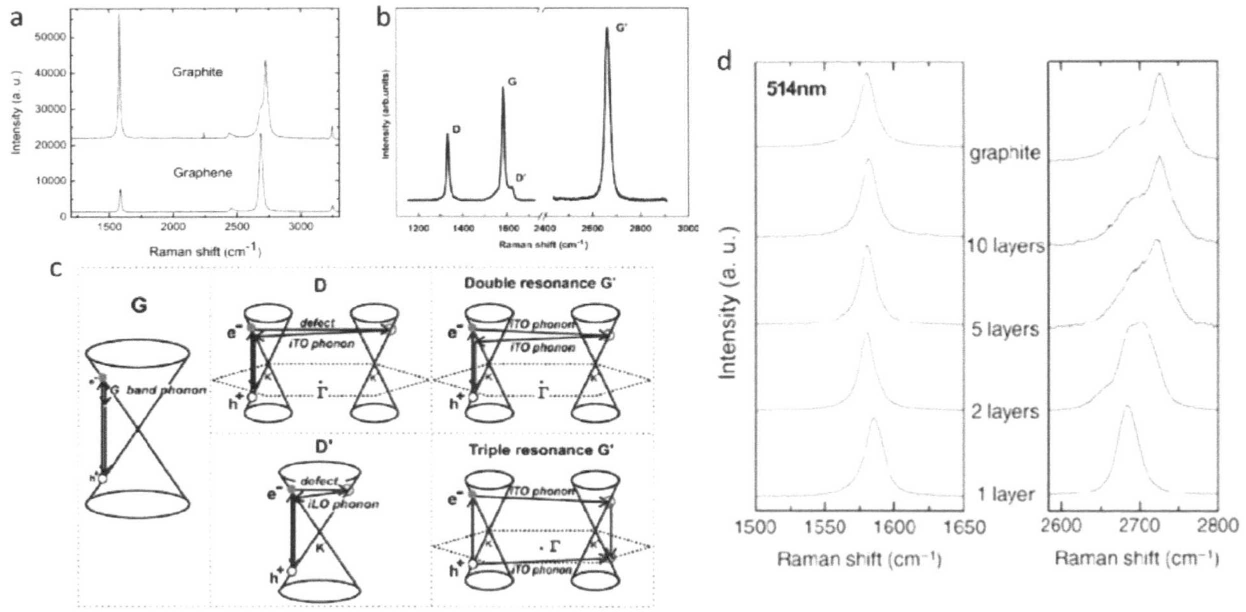

Figure 2.8 (a) Raman spectra of bulk graphite and graphene with 514 nm laser wavelength. Spectra are scaled for having a similar height as the G′ peak. Reprinted (figure) with permission from Ref. [57]. Copyright (2006) by the American Physical Society. (b) Raman spectrum of the graphene edge with the laser excitation of 2.41 eV [56]; the extra D peak is caused by defects. (c) First-order process of the G peak (left), double resonance of the D peak (centre top), D′ peak (centre bottom) and G′ peak (right top), and triple resonance of the G′ peak (right bottom). Reprinted from Ref. [56], Copyright (2009), with permission from Elsevier. (d) The G peak (left) and G′ peak (right) in Raman spectra of graphene with different layer numbers under the illumination of 541 nm laser wavelength. Reprinted from Ref. [62], Copyright (2007), with permission from Elsevier.

different phonon–electron scatterings due to the double splitting of electronic band structures that we have seen before [53]. A further increase in the layer number contributes to the continuous decrease of the relative intensity of the G' peak with respect to the G peak. From the peak broadness, peak shapes and relative intensities, Raman spectroscopy can easily distinguish the single layer and bilayer from other few layers of graphene. However, when the layer number is more than five layers, it becomes quite hard for Raman spectroscopy to tell the difference from graphite [53]. Note that the approach to identify the layer number based on Raman spectroscopy is only applicable for graphene with AB (Bernal) stacking.

Except for the graphene layer number, many factors can affect the Raman signals, such as laser excitation wavelength [59], laser power [60], temperature [61] and the doping level of graphene [62].

2.4.2 Vibrational Property of TMDs

The inter-layer interactions of van der Waals structures facilitate the basic lattice vibrations to present distinct features in layer number evolution of phonons, inter-layer shear, layer-breathing phonons, etc. Due to the versatility of Raman spectroscopy, all these phonon-related properties could be clearly elucidated by Raman spectra. In bulk TMDs, the lattice vibrations obey the D_{6h} point group symmetry and the corresponding modes at the Γ point could be described as $\Gamma = A_{1g} + 2A_{2u} + E_{1g} + 2E_{1u} + 2E_{2g} + E_{2u} + 2B_{2g} + B_{1u}$, where 18 phonon modes are included since the "E" modes are doubly degenerate. Among these modes, A_{1g}, E_{1g} and E_{2g} are Raman-active, A_{2u} and one E_{1u} are IR-active, B_{2g}, B_{1u} and E_{2u} are silent, and A_{2u} and the other E_{1u} are acoustic modes. When the TMDs are thinned down to a bilayer, the point group summery is transferred to D_{3d} and the 18 vibrational modes are expressed as $\Gamma = 3A_{1g} + 3A_{2u} + 3E_g + 3E_u$ due to the inversion symmetry, where A_{1g} and E_g are Raman-active, while others are IR-active or acoustic modes. For a TMD monolayer, the symmetry is reduced and belongs to D_{3h}. There are only nine phonon modes, which can be described as $\Gamma = A_1' + 2A_2'' + 2E' + E''$, where Raman-active modes include E'(both Raman- and IR-active) and A_1' and E'' (Fig. 2.9a) [63]. Evidently, A_{1g}, E_{1g} and E_{2g} are Raman-active in bulk TMDs. Three Raman peaks located at ~33.5 cm^{-1}, 383.6 cm^{-1} and 408.7 cm^{-1} are observed in bulk MoS_2 and can be assigned to

E_{2g}^2, E_{2g}^1 and A_{1g}, respectively [64]. E_{1g} is not observed due to its Raman tensor. Intensive research efforts have been done on monitoring the thickness of TMDs via Raman spectroscopy. As discussed in Fig. 2.9a, the denotations of the photon modes in bulk and monolayer TMDs are different. When the Raman spectra are used to evaluate the thickness of the material from bulk to monolayer, the high-frequency modes E_{2g}^1 (in-plane vibrational mode) and A_{1g} (out-of-plane mode) are commonly monitored as thickness indicators. The layer number identification is favoured by the relative frequency difference between A_{1g} and E_{2g}^1 ($\Delta\omega = A_{1g} - E_{2g}^1$). As demonstrated by Raman spectra of MoS_2 [63, 65], E_{2g}^1 presents a blue shift, while A_{1g} red-shifts when the material thins down from bulk to monolayer (Fig. 2.9b). For thinner samples (layer number, $N < 5$), the relationship between $\Delta\omega$ and layer number can be described well by $\Delta\omega = 2.58 - 8.4/N$ (Fig. 2.9b) [63]. However, this description is not suitable for the thickness materials. The shift of the E_{2g}^1 mode with thickness may assign to the long-range Coulombic interactions or structure changes induced by stacking, while the shift of A_{1g} is caused by the varying restoring forces at different thicknesses [65, 66]. Similar shift trends of the phonon modes have been also observed in $MoSe_2$ [67], WS_2 [68] and WSe_2 [68]. Due to the heavier atomic mass of tungsten, the shifts of the frequencies are less significant.

The determination of the thickness of TMDs by monitoring $\Delta\omega$ is restricted to samples with $N < 5$. The more precise and pertinent approach is to monitor the phonon modes at low frequencies (<50 cm^{-1}). Take MoS_2 as an example. Two obvious modes, E_{2g}^2 (inter-layer shear mode C) and B_{2g}^2 (layer-breathing mode LB) are observed [64, 69]. These phonon modes originate from the inter-layer interactions. As illustrated in Fig. 2.9a, the C modes are the oscillations of the adjacent layers along the plane direction and the LB modes are the oscillations of the layers perpendicular to the slab. The LB and C modes are absent in TMD monolayers since there is no inter-layer interaction for one layer. Starting from the bilayer, the frequencies of LB modes gradually increase with increasing layer numbers, while the frequencies of C modes present the opposite trend. Several sets of C (defined as S in Fig. 2.9c) and LB modes are observed in MoS_2 (Fig. 2.9c) [70]. The LB mode is absent in bulk MoS_2 since the LB mode is not Raman-active at its bulk counterpart.

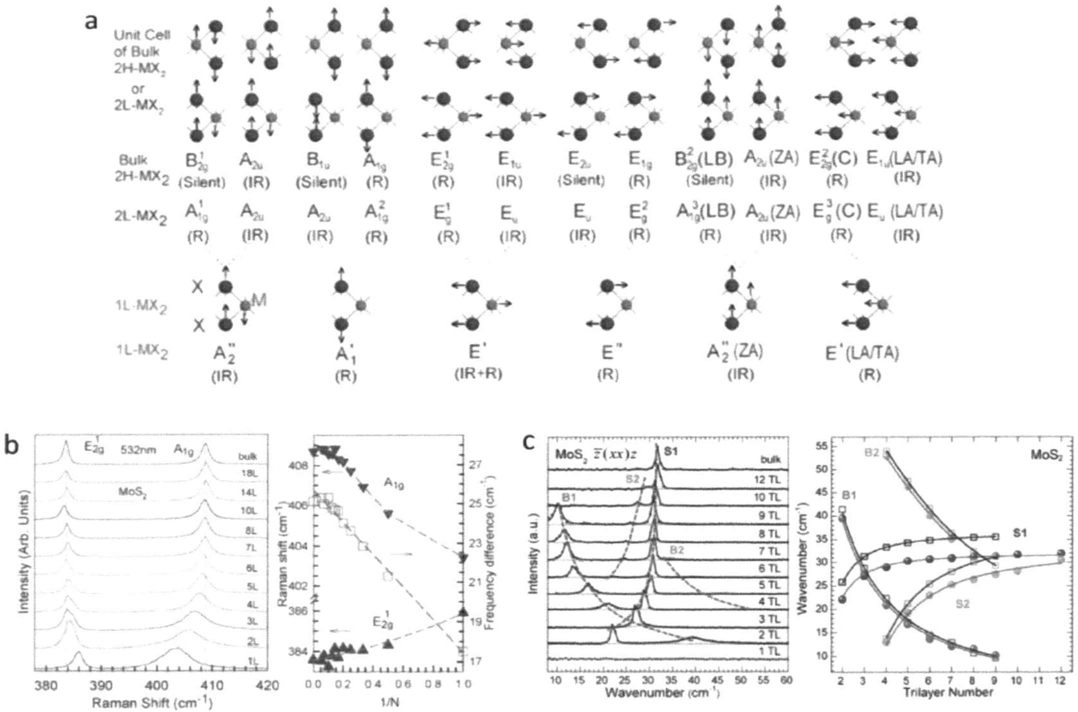

Figure 2.9 (a) Schematic of each optical vibration mode for TMD monolayer, bilayer and bulk. The R, IR and silent stands for Raman-active, IR-active and Raman- and IR-inactive, respectively. The dashed lines indicate that the corresponding in-plane and out-of-plane modes of the bilayer and bulk merge into one monolayer. (b) From left to right: Raman spectra of MoS_2 with different thicknesses; frequency of A_{1g} and E_{2g}^2 and the frequency difference of these two modes as a function of $1/N$; N means the layer numbers. (a, b) Reprinted from Ref. [63] with permission from The Royal Society of Chemistry. (c) From left to right: Raman spectra of MoS_2 with different thicknesses at low frequency collected using $\bar{z}(xx)z$ polarisation; frequencies of breathing modes and inter-layer shear modes as a function of layer number. Reprinted with permission from Ref. [70]. Copyright (2013) American Chemical Society.

The relationship of the phonon modes at low frequency and the layer numbers can be employed to evaluate the thickness of MoS_2 and the applicability is extended to 18 layers [64]. Similar results are also observed in tungsten dichalcogenides [70].

2.4.3 Vibrational Property of BP

The crystal structure of BP is puckered and belongs to the D_{2h}^{18} point group. The primitive cell, half of the unit cell, contains four atoms. Thus the phonon modes at the Γ point can be described as $2A_g + B_{1g} + B_{2g} + 2B_{3g} + A_u + 2B_{1u} + 2B_{2u} + B_{3u}$, where 12 phonon modes are included. Among these modes, B_{1u}, B_{2u} and B_{3u} are acoustic modes and all the rest even-parity modes are Raman-active modes [71]. Therefore, Raman spectroscopy can be used as an effective tool to investigate the vibrational, electronic and structural properties of phosphorene. Generally, three prominent Raman peaks at around ~362.3, 438.9 and 467.3 cm^{-1}, ascribed to the irreducible representations of A_g^1, B_{2g} and A_g^2 phonon modes [39], respectively, can be observed at the Raman spectrum (Fig. 2.10a) and used to identify phosphorene. The vibration of the A_g^1, B_{2g} and A_g^2 modes oscillate along the z (out-of-plane), y (zigzag) and x (armchair) directions, respectively (Fig. 2.10b). Different from TMDs, the Raman spectra are not usually regarded as the unambiguous signature of the layer number of phosphorene. One of the most important reasons is that the frequency of the phonon modes is not significantly dependent on the layer number [72]. Especially, the shifts of A_g^1 and B_{2g} are negligible (<1 cm^{-1}). A_g^2 presents the most significant shift of ~1.9 cm^{-1} [73]. This is consistent with the theoretical prediction where the lattice parameter along y and z directions insignificantly depends on the layer number, while the lattice parameter along the x direction significantly changes with the variation of the layer number [9]. In addition, the broadening of the width of the Raman peaks is observed in phosphorene few layers with decreasing of the layer number. It reaches the maximum at the trilayer and shrinks back at the monolayer. This is due to the larger degree of freedom of P atoms at the primitive cell in phosphorene few layers where the primitive cells consist of $4n$ atoms. This is larger than the four-atom primitive cell of phosphorene monolayer and bulk. The large number of atoms in a unit cell can induce Davydov splitting and give rise to the broadening of the photon modes [73].

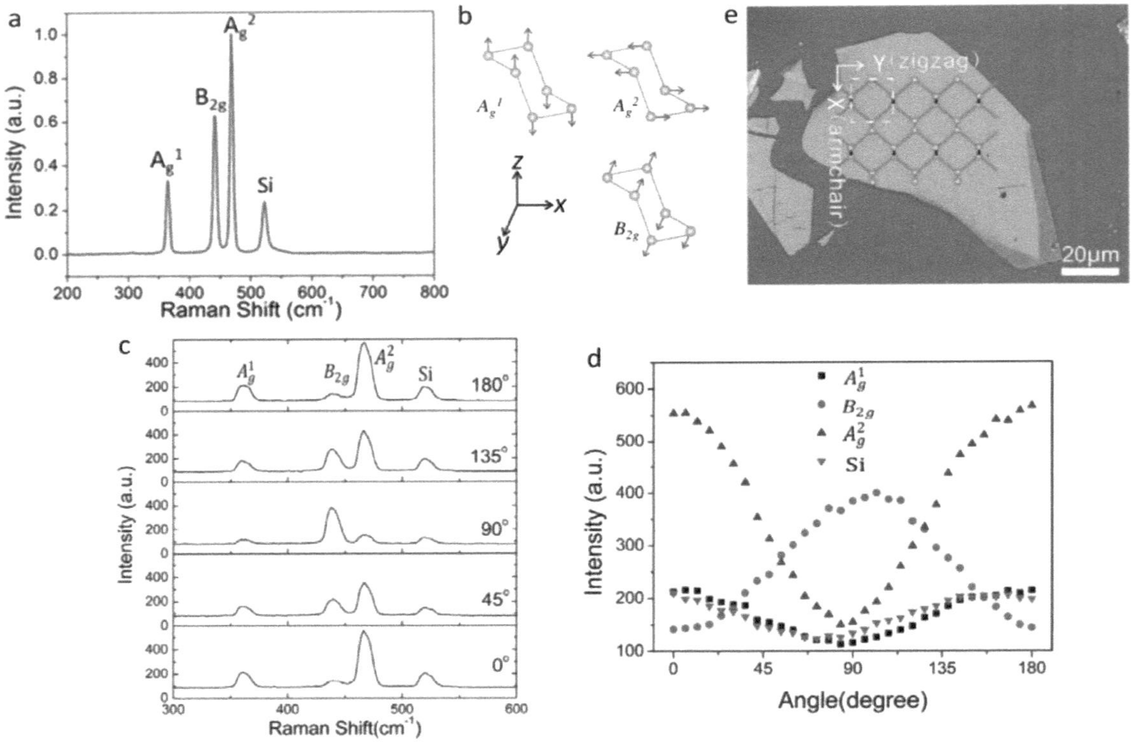

Figure 2.10 (a) Raman spectrum of a typical phosphorene sample. Reproduced from Ref. [39] with permission from John Wiley and Sons. (b) Schematic illustration of the vibration modes. (c) Raman spectra of a phosphorene sample collected at different polarisation angles. (d) Intensities of the Raman peaks as a function of polarisation angle. (e) Determination of crystalline orientation via polarisation-dependent Raman spectra. (b–e) Reprinted with permission from Ref. [74]. Copyright (2014) American Chemical Society.

Despite variation of the Raman peaks not showing very significant dependence on the layer number, the intensity of the Raman peaks exhibits strong dependence on the polarisation angle (Fig. 2.10c) due to anisotropy of phosphorene. Peak intensities of B_{2g} and A_g^2 display opposite variation with changes in the polarisation angle and the angular period is 180° (Fig. 2.10d). However, the intensity of A_g^1 is not sensitive to the angle [74, 75]. As shown in Fig. 2.5b, the vibration of A_g^1 is out of plane. Therefore, it does not vary with the laser polarisation. When the laser polarisation is parallel to the armchair direction where the A_g^2 mode oscillates along, the intensity of A_g^2 reaches its maximum. When the laser polarisation is parallel to the zigzag direction where the B_{2g} mode oscillates along, the intensity of B_{2g} reaches its maximum. Accordingly, polarisation-resolved Raman spectroscopy can be used as a fast and precise way to determine the crystalline orientation of phosphorene (Fig. 2.10e), which is generally realised by a high-resolution electron microscope.

2.5 Carrier Dynamics

Given the prominent optical properties of 2D TMDs, they are expected to present superior performance in optoelectronics with high efficiencies and fast response. To realise the photo-electrical devices with a rapid response rate, it is a prerequisite to understand the ultra-fast optical phenomena of the semiconductors at an ultra-fast timescale. Good performance of TMDs in opto-electronics critically depends on efficient light absorption and excited carrier relaxation dynamics. Taking full advantage of the highly temporal resolution of the ultra-fast (femtosecond) pulse laser techniques, ultra-fast spectroscopies have been devised to directly monitor the carrier dynamics where the decay time of the carriers in each path (carrier lifetime) could be evaluated, while the fundamental mechanisms of the energy loss from the interactions of electron–exciton, exciton–exciton, electron–phonon, etc., could be comprehensively determined. The general ultra-fast spectroscopies include time-resolved photo-luminescence (TRPL) spectroscopy and pump-probe spectroscopies (e.g., transient absorption/reflection spectroscopies).

2.5.1 Carrier Dynamics in Graphene

Transient absorption spectroscopy is another ultra-fast technique usually employed to monitor the carrier dynamics in semiconductors. As compared to TRPL spectroscopy, the characterisations taken by transient absorption spectroscopy would not be effectively affected by the low quantum yield of the materials [76]. In addition to the measurement of the carrier lifetime, it is quite straightforward to extract other crucial physical parameters such as the mean free path, mobility, diffusion length, exciton diffusion coefficient, etc., from analysis of the experimental spectra.

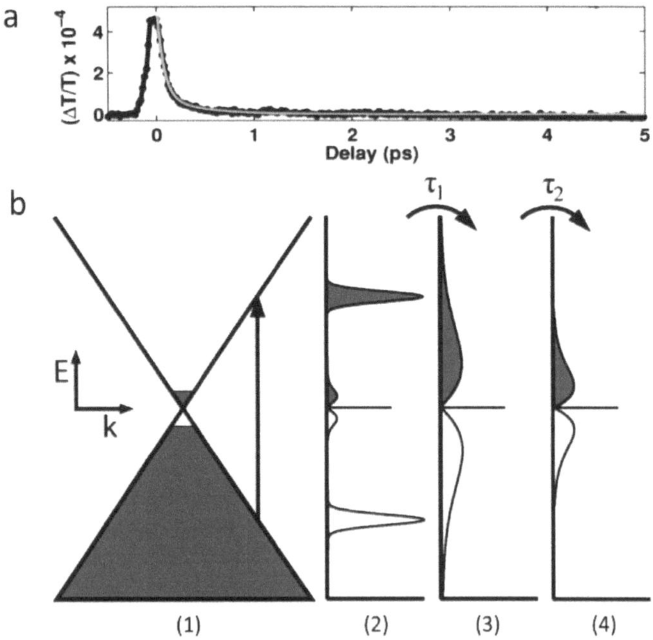

Figure 2.11 (a) Transmission measured using optical pump-probe spectroscopy. The dark dots are experimental results, and the light solid line is the analytic fits using exponentials with timescales τ_1 and τ_2 [77]. (b) (1) Energy dispersion of graphene in the vicinity of the Dirac point; the arrow represents the optical excitation. (2) The non-equilibrium distribution of electrons and holes after photo-excitation. (3) The equilibrium Fermi–Dirac distribution of carriers at high temperature after carrier–carrier scattering on the timescale τ_1. (4) Decay of the hot distribution of carriers after intra-band carrier–phonon scattering on the timescale τ_2. Reproduced from Ref. [77], with the permission of AIP Publishing.

The ultra-fast relaxation and recombination dynamics of excited carriers in graphene are substantially studied using optical pump-probe spectroscopy [77–79], optical-pump terahertz-probe spectroscopy [80, 81] and other techniques [82, 83] due to the important role of carrier dynamics in electronics and opto-electronics. For example, separating the excited electron–hole pair is necessary before recombination in solar cells. However, despite substantial efforts, graphene's carrier dynamics are still not fully understood. For graphene synthesised via different approaches, the carrier dynamics in a timescale is slightly different. Here, the graphene samples used are epitaxially grown on the C-terminated face of semi-insulating 6H-SiC by thermal sublimation of Si [84]. Owing to different stacking arrangement of single layers in epitaxial graphene compared to graphite, the phonons and electrons in different layers of epitaxially grown multi-layer graphene are nearly uncoupled; thus its electronic properties are similar to the graphene single layer [85].

Using optical pump-probe spectroscopy, the differential transmission $\Delta T/T = (T - T_0)/T_0$ is shown in Fig. 2.11a, where T and T_0 are sample transmissions with and without pump excitation. The pump pulses at 780 nm centre wavelength are used to excite electron–hole pairs, while the probe pluses at the same wavelength are used to probe the differential transmission of samples at various delays of the optical probe with respect to the optical pump [77]. From Fig. 2.11a, the transmission has a sharp increase after photo-excitation, and then it gets through two different relaxation timescales, τ_1 around 100 fs and τ_2 around 1 ps, before recovery [77]. Figure 2.11b gives us a simple model with the intra-band carrier–carrier scattering and carrier–phonon scattering to explain what happens after photo-excitation. Photo-excitation of charge carriers by the optical pump blocks the corresponding inter-band transition due to the increase of the filled conduction band states and empty valence band states; this blocking contributes to the initial rapid increase of the transmission in Fig. 2.11a. Shortly after photo-generation, these non-equilibrium-distributed electrons thermalise to form a hot Fermi–Dirac distribution through intra-band carrier–carrier scattering with a temperature T_e much higher than the lattice temperature on a fast timescale τ_1 [86, 87]. The new carrier distribution could possibly block some inter-band transitions in the range of $K_B T_e$. Next, subsequent cooling through intra-band carrier–phonon scattering results in the decay of hot distribution and new thermalisation on the slow time τ_2 [77]. Finally, the electron–hole recombination dominates the process and leads to the equilibrium state, which is the same as the state before photo-excitation. The timescale of carrier inter-band recombination in graphene is carrier-density-dependent, demonstrated by using the optical-pump terahertz-probe spectroscopy [80]. The decay time is highly sensitive to the photon energy [79], and crystalline disorder [77], higher photon energy and less crystal disorder give rise to a low decay time.

2.5.2 Carrier Dynamics in TMDs

Since the excitons in monolayer TMDs dominate their optical properties, it is critical to investigate the corresponding excitonic dynamics. Among the ultra-fast spectroscopy techniques, TRPL is the most frequently used tool to probe this property. As monitored by the TRPL spectra at different temperatures, the PL decay is very fast (~5 ps) at low temperature, while an additional slow recombination process emerges at higher temperature (Fig. 2.12a) [88]. The slow decay process may originate from the exciton–phonon scattering. Besides the classic excitonic behaviour as a traditional direct-bandgap semiconductor, the carriers in 2D TMDs present strong Coulomb interactions due to the co-existence of reduced dielectric screening and quantum confinement effects. The strong Coulomb interactions could facilitate tightly bound excitons and additionally lead to higher-order excitonic states such as the observation of charged excitons (trions) and many-body states. The biexciton, a representative four-body state, is observed and identified by low-temperature PL and TRPL spectroscopies [89]. According to the excitation fluence-dependent analysis, the P_0 peak in Fig. 2.12b presents a different trend than other peaks. The intensity of P_0 grows super-linearly with increasing excitation densities (Fig. 2.12c), while the strength of emission from P_0 also presents a super-linear relationship with respect to exciton emission (X in Fig. 2.12b), as shown in Fig. 2.12d. Consequently, the P_0 peak is identified as the emission from the biexciton (XX) states. As further evidence, TRPL spectra are carried out to characterise the dynamics of the biexciton state (Fig. 2.12e). The lifetime of the biexciton is comparable to those of the exciton and the trion (X^-). It is around tens of picoseconds. However, the lifetime of the emissions from defect-related features (P_1 and P_2) is much longer (>100 ps). Therefore, the emission of P_o is excluded from the defect-related states. The binding energy of the biexciton is defined as $\Delta_{XX} = 2E_X - E_{XX}$, where E_X/E_{XX} is the energy of the free exciton/biexciton. During the decay process, if the radiative decay of the biexciton is assumed to produce an exciton, the energy of the biexciton can be expressed to be $E_{XX} = \hbar\omega_{XX} + E_X = \hbar\omega_{XX} + \hbar\omega_X$, where $\hbar\omega$ is the energy of the radiative emission and can be extracted from the spectra. Therefore, the biexciton binding energy ($\Delta_{XX} = \hbar\omega_X - \hbar\omega_{XX}$) is calculated to be 52 meV. This value is much higher than those observed in conventional quantum wells.

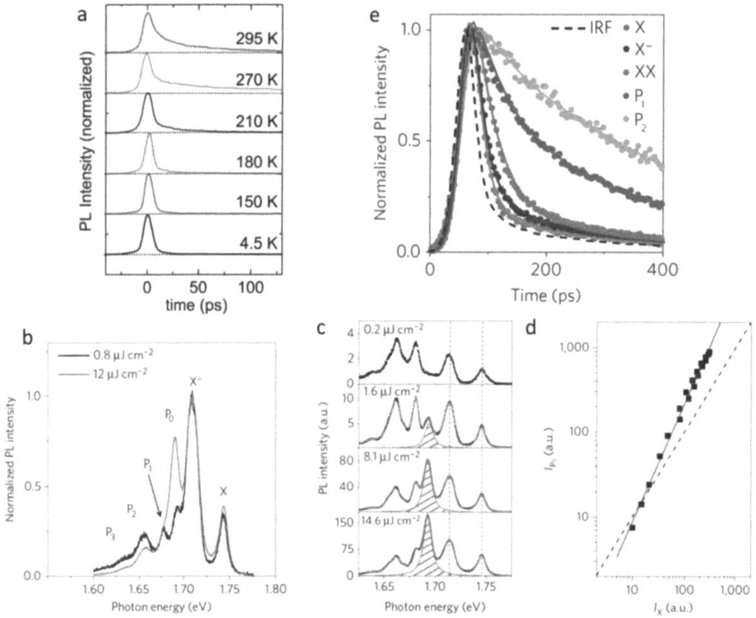

Figure 2.12 (a) TRPL spectra of MoS_2 monolayers measured at different temperatures. Reproduced from Ref. [88], with the permission of AIP Publishing. (b) PL spectra of monolayer WSe_2 measured at 50 K and excited using fluences of 0.8 and 12 μJ cm^{-2}. (c) PL spectra of monolayer WSe_2 at 10 K and excited with different fluences. The identified peaks are the same as (b). Only the P_o peak changes with fluences. (d) P_o emission strength as a function of exciton emission (X peak) strength. (e) TRPL spectra of different peaks labeled in (b). (b–e) Reprinted with permission from Springer Customer Service Centre GmbH: Springer Nature, *Nature Physics*, Ref. [89], Copyright (2015).

2.5.3 Carrier Dynamics in BP

The photo-excitated transient behaviours and dynamical evolution of photo-carriers in BP was studied by means of transient reflection spectroscopy [90]. Benefits from the angle-resolved setup and the dynamics of anisotropic properties in BP are revealed. The conditions in a semiconductor under excitation of an ultra-fast pulse could be used to simulate the working environment of opto-electronic devices operating upon a dynamic field or a high field. The anisotropy of reflectivity of BP is found to be enhanced under the pump-induced quasi-equilibrium state. This indicates that the anisotropy of the transient conductivity is enhanced at the hot-carrier-dominated regime. At a fixed pump polarisation, the probe polarisation-dependent transient reflection spectra are recorded. Obvious polarisation dependence is observed. The sinusoidal relationship between $\Delta R/R$ and probe polarisation at a fixed delay time is extracted from the spectra (Fig. 2.13a). This is attributed to the anisotropic response of the optical conductivity of BP. The change of the optical conductivity, $\Delta\sigma$, could be calculated from the following equation:

$$\Delta R/R = 4\sqrt{\varepsilon_1}\,\Delta\mathrm{Re}\sigma/\varepsilon_o c(\varepsilon_2-\varepsilon_1)+8\sqrt{\varepsilon_1\varepsilon_2}\,(\Delta\mathrm{Im}\sigma)\mathrm{Im}\sigma/\varepsilon_o^2 c^2(\varepsilon_2-\varepsilon_1)^2.$$

The parameter meanings and detailed calculation process can be found in Ref. [90]. After calculation, the dynamical evolution of $\mathrm{Re}(\sigma_{xx})/\mathrm{Re}(\sigma_{yy})$ is shown in Fig. 2.13b. Evidently, the anisotropy is more significant after pumping and recovers to the initial state with increasing time delay. The enhancement is more clearly shown by the dynamic evolution of the $\mathrm{Re}(\sigma)$ ellipse at different delay times (Fig. 2.13c).

2.6 Formidable Interactions between Lasers and 2D Materials

2.6.1 Laser Micro-/NanoManufacturing

Besides the moderate light–matter interactions, formidable interactions would also be triggered in 2D materials when lasers with subversive energy are employed. Techniques based on focused laser beams were designed as a straightforward and effective

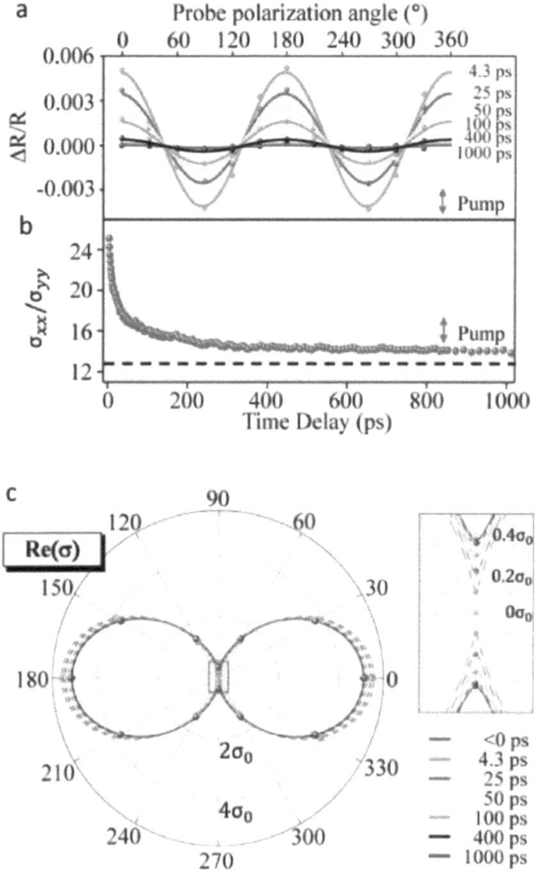

Figure 2.13 (a) $\Delta R/R$ as a function of probe polarisation at different decays. The experimental data is fitted by $a \cos^2 \alpha + b \sin^2 \alpha$. (b) Dynamic evolution of $\text{Re}(\sigma_{xx})/\text{Re}(\sigma_{yy})$. (c) Dynamic evolution of the $\text{Re}(\sigma)$ ellipse at different delay times. Reprinted with permission from Ref. [90]. Copyright (2015) American Chemical Society.

approach to "on demand" fabrication of TMD monolayers [91]. The setup is schematically illustrated in Fig. 2.14a. After scanning the focused laser beam on the multi-layered sample with an optimised power density and scanning speed, a uniform monolayer region could be created. Based on this technique, direct thinning and micro-patterning could be simultaneously achieved. The sample can absorb the photons and converts the energy to thermal energy. The generated heat facilitates the sublimation of the upper layers

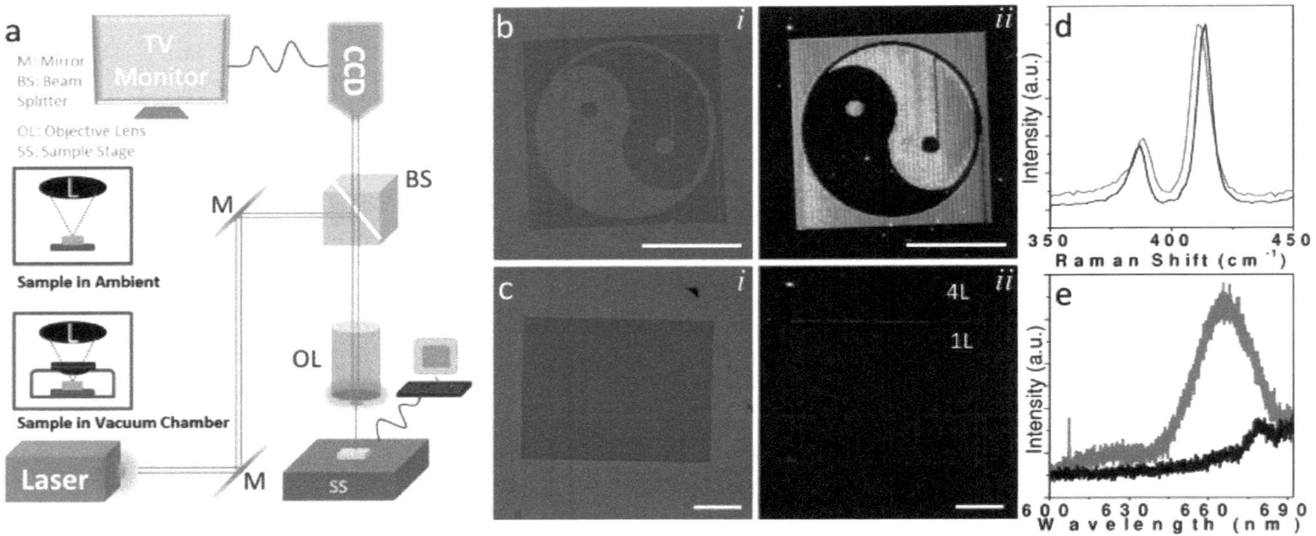

Figure 2.14 (a) Schematic illustration of the laser micro-/nanomanufacturing setup. (b, c) Demonstration of micro-patterning in a MoS_2 film. Scale bars = 20 µm. (d) Raman and (e) PL spectra collected on the as-grown pristine few-layer region and the obtained monolayer region. Reprinted with permission from Ref. [91]. Copyright (2014) American Chemical Society.

to realise direct thinning. The sample is controlled to move with respect to the laser beam in a programmable manner to achieve the patterning (Fig. 2.14b). In such a way, versatile functional junctions and components could be fabricated on a MoS_2 thin film based on layer-dependent bandgaps. For instance, a monolayer domain surrounded by a multi-layer region can be fabricated (Fig. 2.14c). The formation of the monolayer domain is identified by the Raman and PL spectra (Fig. 2.14d,e). This on-demand method has been demonstrated to be applicable to directly thin and other TMD patterns (WSe_2) [92]. In addition to thinning and patterning, this method possesses the capability to modify or alter the property of the samples.

2.6.2 Photo-Chemical Reaction

Nowadays, many ways to synthesise graphene, varying from physical [20] and chemical[93] to exfoliation of graphite to chemical vapour deposition [94], have been developed. One common approach for graphene is the chemical reduction of graphene oxide (GO) which is obtained by treating graphite with strong oxidisers [93]. GO is easy to disperse in organic solvents and water due to the presence of oxygen functionalities such as carboxylic and carbonyl groups; this property provides great convenience when trying to mix the materials with other polymers to improve the electronic and mechanic performances [95, 96]. On the other hand, GO is described as an electrical insulator owing to destruction of the graphene framework. To recover electronic conductivity, a chemical reduction of GO is needed [97]. Howeverthe graphene obtained using the chemical reduction of GO still retains many chemical and physics defects, leading to low performance in some applications.

The thermal instability of GO due to the oxygen functional groups motivates people to use a focused laser beam to do chemical reduction and even micro-structuring [98]. When a focused laser beam is illuminated on the GO film, the energy absorbed from the incident laser will convert to local heat and increase the temperature of the irradiated area; the raised temperature results in localised oxidation of GO to CO or CO_2 if the material is in ambient air [98]. Then the patterned GO can be used for other applications or be subsequently reduced to patterned graphene by thermal annealing at high temperature. However, when focused laser beam illumination happens in an inert atmosphere, such as N_2, the oxidation burning will not happen due to the protection of an inert environment [98]. As a result, the Raman spectra don't show much significant change between the modified and unmodified regions, implying no big change in the lattice structure. However, the laser-modified region displays much improved conductivity of about 1.1 S/m compared to the insulating GO before laser modification [98]. The enhancement in conductivity indicates a reduction of GO to graphene after laser irradiation, originating possibly from the desorption of oxygen functionalities and the reconstruction of the surface [98]. On the other hand, for the reduced GO or graphene layer, the laser patterning doesn't work even in ambient air owing to thermal stability [98].

The oxygen substitution of chalcogen vacancies has been realised using the focused laser beam technique. The substitution process is controlled by the photo-chemical reaction between the WSe_2 monolayers and oxygen in the atmosphere. Isoelectronic substitutional oxygen defects in TMDs are theoretically demonstrated to be electrically neutral [92]. Therefore, passivation of the chalcogen vacancies using oxygen is a feasible approach to improve the quality of TMDs. The degree of substitution could be controlled by altering the oxygen concentration of the environment. X-ray photoelectron spectroscopy (XPS) scans clearly indicate that more oxygen can be introduced into the WSe_2 monolayers in the environment with a higher oxygen concentration (Fig. 2.15a). In the W 4f scans, the two doublets are ascribed to W^{4+} in the W-Se environment and W^{5+}/W^{6+} in the W-O environment. The oxidisation ratio is defined as $I(W^{5+}/W^{6+})/(I(W^{4+}) + I(W^{5+}/W^{6+}))$. The ratio gradually increases when the WSe_2 monolayers are modified by a focused laser beam in an environment with a higher oxygen concentration. The stoichiometric ratio of W^{4+} and Se could be tailored from ~1:1.8 of the pristine sample to be ~1:2.0 after modification of the focused laser beam.

The photo-oxidation of BP via a laser-induced photo-chemical reaction could control the oxidation process and render the oxidation to be favourable. The phosphorene oxides or sub-oxides (P_4O_n) are a protective layer to preserve the phosphorene underneath or a functional material of its own [99, 100]. The crystal structures of representative sub-oxides are shown in Fig. 2.15b. The electronic band structure of phosphorene significantly changes with the degree

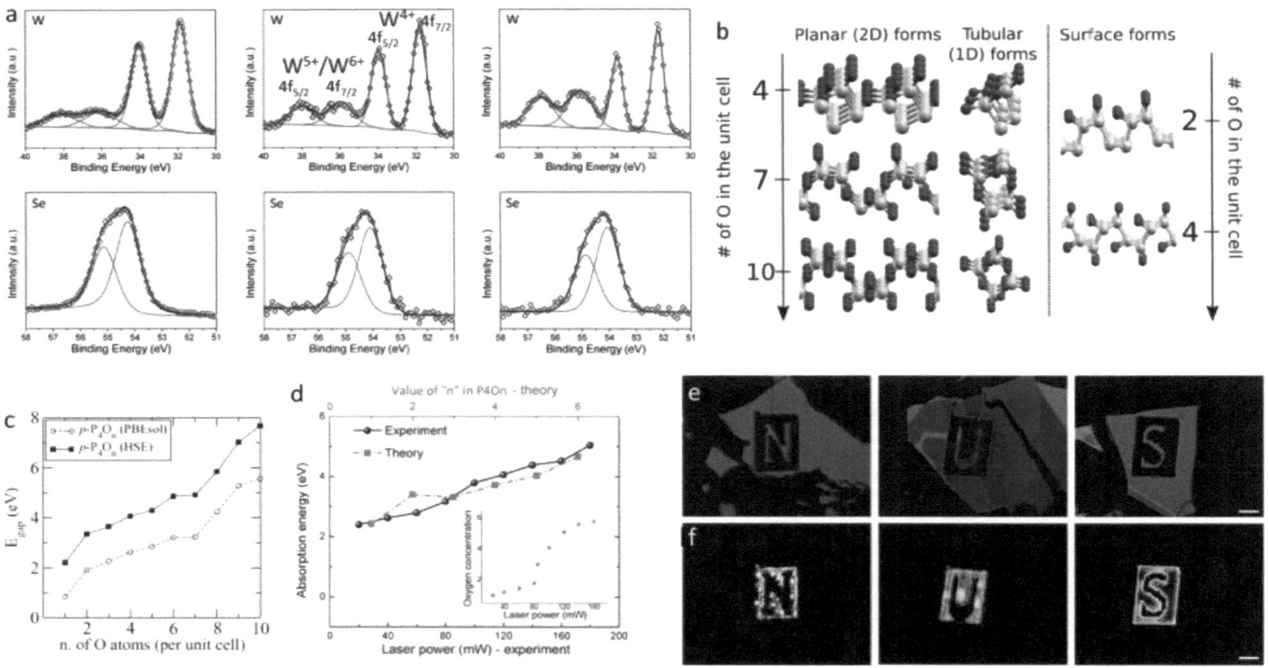

Figure 2.15 (a, left to right) XPS scans of W and Se for a pristine WSe_2 monolayer, sample modified in an environment with a low oxygen concentration and sample modified in an environment with a higher oxygen concentration. Reprinted with permission from Ref. [92]. Copyright (2015) American Chemical Society. (b) Crystal structures of different forms of phosphorene oxides. (c) Calculated bandgap as a function of oxygen concentration. Reprinted (figure) with permission from Ref. [102]. Copyright (2015) by the American Physical Society. (d) Bandgap as a function of laser power (experimental result) and oxygen concentration (theoretical calculation). Inset: Calibration map of oxygen concentration as a function of laser power. (e) Optical and (f) fluorescent images of the micro-patterns on BP. Reprinted with permission from Ref. [100]. Copyright (2015) American Chemical Society.

of oxidation. The bandgap of P_4O_n is predicted to monotonically increase with oxygen concentration (Fig. 2.15c). It changes from ~0.9 eV (n = 1) to ~5.5 eV (n = 10) under the PBEsol calculation method. The wide range of the tunable bandgap extends the spectral regime of the light–matter interactions in phosphorene from visible to deep UV. The broadening of the bandgap arises from the increase of the ionic character of the bonds and the induced localisation of the wave function [101]. Ordered and amorphous domains might be formed depending on the fabrication process. A photo-oxidation technique based on a focused laser beam has been devised to controllably fabricate phosphorene sub-oxides [100]. Upon this technique, sub-oxides can be locally created on the phosphorene with well-defined micro-patterns (Fig. 2.15e). The bandgap of the sub-oxides could be flexibly engineered from ~2.3 eV to ~5.1 eV by controlling the laser power, consistent with the theoretical calculations (Fig. 2.15d). The fabricated sub-oxides are observed to be photo-active in the visible regime and present fluorescence emissions (Fig. 2.15f).

2.7 Conclusion

Light–matter interactions are generally considered as the basis of optoelectronics in condensed matter physics. In this chapter, we provided insight into the optical and photo-electrical properties of 2D materials. Firstly, we presented the electronic band structure of 2D materials with a change in thickness. This feature facilitates strong light–matter interactions in the visible to far-IR frequencies. Subsequently, we described the optical phenomena triggered by light–matter interactions. The description of these phenomena is expanded by discussing the experimentally observed fundamental optical properties, vibrational properties and carrier dynamics. Finally, we introduced the destructive aspect of light–matter interactions in 2D materials, where violent interactions are facilitated by light with high power.

References

1. Novoselov, K. S., Geim, A. K., Morozov, S. V., Jiang, D., Katsnelson, M. I., Grigorieva, I. V., Dubonos, S. V. and Firsov, A. A. (2005). Two-

dimensional gas of massless Dirac fermions in graphene, *Nature*, **438**, pp. 197–200.

2. Castro Neto, A. H., Guinea, F., Peres, N. M. R., Novoselov, K. S. and Geim, A. K. (2009). The electronic properties of graphene, *Rev. Mod. Phys.*, **81**, pp. 109–162.
3. Novoselov, K. S., Jiang, D., Schedin, F., Booth, T. J., Khotkevich, V. V., Morozov, S. V. and Geim, A. K. (2005). Two-dimensional atomic crystals, *Proc. Natl. Acad. Sci. U. S. A.*, **102**, pp. 10451–10453.
4. Wang, Q. H., Kalantar-Zadeh, K., Kis, A., Coleman, J. N. and Strano, M. S. (2012). Electronics and optoelectronics of two-dimensional transition metal dichalcogenides, *Nat. Nanotechnol.*, **7**, pp. 699–712.
5. Li, L., Yu, Y., Ye, G. J., Ge, Q., Ou, X., Wu, H., Feng, D., Chen, X. H. and Zhang, Y. (2014). Black phosphorus field-effect transistors, *Nat. Nanotechnol.*, **9**, pp. 372–377.
6. Liu, H., Neal, A. T., Zhu, Z., Luo, Z., Xu, X., Tománek, D. and Ye, P. D. (2014). Phosphorene: an unexplored 2D semiconductor with a high hole mobility, *ACS Nano*, **8**, pp. 4033–4041.
7. Koenig, S. P., Doganov, R. A., Schmidt, H., Castro Neto, A. H. and Özyilmaz, B. (2014). Electric field effect in ultrathin black phosphorus, *Appl. Phys. Lett.*, **104**, p. 103106.
8. Rodin, A. S., Carvalho, A. and Castro Neto, A. H. (2014). Strain-induced gap modification in black phosphorus, *Phys. Rev. Lett.*, **112**, p. 176801.
9. Qiao, J., Kong, X., Hu, Z.-X., Yang, F. and Ji, W. (2014). High-mobility transport anisotropy and linear dichroism in few-layer black phosphorus, *Nat. Commun.*, **5**, p. 4475.
10. Churchill, H. O. H. and Jarillo-Herrero, P. (2014). Two-dimensional crystals: phosphorus joins the family, *Nat. Nanotechnol.*, **9**, pp. 330–331.
11. Xia, F., Wang, H. and Jia, Y. (2014). Rediscovering black phosphorus as an anisotropic layered material for optoelectronics and electronics, *Nat. Commun.*, **5**, p. 4458.
12. Liu, H., Du, Y., Deng, Y. and Ye, P. D. (2015). Semiconducting black phosphorus: synthesis, transport properties and electronic applications, *Chem. Soc. Rev.*, **44**, pp. 2732–2743.
13. Ling, X., Wang, H., Huang, S., Xia, F. and Dresselhaus, M. S. (2015). The renaissance of black phosphorus, *Proc. Natl. Acad. Sci. U. S. A.*, **112**, pp. 4523–4530.

14. Cao, L., White, J. S., Park, J.-S., Schuller, J. A., Clemens, B. M. and Brongersma, M. L. (2009). Engineering light absorption in semiconductor nanowire devices, *Nat. Mater.*, **8**, pp. 643–647.

15. Pettersson, H., Trägårdh, J., Persson, A. I., Landin, L., Hessman, D. and Samuelson, L. (2006). Infrared photodetectors in heterostructure nanowires, *Nano Lett.*, **6**, pp. 229–232.

16. Soci, C., Zhang, A., Xiang, B., Dayeh, S. A., Aplin, D. P. R., Park, J., Bao, X. Y., Lo, Y. H. and Wang, D. (2007). ZnO nanowire UV photodetectors with high internal gain, *Nano Lett.*, **7**, pp. 1003–1009.

17. Lu, J., Liu, H., Tok, E. S. and Sow, C.-H. (2016). Interactions between lasers and two-dimensional transition metal dichalcogenides, *Chem. Soc. Rev.*, **45**, pp. 2494–2515.

18. Schäffel, F. (2013). The atomic structure of graphene and its few-layer counterparts, in *Graphene* (Elsevier), pp. 5–59.

19. Warner, J. H., Schäffel, F., Bachmatiuk, A. and Rümmeli, M. H. (2013). Properties of graphene, in *Graphene* (Elsevier), pp. 61–127.

20. Novoselov, K. S., Geim, A. K., Morozov, S. V., Jiang, D., Zhang, Y., Dubonos, S. V., Grigorieva, I. V. and Firsov, A. A. (2004). Electric field effect in atomically thin carbon films, *Science*, **306**, pp. 666–669.

21. Haering, R. R. (1958). Band structure of rhombohedral graphite, *Can. J. Phys.*, **36**, pp. 352–362.

22. Wallace, P. R. (1947). The band theory of graphite, *Phys. Rev.*, **71**, pp. 622–634.

23. Morozov, S. V., Novoselov, K. S., Katsnelson, M. I., Schedin, F., Elias, D. C., Jaszczak, J. A. and Geim, A. K. (2008). Giant intrinsic carrier mobilities in graphene and its bilayer, *Phys. Rev. Lett.*, **100**, p. 016602.

24. Latil, S. and Henrard, L. (2006). Charge carriers in few-layer graphene films, *Phys. Rev. Lett.*, **97**, p. 036803.

25. Ohta, T., Bostwick, A., McChesney, J. L., Seyller, T., Horn, K. and Rotenberg, E. (2007). Interlayer interaction and electronic screening in multilayer graphene investigated with angle-resolved photoemission spectroscopy, *Phys. Rev. Lett.*, **98**, p. 206802.

26. Ohta, T., Bostwick, A., Seyller, T., Horn, K. and Rotenberg, E. (2006). Controlling the Electronic structure of bilayer graphene, *Science*, **313**, pp. 951–954.

27. Cao, T., Wang, G., Han, W., Ye, H., Zhu, C., Shi, J., Niu, Q., Tan, P., Wang, E., Liu, B. and Feng, J. (2012). Valley-selective circular dichroism of monolayer molybdenum disulphide, *Nat. Commun.*, **3**, p. 887.

28. Zeng, H. and Cui, X. (2015). An optical spectroscopic study on two-dimensional group-VI transition metal dichalcogenides, *Chem. Soc. Rev.*, **44**, pp. 2629–2642.

29. Mak, K. F., Lee, C., Hone, J., Shan, J. and Heinz, T. F. (2010). Atomically thin MoS_2: a new direct-gap semiconductor, *Phys. Rev. Lett.*, **105**, p. 136805.

30. Splendiani, A., Sun, L., Zhang, Y., Li, T., Kim, J., Chim, C.-Y., Galli, G. and Wang, F. (2010). Emerging photoluminescence in monolayer MoS_2, *Nano Lett.*, **10**, pp. 1271–1275.

31. Zeng, H., Liu, G.-B., Dai, J., Yan, Y., Zhu, B., He, R., Xie, L., Xu, S., Chen, X., Yao, W. and Cui, X. (2013). Optical signature of symmetry variations and spin-valley coupling in atomically thin tungsten dichalcogenides, *Sci. Rep.*, **3**, p. 1608.

32. Schmidt, H., Giustiniano, F. and Eda, G. (2015). Electronic transport properties of transition metal dichalcogenide field-effect devices: surface and interface effects, *Chem. Soc. Rev.*, **44**, pp. 7715-7736.

33. Xiao, D., Liu, G.-B., Feng, W., Xu, X. and Yao, W. (2012). Coupled spin and valley physics in monolayers of MoS_2 and other group-VI dichalcogenides, *Phys. Rev. Lett.*, **108**, p. 196802.

34. Kuc, A., Zibouche, N. and Heine, T. (2011). Influence of quantum confinement on the electronic structure of the transition metal sulfide TS_2, *Phys. Rev. B*, **83**, p. 245213.

35. Jin, W., Yeh, P.-C., Zaki, N., Zhang, D., Sadowski, J. T., Al-Mahboob, A., van der Zande, A. M., Chenet, D. A., Dadap, J. I., Herman, I. P., Sutter, P., Hone, J. and Osgood, R. M. (2013). Direct measurement of the thickness-dependent electronic band structure of MoS_2 using angle-resolved photoemission spectroscopy, *Phys. Rev. Lett.*, **111**, p. 106801.

36. Li, P. and Appelbaum, I. (2014). Electrons and holes in phosphorene, *Phys. Rev. B*, **90**, p. 115439.

37. Tran, V., Soklaski, R., Liang, Y. and Yang, L. (2014). Layer-controlled band gap and anisotropic excitons in few-layer black phosphorus, *Phys. Rev. B*, **89**, p. 235319.

38. Liang, L., Wang, J., Lin, W., Sumpter, B. G., Meunier, V. and Pan, M. (2014). Electronic bandgap and edge reconstruction in phosphorene materials, *Nano Lett.*, **14**, pp. 6400–6406.

39. Lu, J., Carvalho, A., Wu, J., Liu, H., Tok, E. S., Neto, A. H. C., Özyilmaz, B. and Sow, C. H. (2016). Enhanced photoresponse from phosphorene–phosphorene-suboxide junction fashioned by focused laser micromachining, *Adv. Mater.*, **28**, pp. 4090–4096.

40. Low, T., Rodin, A. S., Carvalho, A., Jiang, Y., Wang, H., Xia, F. and Castro Neto, A. H. (2014). Tunable optical properties of multilayer black phosphorus thin films, *Phys. Rev. B*, **90**, p. 075434.

41. Ni, Z., Wang, H., Kasim, J., Fan, H., Yu, T., Wu, Y., Feng, Y. and Shen, Z. (2007). Graphene thickness determination using reflection and contrast spectroscopy, *Nano Lett.*, **7**, pp. 2758–2763.

42. Nair, R. R., Blake, P., Grigorenko, A. N., Novoselov, K. S., Booth, T. J., Stauber, T., Peres, N. M. R. and Geim, A. K. (2008). Fine structure constant defines visual transparency of graphene, *Science*, **320**, pp. 1308–1308.

43. Kim, K. S., Zhao, Y., Jang, H., Lee, S. Y., Kim, J. M., Kim, K. S., Ahn, J.-H., Kim, P., Choi, J.-Y. and Hong, B. H. (2009). Large-scale pattern growth of graphene films for stretchable transparent electrodes, *Nature*, **457**, pp. 706–710.

44. Carvalho, A., Ribeiro, R. M. and Castro Neto, A. H. (2013). Band nesting and the optical response of two-dimensional semiconducting transition metal dichalcogenides, *Phys. Rev. B*, **88**, p. 115205.

45. Eda, G., Yamaguchi, H., Voiry, D., Fujita, T., Chen, M. and Chhowalla, M. (2011). Photoluminescence from chemically exfoliated MoS_2, *Nano Lett.*, **11**, pp. 5111–5116.

46. Kozawa, D., Kumar, R., Carvalho, A., Kumar Amara, K., Zhao, W., Wang, S., Toh, M., Ribeiro, R. M., Castro Neto, A. H., Matsuda, K. and Eda, G. (2014). Photocarrier relaxation pathway in two-dimensional semiconducting transition metal dichalcogenides, *Nat. Commun.*, **5**, p. 4543.

47. Akahama, Y., Endo, S. and Narita, S.-I. (1983). Electrical properties of black phosphorus single crystals, *J. Phys. Soc. Jpn.*, **52**, pp. 2148–2155.

48. Rodin, A. S., Carvalho, A. and Castro Neto, A. H. (2014). Excitons in anisotropic two-dimensional semiconducting crystals, *Phys. Rev. B*, **90**, p. 075429.

49. Lin, J.-H., Zhang, H. and Cheng, X.-L. (2015). First-principle study on the optical response of phosphorene, *Front. Phys.*, **10**, pp. 1–9.

50. Thilagam, A. (1997). Two-dimensional charged-exciton complexes, *Phys. Rev. B*, **55**, pp. 7804–7808.

51. Yang, J., Xu, R., Pei, J., Myint, Y. W., Wang, F., Wang, Z., Zhang, S., Yu, Z. and Lu, Y. (2015). Optical tuning of exciton and trion emissions in monolayer phosphorene, *Light Sci. Appl.*, **4**, p. e312.

52. Xu, R., Zhang, S., Wang, F., Yang, J., Wang, Z., Pei, J., Myint, Y. W., Xing, B., Yu, Z., Fu, L., Qin, Q. and Lu, Y. (2016). Extraordinarily bound quasi-

one-dimensional trions in two-dimensional phosphorene atomic semiconductors, *ACS Nano*, **10**, pp. 2046–2053.

53. Ferrari, A. C., Meyer, J. C., Scardaci, V., Casiraghi, C., Lazzeri, M., Mauri, F., Piscanec, S., Jiang, D., Novoselov, K. S., Roth, S. and Geim, A. K. (2006). Raman spectrum of graphene and graphene layers, *Phys. Rev. Lett.*, **97**, p. 187401.
54. Tuinstra, F. and Koenig, J. L. (1970). Raman spectrum of graphite, *J. Chem. Phys.*, **53**, pp. 1126–1130.
55. Thomsen, C. and Reich, S. (2000). Double resonant Raman scattering in graphite, *Phys. Rev. Lett.*, **85**, p. 5214.
56. Malard, L., Pimenta, M., Dresselhaus, G. and Dresselhaus, M. (2009). Raman spectroscopy in graphene, *Phys. Rep.*, **473**, pp. 51–87.
57. Ferrari, A., Meyer, J., Scardaci, V., Casiraghi, C., Lazzeri, M., Mauri, F., Piscanec, S., Jiang, D., Novoselov, K. and Roth, S. (2006). Raman spectrum of graphene and graphene layers, *Phys. Rev. Lett.*, **97**, p. 187401.
58. Dresselhaus, M. S., Dresselhaus, G., Saito, R. and Jorio, A. (2005). Raman spectroscopy of carbon nanotubes, *Phys. Rep.*, **409**, pp. 47–99.
59. Mafra, D., Samsonidze, G., Malard, L., Elias, D., Brant, J., Plentz, F., Alves, E. and Pimenta, M. (2007). Determination of LA and TO phonon dispersion relations of graphene near the Dirac point by double resonance Raman scattering, *Phys. Rev. B*, **76**, p. 233407.
60. Ferrari, A. C. and Robertson, J. (2004). Raman spectroscopy of amorphous, nanostructured, diamond–like carbon, and nanodiamond, *Philos. Trans. R. Soc. London, Ser. A*, **362**, pp. 2477–2512.
61. Late, D. J., Maitra, U., Panchakarla, L., Waghmare, U. V. and Rao, C. (2011). Temperature effects on the Raman spectra of graphenes: dependence on the number of layers and doping, *J. Phys.: Condens. Matter*, **23**, p. 055303.
62. Ferrari, A. C. (2007). Raman spectroscopy of graphene and graphite: disorder, electron–phonon coupling, doping and nonadiabatic effects, *Solid State Commun.*, **143**, pp. 47–57.
63. Zhang, X., Qiao, X.-F., Shi, W., Wu, J.-B., Jiang, D.-S. and Tan, P.-H. (2015). Phonon and Raman scattering of two-dimensional transition metal dichalcogenides from monolayer, multilayer to bulk material, *Chem. Soc. Rev.*, **44**, pp. 2757–2785.
64. Zhang, X., Han, W. P., Wu, J. B., Milana, S., Lu, Y., Li, Q. Q., Ferrari, A. C. and Tan, P. H. (2013). Raman spectroscopy of shear and layer breathing modes in multilayer MoS_2, *Phys. Rev. B*, **87**, p. 115413.

65. Lee, C., Yan, H., Brus, L. E., Heinz, T. F., Hone, J. and Ryu, S. (2010). Anomalous lattice vibrations of single- and few-layer MoS_2, *ACS Nano*, **4**, pp. 2695–2700.

66. Li, H., Zhang, Q., Yap, C. C. R., Tay, B. K., Edwin, T. H. T., Olivier, A. and Baillargeat, D. (2012). From bulk to monolayer MoS_2: evolution of Raman scattering, *Adv. Funct. Mater.*, **22**, pp. 1385–1390.

67. Tonndorf, P., Schmidt, R., Böttger, P., Zhang, X., Börner, J., Liebig, A., Albrecht, M., Kloc, C., Gordan, O., Zahn, D. R. T., Michaelis de Vasconcellos, S. and Bratschitsch, R. (2013). Photoluminescence emission and Raman response of monolayer MoS_2, $MoSe_2$, and WSe_2, *Opt. Express*, **21**, pp. 4908–4916.

68. Zhao, W., Ghorannevis, Z., Amara, K. K., Pang, J. R., Toh, M., Zhang, X., Kloc, C., Tan, P. H. and Eda, G. (2013). Lattice dynamics in mono- and few-layer sheets of WS_2 and WSe_2, *Nanoscale*, **5**, pp. 9677–9683.

69. Zeng, H., Zhu, B., Liu, K., Fan, J., Cui, X. and Zhang, Q. M. (2012). Low-frequency Raman modes and electronic excitations in atomically thin MoS_2 films, *Phys. Rev. B*, **86**, p. 241301.

70. Zhao, Y., Luo, X., Li, H., Zhang, J., Araujo, P. T., Gan, C. K., Wu, J., Zhang, H., Quek, S. Y., Dresselhaus, M. S. and Xiong, Q. (2013). Interlayer breathing and shear modes in few-trilayer MoS_2 and WSe_2, *Nano Lett.*, **13**, pp. 1007–1015.

71. Sugai, S., Ueda, T. and Murase, K. (1981). Pressure dependence of the lattice vibration in the orthorhombic and rhombohedral structures of black phosphorus, *J. Phys. Soc. Jpn.*, **50**, pp. 3356–3361.

72. Lu, W., Nan, H., Hong, J., Chen, Y., Zhu, C., Liang, Z., Ma, X., Ni, Z., Jin, C. and Zhang, Z. (2014). Plasma-assisted fabrication of monolayer phosphorene and its Raman characterization, *Nano Res.*, **7**, pp. 853–859.

73. Favron, A., Gaufres, E., Fossard, F., Phaneuf-Lheureux, A.-L., Tang, N. Y. W., Levesque, P. L., Loiseau, A., Leonelli, R., Francoeur, S. and Martel, R. (2015). Photooxidation and quantum confinement effects in exfoliated black phosphorus, *Nat. Mater.*, **14**, pp. 826–832.

74. Zhang, S., Yang, J., Xu, R., Wang, F., Li, W., Ghufran, M., Zhang, Y.-W., Yu, Z., Zhang, G., Qin, Q. and Lu, Y. (2014). Extraordinary photoluminescence and strong temperature/angle-dependent Raman responses in few-layer phosphorene, *ACS Nano*, **8**, pp. 9590–9596.

75. Wu, J., Koon, G. K. W., Xiang, D., Han, C., Toh, C. T., Kulkarni, E. S., Verzhbitskiy, I., Carvalho, A., Rodin, A. S., Koenig, S. P., Eda, G., Chen, W., Neto, A. H. C. and Özyilmaz, B. (2015). Colossal ultraviolet

photoresponsivity of few-layer black phosphorus, *ACS Nano*, **9**, pp. 8070–8077.

76. Titova, L. V., Hoang, T. B., Jackson, H. E., Smith, L. M., Yarrison-Rice, J. M., Kim, Y., Joyce, H. J., Tan, H. H. and Jagadish, C. (2006). Temperature dependence of photoluminescence from single core-shell GaAs--AlGaAs nanowires, *Appl. Phys. Lett.*, **89**, p. 173126.

77. Dawlaty, J. M., Shivaraman, S., Chandrashekhar, M., Rana, F. and Spencer, M. G. (2008). Measurement of ultrafast carrier dynamics in epitaxial graphene, *Appl. Phys. Lett.*, **92**, p. 042116.

78. Sun, D., Wu, Z.-K., Divin, C., Li, X., Berger, C., de Heer, W. A., First, P. N. and Norris, T. B. (2008). Ultrafast relaxation of excited Dirac fermions in epitaxial graphene using optical differential transmission spectroscopy, *Phys. Rev. Lett.*, **101**, p. 157402.

79. Breusing, M., Ropers, C. and Elsaesser, T. (2009). Ultrafast carrier dynamics in graphite, *Phys. Rev. Lett.*, **102**, p. 086809.

80. George, P. A., Strait, J., Dawlaty, J., Shivaraman, S., Chandrashekhar, M., Rana, F. and Spencer, M. G. (2008). Ultrafast optical-pump terahertz-probe spectroscopy of the carrier relaxation and recombination dynamics in epitaxial graphene, *Nano Lett.*, **8**, pp. 4248–4251.

81. Choi, H., Borondics, F., Siegel, D. A., Zhou, S. Y., Martin, M. C., Lanzara, A. and Kaindl, R. A. (2009). Broadband electromagnetic response and ultrafast dynamics of few-layer epitaxial graphene, *Appl. Phys. Lett.*, **94**, p. 172102.

82. Huang, L., Hartland, G. V., Chu, L.-Q., Feenstra, R. M., Lian, C., Tahy, K. and Xing, H. (2010). Ultrafast transient absorption microscopy studies of carrier dynamics in epitaxial graphene, *Nano Lett.*, **10**, pp. 1308–1313.

83. Johannsen, J. C., Ulstrup, S., Cilento, F., Crepaldi, A., Zacchigna, M., Cacho, C., Turcu, I. E., Springate, E., Fromm, F. and Raidel, C. (2013). Direct view of hot carrier dynamics in graphene, *Phys. Rev. Lett.*, **111**, p. 027403.

84. Berger, C., Song, Z., Li, X., Wu, X., Brown, N., Naud, C., Mayou, D., Li, T., Hass, J. and Marchenkov, A. N. (2006). Electronic confinement and coherence in patterned epitaxial graphene, *Science*, **312**, pp. 1191–1196.

85. Hass, J., Varchon, F., Millán-Otoya, J., Sprinkle, M., de Heer, W., Berger, C., First, P., Magaud, L. and Conrad, E. (2007). Rotational stacking and its electronic effects on graphene films grown on 4H-SiC (000-1), arXiv:0706.2134.

86. Moos, G., Gahl, C., Fasel, R., Wolf, M. and Hertel, T. (2001). Anisotropy of quasiparticle lifetimes and the role of disorder in graphite from ultrafast time-resolved photoemission spectroscopy, *Phys. Rev. Lett.*, **87**, p. 267402.

87. Kampfrath, T., Perfetti, L., Schapper, F., Frischkorn, C. and Wolf, M. (2005). Strongly coupled optical phonons in the ultrafast dynamics of the electronic energy and current relaxation in graphite, *Phys. Rev. Lett.*, **95**, p. 187403.

88. Korn, T., Heydrich, S., Hirmer, M., Schmutzler, J. and Schüller, C. (2011). Low-temperature photocarrier dynamics in monolayer MoS_2, *Appl. Phys. Lett.*, **99**, p. 102109.

89. You, Y., Zhang, X.-X., Berkelbach, T. C., Hybertsen, M. S., Reichman, D. R. and Heinz, T. F. (2015). Observation of biexcitons in monolayer WSe_2, *Nat. Phys.*, **11**, pp. 477–481.

90. Ge, S., Li, C., Zhang, Z., Zhang, C., Zhang, Y., Qiu, J., Wang, Q., Liu, J., Jia, S., Feng, J. and Sun, D. (2015). Dynamical evolution of anisotropic response in black phosphorus under ultrafast photoexcitation, *Nano Lett.*, **15**, pp. 4650–4656.

91. Lu, J., Lu, J. H., Liu, H., Liu, B., Chan, K. X., Lin, J., Chen, W., Loh, K. P. and Sow, C. H. (2014). Improved photoelectrical properties of MoS_2 films after laser micromachining, *ACS Nano*, **8**, pp. 6334–6343.

92. Lu, J., Carvalho, A., Chan, X. K., Liu, H., Liu, B., Tok, E. S., Loh, K. P., Castro Neto, A. H. and Sow, C. H. (2015). Atomic healing of defects in transition metal dichalcogenides, *Nano Lett.*, **15**, pp. 3524–3532.

93. Park, S. and Ruoff, R. S. (2009). Chemical methods for the production of graphenes, *Nat. Nanotechnol.*, **4**, pp. 217–224.

94. Reina, A., Jia, X., Ho, J., Nezich, D., Son, H., Bulovic, V., Dresselhaus, M. S. and Kong, J. (2008). Large area, few-layer graphene films on arbitrary substrates by chemical vapor deposition, *Nano Lett.*, **9**, pp. 30–35.

95. Stankovich, S., Dikin, D. A., Piner, R. D., Kohlhaas, K. A., Kleinhammes, A., Jia, Y., Wu, Y., Nguyen, S. T. and Ruoff, R. S. (2007). Synthesis of graphene-based nanosheets via chemical reduction of exfoliated graphite oxide, *Carbon*, **45**, pp. 1558–1565.

96. Li, D., Muller, M. B., Gilje, S., Kaner, R. B. and Wallace, G. G. (2008). Processable aqueous dispersions of graphene nanosheets, *Nat. Nanotechnol.*, **3**, pp. 101–105.

97. Park, S. and Ruoff, R. S. (2009). Chemical methods for the production of graphenes, *Nat. Nanotechnol.*, **4**, pp. 217–224.

98. Zhou, Y., Bao, Q., Varghese, B., Tang, L. A. L., Tan, C. K., Sow, C. H. and Loh, K. P. (2010). Microstructuring of graphene oxide nanosheets using direct laser writing, *Adv. Mater.*, **22**, pp. 67–71.

99. Edmonds, M. T., Tadich, A., Carvalho, A., Ziletti, A., O'Donnell, K. M., Koenig, S. P., Coker, D. F., Özyilmaz, B., Neto, A. H. C. and Fuhrer, M. S. (2015). Creating a stable oxide at the surface of black phosphorus, *ACS Appl. Mater. Interfaces*, **7**, pp. 14557–14562.

100. Lu, J., Wu, J., Carvalho, A., Ziletti, A., Liu, H., Tan, J., Chen, Y., Castro Neto, A. H., Özyilmaz, B. and Sow, C. H. (2015). Bandgap engineering of phosphorene by laser oxidation toward functional 2D materials, *ACS Nano*, **9**, pp. 10411–10421.

101. Ziletti, A., Carvalho, A., Campbell, D. K., Coker, D. F. and Castro Neto, A. H. (2015). Oxygen defects in phosphorene, *Phys. Rev. Lett.*, **114**, p. 046801.

102. Ziletti, A., Carvalho, A., Trevisanutto, P. E., Campbell, D. K., Coker, D. F. and Castro Neto, A. H. (2015). Phosphorene oxides: bandgap engineering of phosphorene by oxidation, *Phys. Rev. B*, **91**, p. 085407.

Chapter 3

Atomically Thin Optical Lenses and Gratings

Jiong Yang[a,b] **and Yuerui Lu**[a,c]
[a]*College of Chemistry and Environmental Engineering, Shenzhen University, Shenzhen 518060, China*
[b]*Key Laboratory of Optoelectronic Devices and Systems of Ministry of Education and Guangdong Province, College of Optoelectronic Engineering, Shenzhen University, Shenzhen 518060, China*
[c]*Research School of Engineering, College of Engineering and Computer Science, Australian National University, Canberra ACT 2601, Australia*
yuerui.lu@anu.edu.au; yrlu@szu.edu.cn

3.1 Introduction

Two-dimensional (2D) layered materials have emerged as promising candidates for miniaturisation of optoelectronic devices, such as photo-detectors [1, 2], light emitters [3, 4], absorbers [5, 6], edge non-linear optics [7] and valleytronic devices [8, 9], due to their strong inelastic interactions with light [10, 11]. On the other hand, a miniaturised optical system also requires strong elastic light–matter interactions to control the flow of light [12]. Unfortunately,

Two-Dimensional Materials in Nanophotonics: Developments, Devices, and Applications
Edited by Yuerui Lu

ISBN 978-981-4800-22-8 (Hardcover), 978-0-429-42877-7 (eBook)
www.jennystanford.com

elastic interactions in 2D materials are normally weak because their structures are atomically thin. Considerable effort has been devoted to this issue, but success has been only achieved in the mid- to far-infrared where the plasmonic resonance in graphene can enhance the elastic optical response [13–15]. It remains a great challenge to manipulate the flow of light using atomically thin 2D materials in the important visible and near-infrared spectral regions where 2D materials have most interesting optoelectronic properties.

In this chapter, we will report giant optical path length (OPL) from thin-layer molybdenum disulphide (MoS_2) originated from their high refractive indices. Using such giant OPL to engineer the phase front of optical beams, we will present an atomically thin optical lens consisting of a few layers of MoS_2 less than 6.3 nm thick. Moreover, by taking advantage of the giant elastic scattering efficiency in ultra-thin high-index 2D materials, we will demonstrate high-efficiency gratings based on thin-layer MoS_2.

3.2 Giant Optical Path Lengths

Refractive optical components rely on the OPL to modify the phase front of an optical beam. The OPL is directly related to the geometrical length of the light path. As a result, it is normally expected that the OPL of an ultra-thin 2D material would be too small to have a significant impact on the phase front because of its ultra-thin thicknesses. In this section, we will show a giant OPL of 38 nm from single-layer (1L) MoS_2, which is more than 50 times larger than its physical thickness of 0.67 nm and around 1 order of magnitude larger than the measured OPL of 1L graphene which is measured to be only 4.4 nm, as shown in Fig. 3.1 [16].

In our experiments, 1L to four-layer (4L) MoS_2 flakes on a SiO_2/Si (275 nm thermal oxide on n+-doped silicon) substrate are firstly identified by their optical contrast in an optical microscope. Regions with different colours correspond to MoS_2 flakes with different thicknesses, as illustrated in Fig. 3.1a. Due to their high refractive index, these atomically thin MoS_2 layers have a significant and layer-dependent OPL and this enables their layer numbers to be easily identified by a phase-shifting interferometer (PSI), as shown in Fig. 3.1b,c. The PSI can measure the vertical OPL to an accuracy of around 0.1 nm by analysing the digitised interference pattern obtained during a well-controlled phase shift. The measured OPL of a MoS_2 flake on a SiO_2/Si substrate is determined by

$$OPL_{\mathrm{MoS_2}} = -\frac{\lambda}{2\pi}(\phi_{\mathrm{MoS_2}} - \phi_{\mathrm{SiO_2}}), \qquad (3.1)$$

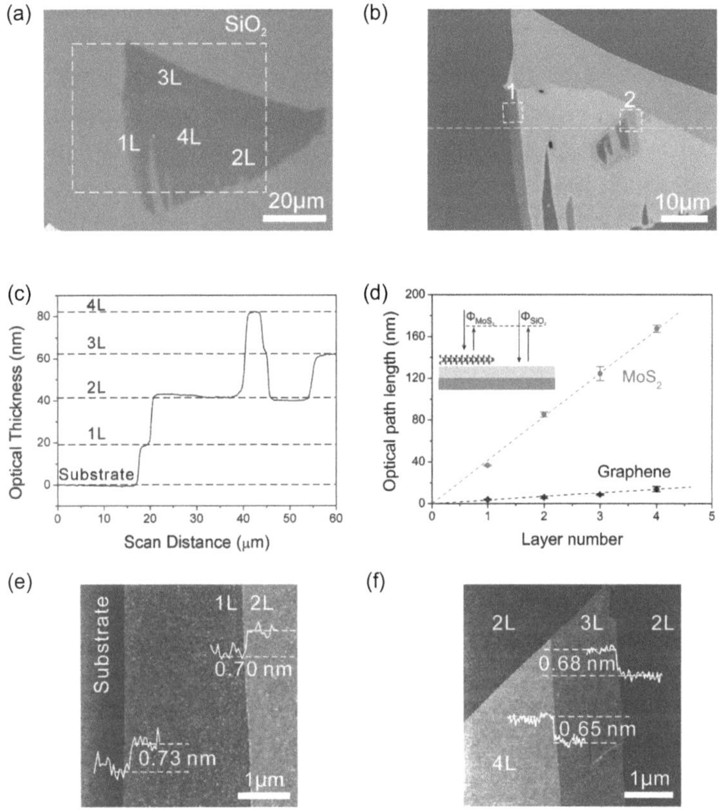

Figure 3.1 Giant OPLs from 1L to 4L MoS_2. (a) Optical microscope image of a mechanically exfoliated MoS_2 sample on a SiO_2/Si substrate. (b) PSI image of the dashed line–enclosed area in (a). (c) PSI-measured OPL versus position for 1L to 4L MoS_2 along the dashed line in (b). (d) Statistical data of the OPL from the PSI for 1L to 4L MoS_2 and graphene samples. Inset: Schematic plot showing the PSI-measured phase shifts of the reflected light from the MoS_2 flake and the SiO_2 substrate. (e, f) AFM image of 1L and 2L, 3L and 4L MoS_2 from dashed line–enclosed area 1 and 2, respectively, in (b). Reprinted with permission from Ref. [16]. Copyright © 2016, Springer Nature.

where λ is the wavelength of the light source and is 535 nm in the experiment, and φ_{MoS2} and φ_{SiO2} are the PSI-measured phase shifts of the light reflected from the MoS_2 flake and the SiO_2 substrate, as shown in the inset of Fig. 3.1d. Multiple samples are characterised and the obtained statistical values of the OPL for 1L to 4L MoS_2 samples are shown in Fig. 3.1d. For each number of layers of MoS_2, at least five different samples are characterised in these statistical measurements. The layer number can be quickly determined by the measured layer-dependent OPL, and the deduced layer number can be further confirmed by corresponding atomic force microscope (AFM) images, as shown in Fig. 3.1e,f. For comparison, we have also performed the same characterisations on mechanically exfoliated graphene samples, and their statistical OPLs are presented in Fig. 3.1d. The measured average and standard deviation error of the OPL from 1L to 4L MoS_2 samples are (38.0 ± 2.8) nm, (85.4 ± 2.2) nm, (124.0 ± 6.6) nm and (162.6 ± 9.0) nm, respectively, while those from 1L to 4L graphene samples are (4.4 ± 0.8) nm, (8.2 ± 2.0) nm, (13.0 ± 3.2) nm and (17.2 ± 3.6) nm, respectively, indicating that MoS_2 has an OPL per layer approximately 1 order of magnitude larger than graphene.

3.3 Origin of the Giant Optical Path Length

3.3.1 High Refractive Index–Enabled Giant Optical Path Length

After careful consideration, we conclude that this giant OPL of thin-layer MoS_2 is created by relatively strong multiple reflections at the air–MoS_2 and MoS_2–SiO_2 interfaces enabled by the high refractive index of thin-layer MoS_2. We consider a simple interface between air and SiO_2, each occupying half infinite space. A layer of 2D material with a real refractive index n_{high} and n_{low} is placed in between the two media, as shown in Fig. 3.2a,b. The high impedance mismatch at these interfaces leads to large reflection coefficients, which will cause strong multiple reflections of light in the 2D material. More reflection rounds of light will happen in the 2D semiconductor with higher refractive index than those with lower refractive index. The amplitude of the reflected light is the summation of the multiple reflections off the interfaces of the thin high-index layer R_i, where i indicates the number of round trips in the 2D material. As the index increases so does the reflectivity of the interfaces, which increases the effective number of transits of the light through the high-index layer and thus the OPL of the reflected light, as illustrated in Fig. 3.2a. The OPL is low for low-index 2D materials, where the small reflection coefficients cause R_i to be small. This situation is illustrated schematically in Fig. 3.2b. We verify this intuition with a numerical calculation, as shown by the dashed line in Fig. 3.2c. Additionally, in the experiment, a Si substrate with a layer of 275 nm thermal SiO_2 on its surface forms a weak Fabry–Perot resonance. As a result of this weak resonant enhancement, the OPL is further enhanced by a factor of around 1.5, as shown by the solid line in Fig. 3.2c. Figure 3.2c also shows the OPL for a few other materials. The OPL of high-index 2D materials, such as MoS_2, is remarkably larger than that of SiO_2, graphene, Au or Si. The wavelength used for these calculations is 535 nm, same as the wavelength of the light source of our PSI system. The refractive indices used for MoS_2 [17], Si, grapheme [18], SiO_2 and Au are $5.3 + 1.3i$, $4.15 + 0.0439i$, $2.6 + 1.3i$, 1.46 and $0.467 + 2.4i$, respectively.

Even more remarkably, from our measured data presented in Fig. 3.1d and calculated results presented in Fig. 3.2d, the OPL of 1L to 4L MoS_2 scales almost linearly with the layer number, offering an exciting opportunity of controlling the OPL using thin-layer MoS_2. When the layer thickness increases by 1 nm, the OPL increases by over 50 nm. Such a rapid change of OPL with thickness allows us to control the phase front of an optical beam very effectively using only an atomically thin structure. The theoretical and numerical predictions in Fig. 3.2d are well-supported by the experimental data, as shown in Fig. 3.1d. The calculated OPL–layer number relationship in Fig. 3.2d deviates slightly from the linear relation obtained in experiments, as shown in Fig. 3.1d, because the refractive indices for different layers are expected to be slightly different, while constant index values are used for all simulations.

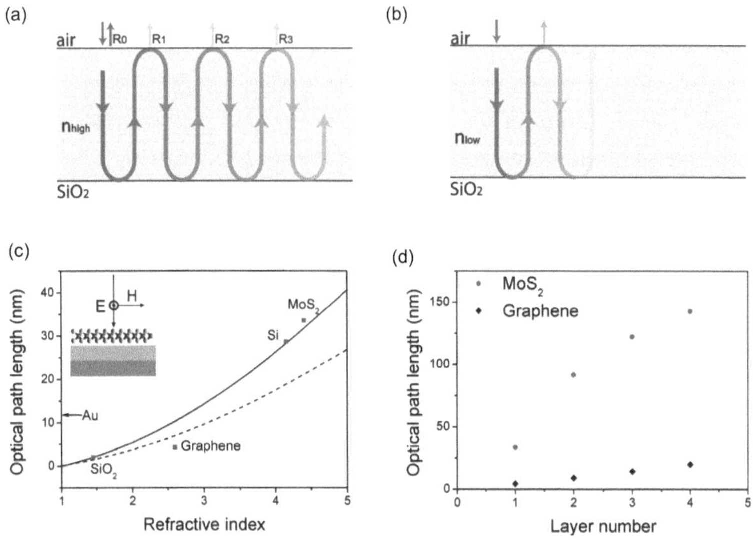

Figure 3.2 High refractive index–enabled giant OPL in ultra-thin film. (a, b) Schematic plots of multiple reflections at the interfaces of ultra-thin 2D materials with a high and a low refractive index, respectively. (c) Simulated OPL for light reflected from a 2D material (0.67 nm in thickness) with different indices on a SiO_2 (275 nm)/Si substrate (solid line) and a SiO_2 substrate with infinite thickness (dashed line). The calculated OPLs of 0.67 nm Au, 0.67 nm SiO_2, 1L (0.34 nm) graphene and 1L (0.67 nm) MoS_2 are represented by markers. (d) Simulated OPL for 1L to 4L MoS_2 and graphene on a SiO_2 (275 nm)/Si substrate, respectively. Reprinted with permission from Ref. [16]. Copyright © 2016, Springer Nature.

3.3.2 Bandwidth of the Giant Optical Path Length

It should be noted that the giant OPL is not a narrow-band effect. The calculated OPL for 1L MoS_2 is above 20 nm at the wavelength ranging from 450 nm to 560 nm with a bandwidth of over 100 nm, as indicated in Fig. 3.3. In the calculation, the wavelength-dependent dielectric constant is obtained from reference. We can adjust the thickness of the SiO_2 layer to shift this operational bandwidth to any desired wavelength range on the basis of requirements in actual applications, and the bandwidth of the giant OPL will remain wide.

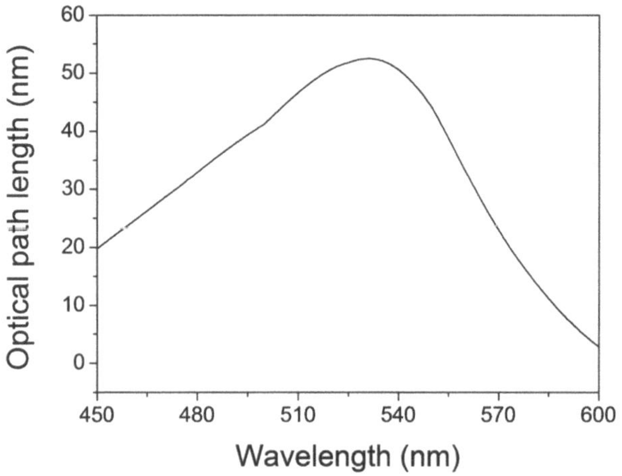

Figure 3.3 Calculated OPL as a function of wavelength for 1L MoS_2 on a SiO_2/Si (275 nm thermal oxide on n^+-doped silicon) substrate. Reprinted with permission from Ref. [16]. Copyright © 2016, Springer Nature.

3.4 Atomically Thin Micro-Lens

In this section, we will demonstrate phase-front engineering by fabricating an atomically thin optical lens based on a few atomic layers of MoS_2. We will start by introducing the fabrication process with a focused gallium ion beam. With the help of the measured OPL versus layer number for thin-layer MoS_2, the focused ion beam (FIB) parameters used in experiments can be well-designed. And then on the basis of the PSI-measured OPL curves along the diameter of the FIB-milled circle, we will calculate the focal length of the fabricated MoS_2 micro-lens. After that, a far-field scanning optical microscope (SOM) system will be introduced to characterise the fabricated MoS_2 micro-lens. A clear focusing effect can be observed from our MoS_2 micro-lens comparing with a planar SiO_2/Si substrate, and the measured focal length matches well with the calculated value.

3.4.1 Fabrication of a MoS_2 Micro-Lens

The fabrication process starts with a flake of uniform nine-layer (9L) MoS_2. Its image taken by the PSI system is shown in Fig. 3.4a, and the

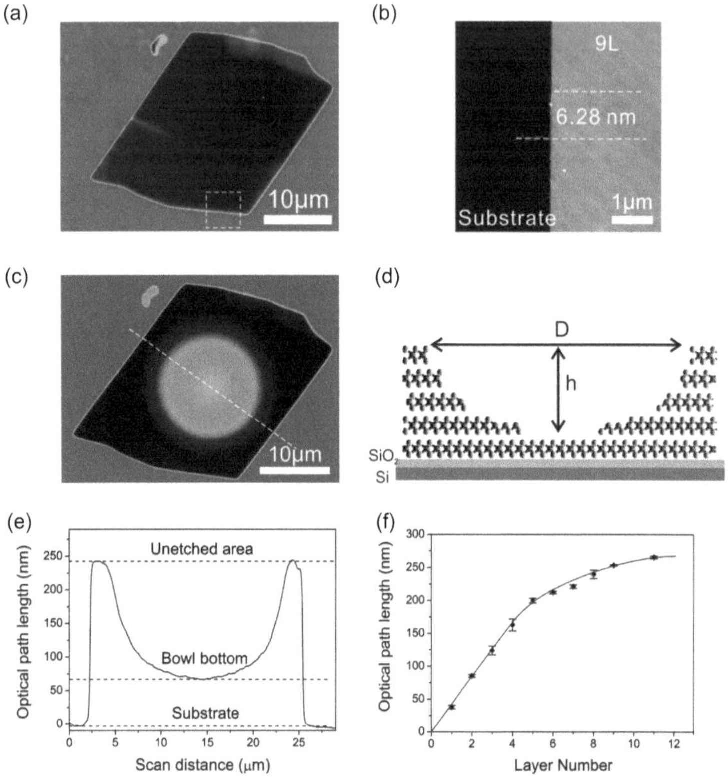

Figure 3.4 Fabrication of atomically thin MoS_2 micro-lens. (a) PSI image of the 9L MoS_2 flake before FIB milling. (b) AFM image of dashed line–enclosed area in (a), showing a thickness of 6.28 nm. (c) PSI image of the same MoS_2 flake after FIB milling. (d) Schematic plot of the atomically thin MoS_2 optical lens. (e) PSI profile of the MoS_2 optical lens along the dashed line in (c). (f) OPL vs. layer number for thin-layer MoS_2. Reprinted with permission from Ref. [16]. Copyright © 2016, Springer Nature.

AFM image of the dashed line–enclosed area in Fig. 3.4a is displayed in Fig. 3.4b, showing a thickness of 6.28 nm. And then a focused gallium ion beam is used in a FEI SEM/FIB dual-beam system to mill a pre-designed bowl-shape structure (20 μm in diameter) into the flake. The FIB in this system has an atomic resolution in the vertical direction and sub-20 nm resolution in the lateral direction. After FIB milling, the image taken with the PSI system is shown in Fig. 3.4c. By comparing Figs. 3.4a and 3.4c, it is clear to see the gradual colour change from outside to inside of the milled area on a MoS_2 flake, corresponding to the gradual thickness change of MoS_2

in the milled area. Figure 3.4d displays a schematic plot of a side view of the atomically thin MoS_2 micro-lens. The gradual change of MoS_2 thickness, from the edge to the centre, leads to a continuous and curved OPL profile for an incident beam, and it serves as an atomically thin reflective concave micro-lens, as shown in Fig. 3.4e. To realise the precise design for this MoS_2 micro-lens in the milling process, the statistical calibration curve measured from experiments between the OPL of MoS_2 flakes and their layer numbers is used, which is shown in Fig. 3.4f. When the layer number is less than 5, the OPL increases with layer number but with a saturation manner when the layer number is more than 10 layers.

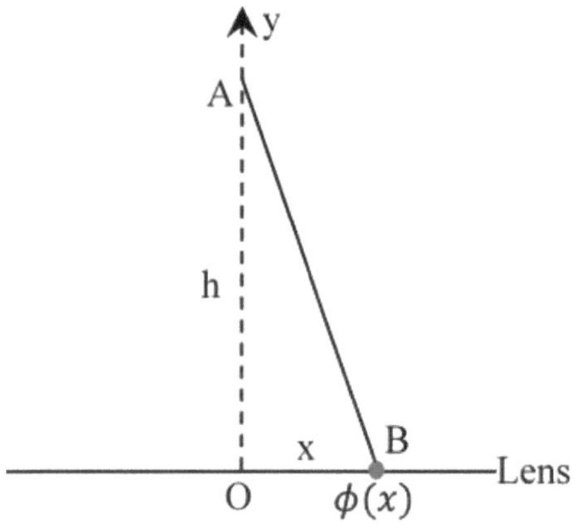

Figure 3.5 Schematic plot of the focal length calculation for an atomically thin MoS_2 micro-lens. Reprinted with permission from Ref. [16]. Copyright © 2016, Springer Nature.

Figure 3.5 shows the schematic cross section of the MoS_2 micro-lens. The phase shift $\phi(x)$ is impinged on the x axis, where point O is the centre of the lens. The focal length of the micro-lens can be extracted on the basis of the phase shift $\phi(x)$ profile measured in experiment. We can evaluate the interference of light along the perpendicular axis y. For each point on the lens plane, the phase shifts can be calculated, including the propagation phase and the phase induced by the MoS_2 micro-lens. For example, for a point B, point A receives light from point B with a phase as $e^{i\phi(x)}e^{-i2\pi(h^2+x^2)/\lambda}$. The minus sign in the exponential term is because the phase-shift profile produces a virtual focus. We can find the focal length by evaluating the maximum value of the following term:

$$I(y)=\int_0^r dx\, e^{i\phi(x)}e^{-i2\pi(h^2+x^2)/\lambda}, \tag{3.2}$$

where r is the radius of the lens. Using the experimentally measured OPL profile, as shown in Fig. 3.4e, the focal length of the MoS_2 micro-lens is calculated to be –248 μm.

3.4.2 Characterisation of a MoS_2 Micro-Lens

A far-field SOM is used to characterise the fabricated MoS_2 micro-lens, as shown in Fig. 3.6. The light path of the Horiba Jobin Yvon T64000 micro-Raman system is exploited to set up the far-field SOM. The far-field SOM uses a green laser at a wavelength of 532 nm, which is focused by an Olympus 10X (numerical aperture [NA] = 0.25; depth of focus = 18 μm) objective lens. In the experiment, the sample can be moved along the z axis in steps of 10 μm by a piezo-electrically-driven stage. If the sample is placed on the focal plane of the objective, light coming from the focal spot on the sample will converge into a light spot at the camera. In these conditions the light path is indicated by the solid line in Fig. 3.6. Otherwise, if the sample is out of the focal plane, the light coming from the sample will be either over-focused into a spot in front of the camera or less-focused, and leave a halo at the camera, as indicated by two sets of dashed lines in Fig. 3.6. The setup offers good collection efficiency for the light coming from (or effectively coming from) a small volume around the focal plane.

The micro-lens is moved along the z axis in steps of 10 μm by a piezo-electrically-driven stage, and the camera records a series of the intensity distributions with the MoS_2 micro-lens positioned at different z values, as illustrated in Fig. 3.7. Our MoS_2 micro-lens is effectively a reflective concave lens, and we can use following lens formula to characterise the relationship between the focal length f, objective distance f_1 and image distance f_2:

$$\frac{1}{f}=\frac{1}{f_1}+\frac{1}{f_2} \tag{3.3}$$

For this MoS_2 micro-lens, the focal length f has a negative value. In our experimental characterisation, f_1 has a negative value and its absolute value is always the distance between the focal plane and the lens, d. With $f_1 = -d$, Eq. 3.3 can be rewritten as

$$\frac{1}{f_2} = \frac{1}{d} - \frac{1}{|f|} \tag{3.4}$$

And on the basis of Eq. 3.4, we can determine f_2 and the corresponding light paths in five cases.

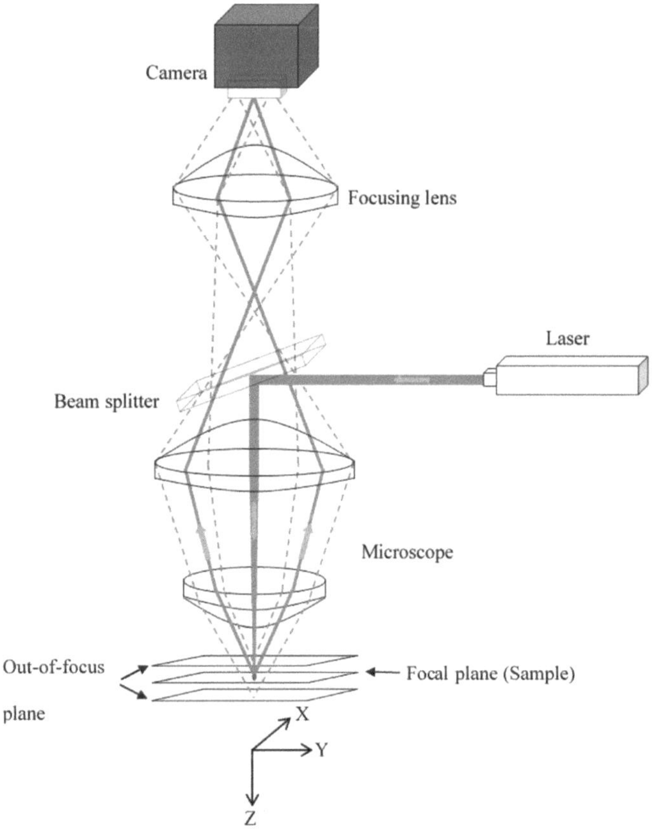

Figure 3.6 Schematic plot of a far-field scanning optical microscope. Reprinted with permission from Ref. [16]. Copyright © 2016, Springer Nature.

When $0 < d < |f|$, from Eq. 3.4 we can get that $f_2 > 0$, which means that a real light spot will be formed on the upper side of the MoS_2 micro-lens, as indicated in Fig. 3.7a, left panel. Under this condition, the camera will observe a circular disk pattern which is effectively coming from the focal plane. The observed intensity pattern distribution by camera at $z = -120$ μm is shown in Fig. 3.7a, right panel.

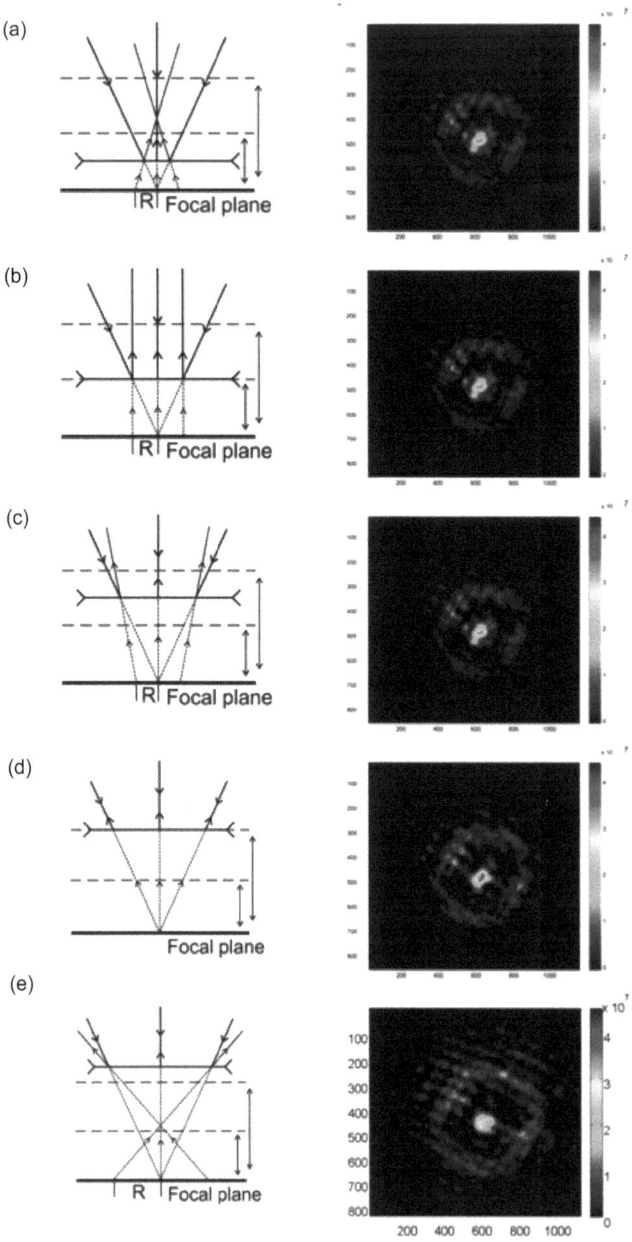

Figure 3.7 Characterisation of the atomically thin MoS_2 micro-lens. (a–e) Schematic plots of the light path (left) and the corresponding recorded intensity distribution pattern images (right) by a camera in SOM, when (a) $d < |f|$, (b) $d = |f|$, (c) $|f| < d < 2|f|$, (d) $d = 2|f|$ and (e) $d > 2|f|$. Reprinted with permission from Ref. [16]. Copyright © 2016, Springer Nature.

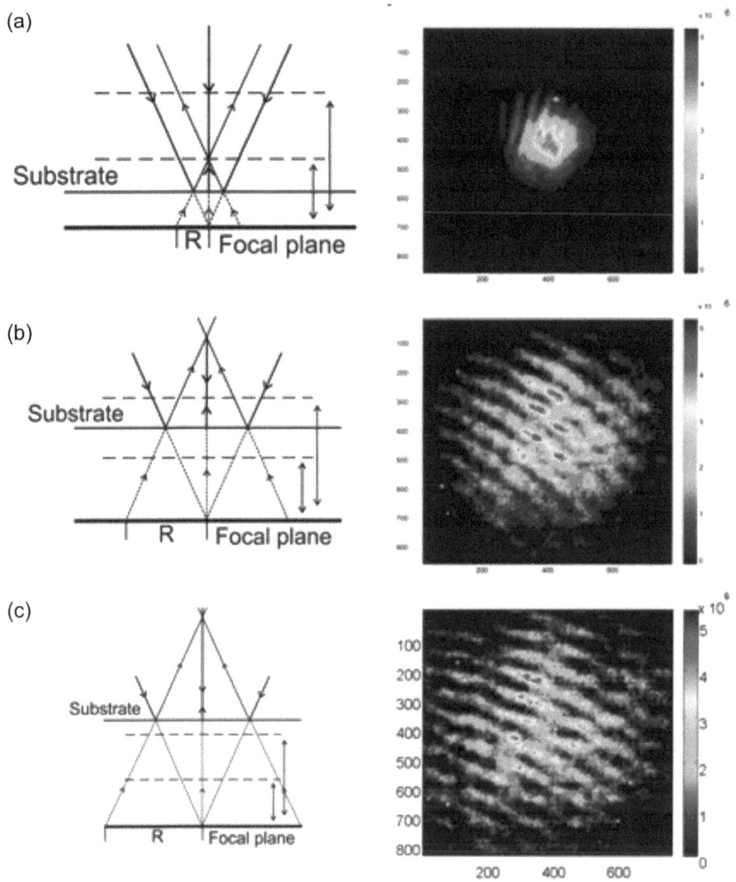

Figure 3.8 Characterisation experiments using a planar SiO_2/Si substrate as a control. (a–c) Schematic plots of the light path (left) and the corresponding camera-recorded optical images (right), when (a) $d < |f|$, (b) $|f| < d < 2|f|$ and (c) $d > 2|f|$. Reprinted with permission from Ref. [16]. Copyright © 2016, Springer Nature.

When $d = |f|$, from Eq. 3.4 we can get that $f_2 = \infty$, which means that the light will be reflected as a group of parallel light, as indicated in Fig. 3.7b, left panel. The observed intensity pattern distribution by the camera at $z = -240$ μm is shown in Fig. 3.7b, right panel.

When $|f| < d < 2|f|$, from Eq. 3.4 we can get that $f_2 < -2|f|$, which means that light will be reflected as indicated in Fig. 3.7c. The observed intensity pattern distribution by the camera at $z = -360$ μm is shown in Fig. 3.7c, right panel.

When $d = 2|f|$, from Eq. 3.4 we can get that $f_2 = -2|f|$, which means that light will be reflected exactly along the same route of the incident light by the micro-lens, as indicated in Fig. 3.7d. The observed intensity pattern distribution by the camera at $z = -480$ μm is shown in Fig. 3.7d, right panel.

When $d > 2|f|$, from Eq. 3.4 we can get that $-2|f| < f_2 < -|f|$, which means that light will be reflected as indicated in Fig. 3.7e. The observed intensity pattern distribution by the camera at $z = -600$ μm is shown in Fig. 3.7e, right panel.

For comparison, a control characterisation experiment on the planar SiO_2/Si substrate using the same procedure has also been conducted. The schematic plots and recorded images are displayed in Fig. 3.8. The SiO_2/Si substrate can be considered a flat mirror. In Fig. 3.8, three conditions are illustrated: $d < |f|$, $|f| < d < 2|f|$ and $d > 2|f|$. In the control measurements, when the distance between the planar substrate and the focal plane d increases, the radius of the circular disk pattern observed by the camera will increase, without any lensing effect. The fringe patterns are the SOM system background noise, which is due to the interference from light beams reflected by various interfaces in the optics.

3.4.3 Focusing Effect of a MoS_2 Micro-Lens

During the characterisation process, the same laser power needs to be kept to ensure that we can compare the recorded intensity distribution images at different heights. But as the stage moves upwards (the z absolute value increases), the laser spot size on the stage (micro-lens plane) will increase and the power density will decrease appropriately following the scaling of $1/z^2$. To make sure that the incident beam has the same power area density at all z values, the following normalisation is necessary for all the recorded images:

$$I_{\text{normalised}} = I_{\text{measured}} \times z^2 \quad (5)$$

where $I_{\text{normalised}}$ is the normalised intensity, I_{measured} is the intensity of the spot in the image, and z is the value of height where the image is taken. Then a 3D dataset can be composed of the series of the images along the z axis. A cross-section profile will be obtained along the x

and z axes to illustrate the distribution of the intensity along these directions.

From the characterisation results, we can conclude the lensing effect of our atomically thin MoS_2 optical lens. When the MoS_2 micro-lens is placed at a distance $2|f|$ above the focal plane, the focused incident light will be exactly re-imaged. Therefore, the camera will record a well-focused light spot. The focal length f of the MoS_2 micro-lens is thus measured to be −240 μm ($2f$ = −480 μm), which matches very well with the simulated value (−248 μm) using the measured OPL profile of the micro-lens. To better represent the data acquired during the experiment, the intensities of the cross section along x and z axes are normalised. At a given distance, a virtual circle is drawn to the centre peak of the spot, and 600 points are chosen evenly on this circle and their intensities are averaged. Then the average intensity is selected to represent the value at this specific radius to the centre. Finally, the data at different heights are assembled, interpolated and plotted in contour, as shown in Fig. 3.9a. From the comparison characterisation by using a planar substrate without the MoS_2 micro-lens, we obtain the intensity distribution shown in Fig. 3.9b. The lensing effect is clearly demonstrated by comparing the difference between Figs. 3.9a and 3.9b.

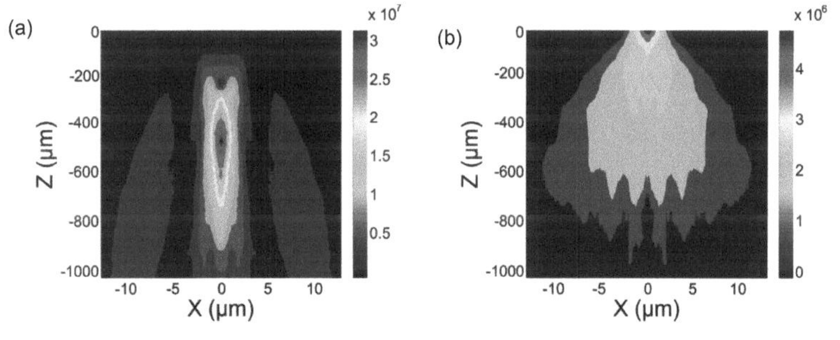

Figure 3.9 Characterisation of atomically thin MoS_2 micro-lens. (a) Intensity distribution pattern of the MoS_2 micro-lens measured by an SOM. (b) Intensity distribution pattern of the planar reference SiO_2/Si substrate measured by the same SOM setup. Reprinted with permission from Ref. [16]. Copyright © 2016, Springer Nature.

We have also calculated the NA of our atomically thin MoS_2 micro-lens. With a focal length of 240 μm, a lens diameter of 20 μm and a lens material refractive index of 4.4, a simple calculation shows that the NA of this micro-lens based on few-layer MoS_2 is ~0.2 by using the equation

$$\mathrm{NA} \approx \frac{n \cdot D}{2f}, \tag{3.5}$$

where n is the refractive index of the medium in which the lens is working, D is the lens diameter and f is the focal length of the lens. This NA value of our MoS_2 micro-lens needs to be further improved for application as ultra-thin optical elements. Several approaches can be considered, including electrically tuning the refractive index of thin-layer MoS_2, as indicated in Ref. [19], and re-designing the lens structure to achieve a higher D/f ratio.

3.5 Atomically Thin Gratings

In this section, we will demonstrate the high-index-enabled high scattering efficiency of thin-layer MoS_2. Firstly, we will calculate the scattering cross section of different materials with different refractive indices but with the same thickness. It can be seen from the calculation that the scattering cross section is highly related to the refractive index, and thin-layer MoS_2 owns a large cross section due to its high refractive index. Then we will make use of its large scattering cross section to fabricate atomically thin gratings from thin-layer MoS_2 with optimised parameters from simulation. And finally, the diffraction efficiency of fabricated MoS_2 gratings will be measured.

3.5.1 High Scattering Efficiency of Thin-Layer MoS_2

The efficiency of light scattering is a critical parameter for advanced light manipulation. Devices which employ photonic bandgaps [20], Anderson localisation [21] and light trapping such as with thin-film solar cells [22] all rely heavily on strong light scattering. Unfortunately, in typical 2D materials, such as graphene, the scattering efficiency is low, making it impossible to rely on collective scattering of nanostructured graphene to achieve functionalities such as gratings. Here, we show that thin-layer MoS_2 films have an extraordinarily high scattering efficiency, enabled by the combination of a high index in a thin structure. The scattering efficiency is determined by the strength of the electric field in the material. Normally, the electric field inside a bulk material, particularly a high-index material, is much weaker than that of incident light because of the impedance mismatch. The boundary condition of Maxwell's equations requires the tangential component of the electric field to be continuous across any interface. Because the layer is thin, this condition indicates that the electrical field inside a 2D material is almost as strong as the tangential component of the incident field. As a result, there is a strong polarisation:

$$P = \varepsilon_0(n^2 - 1)E_0, \tag{3.6}$$

where E_0 is the electric field of s-polarised incident light, n is the index of the material and ε_0 is the electric permittivity of free space. The scattering power is proportional to P^2 and, therefore, scales roughly as n^4. This scaling rule greatly favours high-index materials and is again uniquely available in ultra-thin materials. In contrast, for nanoparticles, the scattering power is proportional to $[(n^2 - 1)/(n^2 + 2)]^2$, which does not increase appreciably with the refractive index [23].

Here we use the finite element method to explicitly calculate the scattering efficiency of 2D ribbons by solving Maxwell's equations. In Fig. 3.10a, the cross section for a 1L MoS_2 (red line) nanoribbon with a 30 nm width is calculated for different incident angles, showing an excellent isotropic response for s-polarised light. In comparison, the black line indicates the scattering cross section for a 10-nm-thick SiO_2 ribbon with a 30 nm width. Figure 3.10b shows the calculated scattering cross section of an infinitely long ribbon (30 nm wide and 0.67 nm thick) in air for s-polarised light incident from the normal direction. The scattering cross section has units of nanometres because the length of the ribbon is considered infinite. From the calculation, the scattering cross section increases by orders of magnitude when the index increases by just a few times. For example, the scattering cross section of a 1L MoS_2 ribbon is around 670 times, 54 times and 18 times of those in 0.67 nm SiO_2, 1L graphene and 0.67 nm of gold, respectively. Metal is generally considered as one of the strongest scattering materials, and it is important to note that MoS_2 even displays much stronger light scattering than gold. Such favourable scaling for high-index materials is uniquely available in ultra-thin materials. The giant scattering efficiency in high-index 2D materials makes it possible to achieve sophisticated light manipulation based on collective scattering by nanostructured patterns.

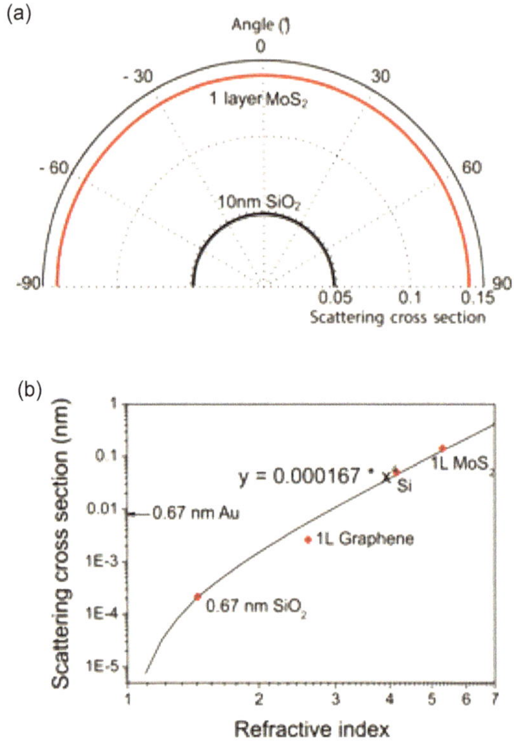

Figure 3.10 Scattering cross section of 1L MoS_2 and several other materials. (a) Isotropic scattering cross section of 1L MoS_2 and 10 nm SiO_2. (b) Calculated scattering cross section vs. refractive index. Reprinted with permission from Ref. [16]. Copyright © 2016, Springer Nature.

3.5.2 Fabrication of Atomically Thin MoS_2 Gratings

To start with the fabrication of atomically thin MoS_2 gratings, thin-layer MoS_2 flakes are transferred onto a SiO_2/Si (275 nm thermal oxide on n+-doped silicon) substrate, including 1L, 2L, 5L, six-layer (6L) and eight-layer (8L) MoS_2. The grating efficiencies are calculated using rigorous coupled wave analysis (RCWA). Three parameters are scanned to search the maximum efficiency: the periodicity, the filling ratio of MoS_2 and the incident angle. The maximum efficiencies and corresponding optimal parameters of these gratings for FIB fabrication are given in Table 3.1.

Table 3.1 Optimal parameters and first-order efficiency for MoS_2, graphene and gold gratings on a SiO_2/Si substrate

Materials	Filling ratio	Periodicity	Incident angle (°)	Efficiency simulation (%)	Efficiency experiments (%)
1L MoS_2	0.52	1.8λ	8.7	0.4	0.3
2L MoS_2	0.39	1.2λ	23.9	1.3	0.8
6L MoS_2	0.39	1.2λ	23.9	7.4	4.4
8L MoS_2	0.44	1.2λ	23.9	10.2	10.1
2 nm SiO_2	0.44	1.2λ	21.0	0.0051	–
1L graphene	0.52	1.9λ	8.0	0.0078	–
1 nm Au	0.49	1.9λ	7.8	0.0519	–

In our simulations, we used the reported refractive index values for MoS_2 [17] and grapheme [18], which are n = 5.3 + 1.3i and n = 2.6 + 1.3i, respectively. The optimal efficiency and the parameters are listed in Table 3.1. With the optimal parameters, FIB will be used to mill thin-layer MoS_2 into atomically thin gratings, as shown in Fig. 3.9. It can be seen that for 1L and 2L, 5L and 6L, and 8L MoS_2 gratings, the FIB-etched area owns different colours compared to the unetched area. From our measurement results, the 1L, 2L, 5L, 6L and 8L MoS_2 are fully etched through and the SiO_2 substrates underneath are over-etched by around 1.5 nm. From the simulated efficiency of 2 nm SiO_2 presented in Table 3.1, the grating contribution from this over-etched SiO_2 is negligible.

We have also experimentally testified the simulation results on 2 nm SiO_2 and 1L graphene grown by chemical vapour deposition (CVD). The AFM scanning results presented in Fig. 3.12 indicate the grating thickness. Figure 3.12a shows the SEM images of a 1L graphene grating, because it cannot be clearly identified with an optical microscope image due to its high transparency. Figure 3.12b shows its AFM image with the height of 0.81 nm, confirming that this is 1L graphene. Figure 3.12c shows the optical microscope image of a grating fabricated from thermal SiO_2 on a Si substrate, and Fig. 3.12d shows its AFM image with a height of 2.17 nm. We have conducted a diffraction efficiency characterisation experiment on both graphene and SiO_2 gratings, and we cannot observe any diffraction pattern, consistent with the simulated results that the scattering efficiency of those two gratings is very low, below the lower limit of our light detection system.

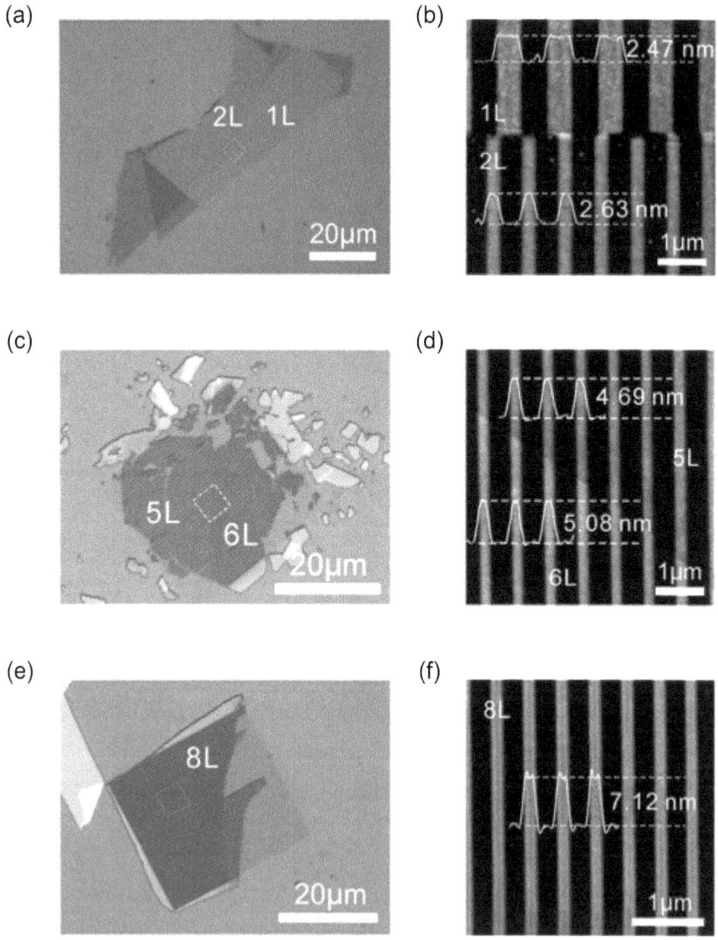

Figure 3.11 Atomically thin MoS_2 gratings. (a, c and e) Optical microscope images of 1L and 2L, 5L and 6L, and 8L MoS_2 gratings. (b, d and f) AFM images of 1L and 2L, 5L and 6L, and 8L MoS_2 gratings. Reprinted with permission from Ref. [16]. Copyright © 2016, Springer Nature.

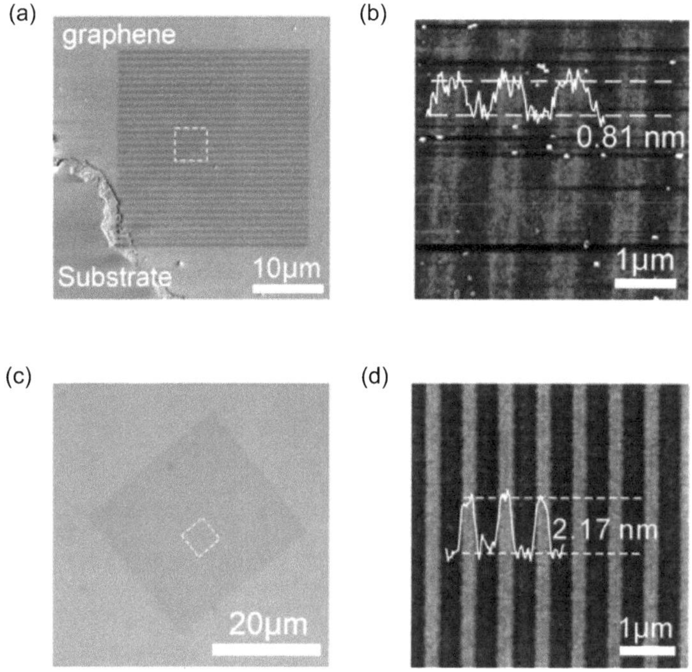

Figure 3.12 Control gratings fabricated from 1L graphene and SiO_2. (a) SEM image of a 1L graphene grating. (b) AFM image of a 1L graphene grating. (c) Optical microscope image of a SiO_2 grating. (d) AFM image of a SiO_2 grating. Reprinted with permission from Ref. [16]. Copyright © 2016, Springer Nature.

3.5.3 Characterisation of Atomically Thin MoS_2 Gratings

The gratings are characterised using an s-polarised green laser at a wavelength of 532 nm. The laser beam has a diameter of around 200 μm, which is large enough to fully cover the grating. The schematic plot of the MoS_2 gratings and their characterisation are shown in Fig. 3.13a. First-order and second-order diffraction beams can be observed during characterisation, and the measured diffraction angles agree with the predictions of the diffraction equation

$$d(\sin \theta_d + \sin \theta_i) = m\lambda, \tag{3.7}$$

where θ_d and θ_i are the diffraction angle and the incident angle, respectively; d is the period of the grating elements; and m is an integer characterising the diffraction order. The power of the first-order diffraction beam is measured, and the grating efficiency η can be determined by

$$\eta = \frac{P_\text{d}}{P_\text{i}} \times \frac{S_\text{b}}{S_\text{g}}, \tag{3.8}$$

where P_d and P_i are the measured powers of the diffracted and incident beams, respectively, and S_b and S_g are the measured areas of the incident beam and the MoS_2 grating, respectively. Taking the 8L MoS_2 grating as an example, the measured grating efficiency is a function of the incident angle, which agrees well with our simulation, as shown in Fig. 3.13b.

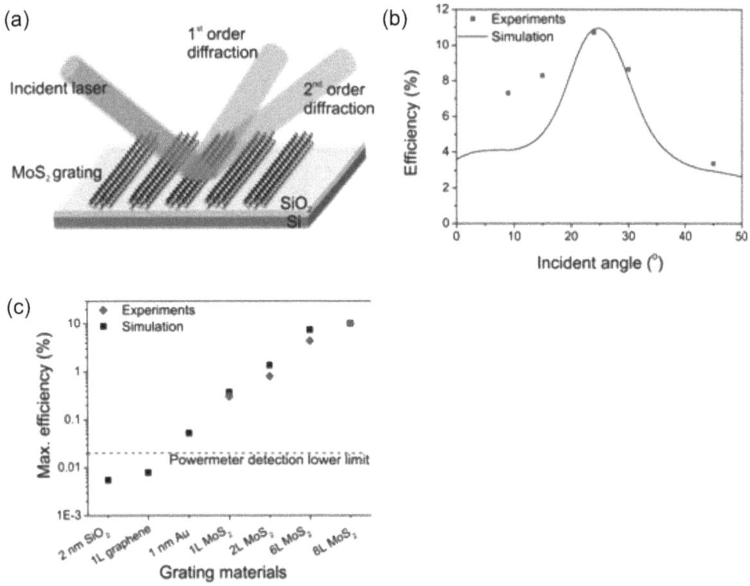

Figure 3.13 Characterisation of atomically thin MoS_2 gratings. (a) Schematic plot of a MoS_2 grating and its characterisation. (b) Simulated and measured efficiency of an 8L MoS_2 grating versus incident angle of the light beam. (c) Comparison of the simulated and measured maximum grating efficiencies for different materials. The dashed line represents the noise level of our light detection system, with the minimum detectable grating efficiency being 0.02%. Reprinted with permission from Ref. [16]. Copyright © 2016, Springer Nature.

The maximum grating efficiencies for the 1L, 2L, 6L and 8L MoS_2 gratings are measured to be 0.3%, 0.8%, 4.4% and 10.1%, respectively, which also agree well with the simulations, as shown in

Fig. 3.13c and Table 3.1. For comparison, a grating from a graphene sheet deposited by large-area CVD is also fabricated, as shown in Fig. 3.12a,b. The intensity of the diffracted beam from the graphene grating is lower than the noise level of our light detection system and thus had a maximum efficiency no greater than 0.02%. From our simulations, the maximum grating efficiency of 1L graphene will be only 0.0078%, which is around 47 times lower than that of a 1L MoS_2 grating.

3.6 Summary

In this chapter, we have shown that high-index 2D materials have extraordinary elastic interactions with light, enabled uniquely by the combination of their ultra-thin nature and high refractive indices. As a result, wavefront shaping [24, 25] and efficient light scattering can be accomplished with atomically thin 2D materials, enabling a new class of optical components entirely based on high-index 2D materials. Moreover, compared to conventional diffractive optical components, the spatial resolution of phase-front shaping is much smaller than the wavelength and is only limited by the nanofabrication resolution, making it possible to eliminate undesired diffractive orders [25]. 2D materials also offer many unique advantages. First, considering the strong tunability of 2D materials, advanced beam steering can be envisioned [25]. Second, a similar giant OPL in another transition metal dichalcogenide (TMD) family MX_2 (M = Mo, W; X = S, Se, Te) semiconductors has also been observed in our experiments, such as WS_2 and WSe_2, as shown in Fig. 3.14. The availability of different functional materials offers rich opportunities for the combination of optical and electronic properties, such as stacked atomically thin heterostructures for 2D optoelectronics. Third, high-quality 2D TMD semiconductors can be deposited directly onto (or transferred to) various substrates with large size by CVD at low cost [26] potentially enabling low-cost flexible optical components. Lastly, 2D optical components represents a significant advantage in manufacturing compared to conventional 3D optical components, because different functionalities can all be achieved in a 2D platform sharing the same fabrication processes and this will greatly facilitate large-scale manufacturing and integration. In summary, the capability of manipulating the flow of light in 2D materials opens an exciting avenue towards unprecedented miniaturisation of optical components and the integration of advanced optical functionalities into compact 2D optical systems.

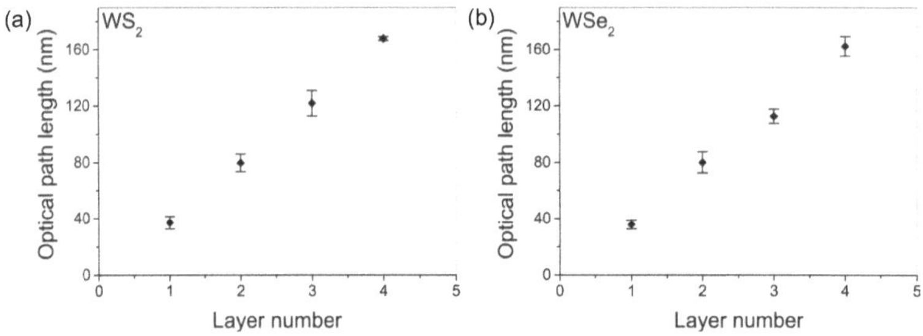

Figure 3.14 Statistical OPL data from 1L to 4L WS_2 (a) and WSe_2 (b), measured by a PSI. At least five samples are measured for each layer number, and all the layer numbers are confirmed by an AFM. Reprinted with permission from Ref. [16]. Copyright © 2016, Springer Nature.

References

1. Lopez-Sanchez, O., Lembke, D., Kayci, M., Radenovic, A. and Kis, A. (2013). Ultrasensitive photodetectors based on monolayer MoS_2, *Nat. Nanotechnol.*, **8**(7), pp. 497–501.
2. Liu, C.-H., Chang, Y.-C., Norris, T. B. and Zhong, Z. (2014). Graphene photodetectors with ultra-broadband and high responsivity at room temperature, *Nat. Nanotechnol.*, **9**(4), pp. 273–278.
3. Splendiani, A., Sun, L., Zhang, Y., Li, T., Kim, J., et al. (2010). Emerging photoluminescence in monolayer MoS_2, *Nano Lett.*, **10**(4), pp. 1271–1275.
4. Sundaram, R. S., Engel, M., Lombardo, A., Krupke, R., Ferrari, A. C., et al. (2013). Electroluminescence in single layer MoS_2, *Nano Lett.*, **13**(4), pp. 1416–1421.
5. Bao, Q., Zhang, H., Wang, Y., Ni, Z., Yan, Y., et al. (2009). Atomic-layer graphene as a saturable absorber for ultrafast pulsed lasers, *Adv. Funct. Mater.*, **19**(19), pp. 3077–3083.
6. Wang, S., Yu, H., Zhang, H., Wang, A., Zhao, M., et al. (2014). Broadband few-layer MoS_2 saturable absorbers, *Adv. Mater.*, **26**(21), pp. 3538–3544.
7. Yin, X., Ye, Z., Chenet, D. A., Ye, Y., O'Brien, K., et al. (2014). Edge nonlinear optics on a MoS_2 atomic monolayer, *Science*, **344**(6183), pp. 488–490.

8. Zeng, H., Dai, J., Yao, W., Xiao, D. and Cui, X. (2012). Valley polarization in MoS_2 monolayers by optical pumping, *Nat. Nanotechnol.*, **7**(8), pp. 490–493.

9. Mak, K. F., He, K., Shan, J. and Heinz, T. F. (2012). Control of valley polarization in monolayer MoS_2 by optical helicity, *Nat. Nanotechnol.*, **7**(8), pp. 494–498.

10. Chen, C.-F., Park, C.-H., Boudouris, B. W., Horng, J., Geng, B., et al. (2011). Controlling inelastic light scattering quantum pathways in graphene, *Nature*, **471**(7340), pp. 617–620.

11. Britnell, L., Ribeiro, R. M., Eckmann, A., Jalil, R., Belle, B. D., et al. (2013). Strong light-matter interactions in heterostructures of atomically thin films, *Science*, **340**(6138), pp. 1311–1314.

12. Weiner, J. and Ho, P.-T. (2008). *Light-Matter Interaction, Fundamentals and Applications*, Vol. 1 (John Wiley & Sons).

13. Lee, S. H., Choi, M., Kim, T.-T., Lee, S., Liu, M., et al. (2012). Switching terahertz waves with gate-controlled active graphene metamaterials, *Nat. Mater.*, **11**(11), pp. 936–941.

14. Ju, L., Geng, B., Horng, J., Girit, C., Martin, M., et al. (2011). Graphene plasmonics for tunable terahertz metamaterials, *Nat. Nanotechnol.*, **6**(10), pp. 630–634.

15. Vakil, A. and Engheta, N. (2011). Transformation optics using graphene, *Science*, **332**(6035), pp. 1291–1294.

16. Yang, J., Wang, Z., Wang, F., Xu, R., Tao, J., Zhang, S., Qin, Q., Luther-Davies, B., Jagadish, C., Yu, Z. and Lu, Y. (2016). Atomically thin optical lenses and gratings, *Light: Sci. Appl.*, **5**, p. e16046.

17. Shen, C.-C., Hsu, Y.-T., Li, L.-J. and Liu, H.-L. (2013). Charge dynamics and electronic structures of monolayer MoS_2 films grown by chemical vapor deposition, *Appl. Phys. Express*, **6**(12), p. 125801.

18. Blake, P., Hill, E. W., Castro Neto, A. H., Novoselov, K. S., Jiang, D., et al. (2007). Making graphene visible, *Appl. Phys. Lett.*, **91**(6), p. 063124.

19. Chih-Chiang, S., Yu-Te, H., Lain-Jong, L. and Hsiang-Lin, L. (2013). Charge dynamics and electronic structures of monolayer MoS_2 films grown by chemical vapor deposition, *Appl. Phys. Express*, **6**(12), p. 125801.

20. Santos, E. J. G. and Kaxiras, E. (2013). Electrically driven tuning of the dielectric constant in MoS_2 layers, *ACS Nano*, **7**(12), pp. 10741–10746.

21. Yablonovitch, E. (1987). Inhibited spontaneous emission in solid-state physics and electronics, *Phys. Rev. Lett.*, **58**(20), pp. 2059–2062.

22. John, S. (1987). Strong localization of photons in certain disordered dielectric superlattices, *Phys. Rev. Lett.*, **58**(23), pp. 2486–2489.
23. Yu, Z., Raman, A. and Fan, S. (2012). Thermodynamic upper bound on broadband light coupling with photonic structures, *Phys. Rev. Lett.*, **109**(17), p. 173901.
24. Jackson, J. D. (1998). *Classical Electrodynamics*, 3rd Edition (John Wiley & Sons).
25. Lin, D., Fan, P., Hasman, E. and Brongersma, M. L. (2014). Dielectric gradient metasurface optical elements, *Science*, **345**(6194), pp. 298–302.
26. Yu, N. and Capasso, F. (2014). Flat optics with designer metasurfaces, *Nat. Mater.*, **13**(2), pp. 139–150.

Chapter 4

Two-Dimensional Layered Materials for Optical Pulse Generation

Diao Li,[a] **Dong Mao**[b*] **and Zhipei Sun**[c]
[a]*State Key Lab Incubation Base of Photoelectric Technology and Functional Materials, International Cooperation Base of Photoelectric Technology and Functional Materials, and Institute of Photonics and Photon-Technology, Northwest University, 710069, Xi'an, China*
[b]*Key Laboratory of Space Applied Physics and Chemistry, School of Science, Northwestern Polytechnical University, 710072, Xi'an, China*
[c]*Department of Electronics and Nanoengineering, Aalto University, Tietotie 3, 02150, Espoo, Finland*
lidiaoxc@163.com; maodong@nwpu.edu.cn

4.1 Introduction

Lasers emitting short (μs, ns) and ultra-short (ps, fs) pulses are powerful optical platforms due to their features of manipulating power, energy and spectra, in contrast to sources operating in the continuous-wave regime. Various applications ranging from scientific research (e.g., spectroscopy, sensing, frequency comb, etc.) to commercial requirements (e.g., materials processing, bio-medical diagnostics, etc.) [1–6] have drawn a large amount of

Two-Dimensional Materials in Nanophotonics: Developments, Devices, and Applications
Edited by Yuerui Lu

ISBN 978-981-4800-22-8 (Hardcover), 978-0-429-42877-7 (eBook)
www.jennystanford.com

research efforts and promoted significant advance in pulsed laser technology. Realisation of pulsed lasers usually relies on a nonlinear optical component named saturable absorber (SA), to convert the continuous wave to periodic pulses via two typical technologies (i.e., mode-locking and *Q*-switching) [2]. Independent to the laser formats, SAs with desirable properties are crucial elements for specific laser parametres. Therefore, a single SA possessing versatile functions is always preferred for the rapidly increased demand of laser sources with enhanced performance, and thus, pushing forward the state-of-the-art laser technology. The current-dominating SA is a semiconductor saturable absorber mirror (SESAM) [2, 3, 7]. Controllable epitaxial growth enables to tailor the SA parametres such as modulation depth, optical response time and operation wavelength, but with a cost of complicated fabrication and commercial ineffectiveness [2, 3]. Nevertheless, the available operation bandwidth of a single SESAM is limited in a few tens of nanometres [2]. Single-walled carbon nanotubes (SWCNTs) are an alternative SA material since their first demonstration in 2003 [8]. Although the fabrication of an SWCNT SA is much easier and cheaper than a SESAM, its broadband operation needs a wide tube diameter distribution [9–23]. Another issue of the SWCNTs is the relatively large non-saturable loss induced by the nanotubes that are not in resonance to the laser wavelength [10].

Motivated by saturable absorption of a 1D SWCNT [9, 10], 2D layered materials (e.g., graphene [24–26], transition metal dichalcogenides (TMDs) [27–29], black phosphorus (BP) [30–32] and topological insulators (TIs) [33]) have been demonstrated for ultra-fast pulse generation. One of the significant differences of these studied 2D materials is their bandgap structures. Graphene features a zero bandgap due to the linear dispersion of its Dirac electrons [34], which induces a fullband [10, 35–38] optical absorption. Other materials with a direct or indirect bandgap in 2D or bulk formats provide multiple properties for electronics and photonics applications. In this chapter, we present a summary of the recent advances in ultra-fast pulsed lasers based on 2D layered materials. We will discuss the fundamentals of saturable absorption of 2D layered materials in Section 4.2, the current state of the art of graphene pulsed lasers in Section 4.3, TMD pulsed lasers in Section 4.4 and other 2D layered material (BP and TIs) pulsed lasers in

Section 4.5. Finally, we will also present our perspective in Section 4.6.

4.2 Saturable Absorption of 2D Layered Materials

4.2.1 Graphene

Graphene behaves as a semi-metal due to its special Dirac energy band structure, that is, the conduction and valence bands touch at a charge neutrality point [39]. The linear electron dispersion in this unique bi-conical structure determines graphene's broadband optical response ranging from the visible to the far-infrared region and the terahertz range, which suggests graphene is suitable for wide-spectral-bandwidth photon manipulation. Figure 4.1a shows a linear optical absorption measurement of graphene over a broad spectral range; the smooth curves of polyvinyl alcohol (PVA) and grapheme-PVA identify a wavelength-independent response [25, 40]. An SA is a crucial optical switching component in pulsed laser technology, which needs to have large non-linear optical absorption at the laser wavelength and ultra-fast optical response [2]. To characterise the non-linear absorption of graphene, Z-scan [41] or balanced twin-detector transmission measurements [42] are commonly used according to the SA format (free space or fibre integration). A typical experimental measurement of the non-linear transmission carried out with separated wavelengths over a wide span at the telecommunication band verifies a power-dependent optical response [25], as depicted in Fig. 4.2b. The transmission change from the initial value to the maximum represents an absorption transition process between linear and non-linear, while the material saturates as the increase of input power. Broadband absorption saturation is demonstrated over 20 nm between 1548 nm and 1568 nm. Ultra-fast optical relaxation time is also important for SAs [10, 35, 37]. Time-resolved spectroscopy and pump probe investigations have disclosed two typical distinct carrier relaxation timescales corresponding to ~100 fs and ~1 ps, which were dominated by carrier-carrier intra-band scattering and electron-hole inter-band recombination processes [43, 44], respectively.

Large optical non-linearity, broad optical responsivity, ultra-fast relaxation time, controllable modulation depth, low saturation intensity, as well as the gradually matured fabrication technology with low cost have created promise for scalable optical applications where novel SAs are desired.

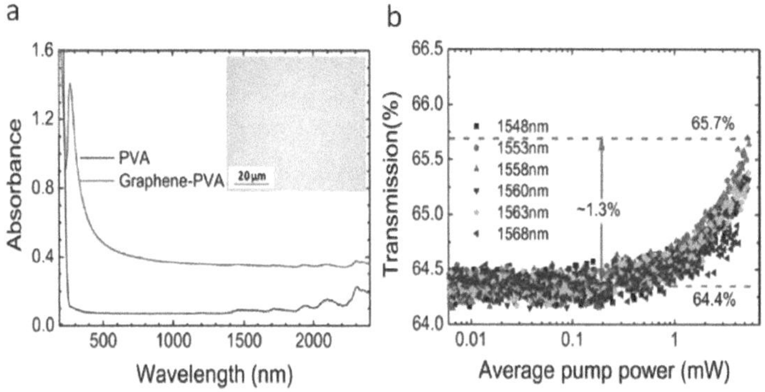

Figure 4.1 (a) Broad linear absorption spectrum of graphene-PVA composite. (b) Optical transmission of grapheme-PVA with non-linear saturation. Reprinted with permission from Ref. [25]. Copyright (2010) American Chemical Society.

To date, various approaches have been exploited to produce graphene-based saturable absorber (GSA) devices for different laser formats. These include mechanical exfoliation [45], liquid-phase exfoliation (LPE) [40, 46] and chemical vapour deposition (CVD) [47], which can provide graphene in flakes, an aqueous dispersion and a large uniform polymer composite [48]. Graphene derivatives (e.g., graphene oxide [GO] [49–51], reduced graphene oxide [RGO] [52], etc.) were also investigated. The fabrication of a GSA for a specified laser is thus selected to obtain optimum absorption parametres. For instance, large-area CVD-grown graphene can be transferred on a glass substrate to form a saturable absorber mirror (SAM) for pulsed laser modulation [53–55]. In fact, GSAs have been targeted for various laser pulse investigations in fibre [24–26], solid-state [53–56], waveguide [57, 58] and semiconductor [59] lasers, these lasers with specific parametres have shown huge potential for applications. Another fascinating graphene-based pulsed laser can be graphene actively modulated lasers [60]. For example, a graphene-based electro-optic modulator with capacitor configuration can be inserted into a fibre laser cavity for active pulse generation [61].

4.2.2 Transition Metal Dichalcogenides

TMDs, a family of layered 2D materials, have captured a great deal of research interest for potential applications in optoelectronic devices [60, 62–66]. TMDs have a chemical formula MX_2, where M is a transition metal (such as Mo, W, Re) and X is a group VI element (such as S, Se, Te). Depending on the combination of transition metal and chalcogen, 40 different types of TMDs have been discovered [67], including MoS_2, WS_2, $MoSe_2$, WSe_2, $MoTe_2$, WTe_2, etc. Similar to graphite, layered TMDs show strong in-plane chemical bounds but display weak van der Waals interaction between layers [63, 68–71]. In monolayer TMDs, the metal atoms are sandwiched between two layers of chalcogen atoms [70], as illustrated in Fig. 4.2 [72]. Unlike the zero-bandgap structure of graphene, TMD materials exhibit a non-zero bandgap from the visible to the near-infrared waveband [73]. Additionally, the band structure of TMD materials varies dramatically with the layer numbers of the nanosheets, for example, the bulk-state MoS_2 crystal is an indirect semiconductor with a bandgap of a 1.2 eV, while monolayer MoS_2 is a direct semiconductor with a bandgap of 1.9 eV. Such difference is attributed to hybridisation between the d orbitals of transition metal atoms and pz orbitals of S atoms [73]. The bandgap and stability of TMDs decrease as the mass of the chalcogen atom increase from S to Se to Te. Thus, $MoTe_2$ and WTe_2 have smaller bandgaps and weaker stability than MoS_2 andWS_2 [74].

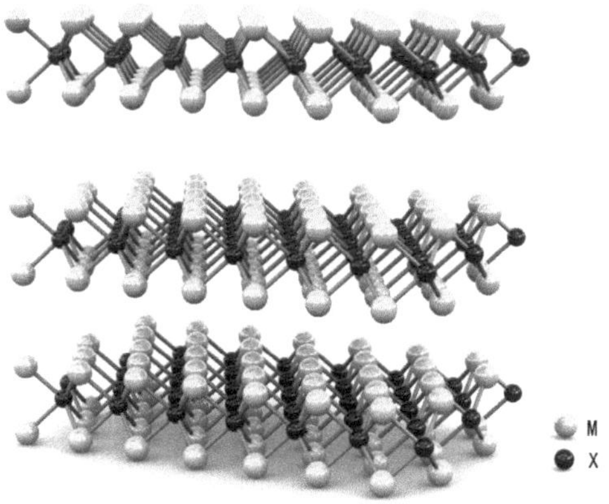

Figure 4.2 3D schematic representation of typical TMD materials with a MX_2 structure. Reprinted by permission from Springer Nature Customer Service Centre GmbH: Springer Nature, *Nature Nanotechnology*, Ref. [72], Copyright (2011).

TMD materials have been used to fabricate solar cells, photodetectors and transistors [75–77]. In the field of non-linear optics, MoS_2 nanosheets were proven to display an intra-band relaxation time of ~30 fs and a carrier lifetime of ~100 ps, and exhibit higher third-order non-linear susceptibility than graphene at 515 and 800 nm [78]. Using an open-aperture Z-scan technique, MoS_2 nanosheets were found to possess a non-linear saturable absorption property that the optical absorbance decreases with the increase of the incident light intensity and becomes saturated above a certain value [78]. Ref. [79] demonstrated that when the ratio between Mo and S ions was decreased from 2 to 1.89, the MoS_2 bandgap can be reduced from 1.08 to 0.26 eV. They have demonstrated a *Q*-switched solid-state laser operating at 1.06, 1.42 and 2.1 with a MoS_2 SA [79]. After that, MoS_2 SAs were widely used in various *Q*-switched and mode-locked lasers [27, 29, 80–88]. Other types of TMDs materials, such as WS_2 [89-93], $MoSe_2$ [94–96], WSe_2[95], $MoTe_2$ [97], WTe_2 [97, 98] and ReS_2 [99] were found to exhibit saturable absorption property and can work as SAs for *Q*-switched and mode-locked lasers.

A number of experiments show that TMD materials exhibit a sub-bandgap saturable absorption property. For example, MoS_2 has a bandgap in the range of 1.2 to 1.8 eV [79], while it displays a saturable absorption property at 1.565 (0.793 eV) [82] and 1.941 μm (0.62 eV) [85, 86]. Various theories have been developed to explain the intriguing experimental results, such as defect-induced bandgap decreasing [79, 90], edge state of the materials [81], co-existence of semiconducting and metallic states [29] and two-photon absorption saturation [74]. Actually, the edge state can be regarded as a special defect located at the edges of the nanosheets. However, an ultimate interpretation of the governing mechanism has not been fully established to explain the SAs of TMDs.

4.2.3 Other 2D Layered Materials

Inspired by successful demonstration of graphene- and TMD-based SAs, there is also great interest to try other 2D materials. For example, BP, a recently re-discovered 2D layered semiconductor composed of only phosphorus atoms, was dramatically focused on for electronics and photonics [100–105]. Theoretical calculations reveal a large bandgap variation range from ~0.3 eV (~ 4 μm) in bulk to 2 eV (~ 600 nm) in monolayer [106]; this interesting property can bridge the energy blank between zero-gap graphene and large-gap TMDs for specific electronic applications where a modest gap is required. For multilayer BP flakes with relatively a low-energy gap, the optical response at the infrared region is also broad like graphene.

Large non-linear optical absorption in a BP-enriched solution and fibre-integrated BP flakes has been recently demonstrated with the Z-scan technique and twin-detector transmission measurements, respectively, which identify BP as a new candidate for SAs [30, 31, 107, 108]. Angle-resolved transient reflection spectroscopy investigation reveals that the ultra-fast carrier relaxation in few-layer BP is in the sub-200 fs timescale [109], proving the potential of BP-based SAs for ultra-fast pulse generation. Particularly, the natural anisotropic arrangement of BP atoms in a puckered plane determines an apparent optical anisotropy (i.e., polarisation-dependent absorption) [30]. Experimental investigations have demonstrated that both linear absorption and non-linear absorption of BP flakes are polarisation-dependent, and the absorption rate can be adjusted by the sample thickness [30]. Figure 4.3a shows a comparison between the polarisation-dependent linear transmittance of two BP flakes with different thicknesses. It is obvious to see a much stronger polarisation-dependent absorption contrast in the thicker film and a significant decrease of transmittance. In non-linear optical transmittance measurements, a BP flake also presents pronounced polarisation dependence, as illustrated in Fig. 4.3b. For a single sample, optical absorption along the zigzag axis is relatively weaker than that along the armchair direction, which is a result of the anisotropic phosphorus atom arrangement [30].

Although BP is the most stable allotrope in the phosphorus elemental family, the chemical reaction of phosphorus atoms with water and oxygen in an air environment has been a significant obstacle for the prospect of electronic and optical devices formed by few-layer BP [110]. Especially in those BP films with only few layers or a monolayer, the chemical reaction-caused performance degradation has to be considered during fabrication. This issue has inspired research efforts to develop modification and protection technology for more stable performance. A few methods have been proposed and confirmed to be effective for BP passivation, for instance, mixing LPE BP flakes with an inert polymer composite (e.g., polycarbonate, PVA and polyimide) or covering few-layer BP by polymethyl methacrylate (PMMA) or an atomic layer deposition-fabricated Al_2O_3 film [108, 111–114].

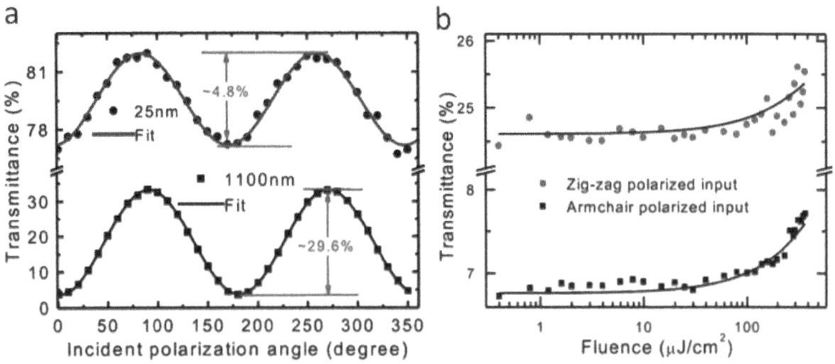

Figure 4.3 (a) Polarisation-dependent transmittance for 25-nm and 1100-nm-thick BP flakes. The polarisation directions corresponding to the maximum and minimum transmittance are linked with the zigzag and armchair axes of BP flakes. (Input laser: 1550 nm continuous wave). (b) Fluence-dependent transmittance of BP flakes measured with ultra-fast pulses at two orthogonally polarised light directions. Adapted from Ref. [30] under a Creative Commons License Deed Attribution 4.0 International (CC BY 4.0), https://creativecommons.org/licenses/by/4.0/.

Besides the 2D semiconductors mentioned before, a group of quantum well materials, that is, TIs were also discovered and demonstrated to present saturable absorption property. Representative samples are the recently investigated SAs [28] of Sb_2Te_3 [115], Bi_2Te_3 [116] and Bi_2Se_3 [117], etc. These materials feature a unique bulk energy gap and zero gap in the surface or edge state [115]. As a result, the TI-based SAs can be in a bulk sample when they are employed in lasers for pulse formation. Similar to the fabrication and characterisation approaches used for graphene, TMDs and BP, TI-based SAs can be used for *Q*-switched or mode-locked pulse generation over a broadband wavelength range, either in fibre or in solid-state lasers [116, 118, 119]. The discovery of TIs expands the family of 2D layered SAs and provides new candidates for future pulsed laser technology. A detailed discussion of the non-linear optical properties of these materials is not given in order to make this chapter concise, as their general performance is comparative with other aforementioned SAs.

4.3 Graphene-Based Pulsed Lasers

4.3.1 Graphene *Q*-switched Lasers

Q-switching is a pulsed laser technology typically for large-energy pulse generation, which has been widely applied in sensing, materials processing and medicine [1, 3, 5]. Physically, the pulse formation in a cavity is dependent on the switching of cavity *Q*-factor defined by the ratio between stored energy and optical loss [120]. The key element of such lasers is an optical *Q*-switcher, which functions as a modulation component in the laser via either active or passive regime. In passive *Q*-switching, an SA with intensity-dependent absorption can operate as an optical modulator to switch the pulse between on and off states. *Q*-switched lasers based on SAs typically release light pulses with short pulse durations (ns to μs), low repetition rate (kilohertz) and high pulse energy (nJ to mJ). These laser parametres are strongly dependent on the interplay of various laser dynamics in terms of gain, loss, modulation depth, etc. The adjustment of these parametres can usually be indirectly achieved by changing the pump power.

The broadband optical absorption, ultra-fast optical response and easy integration enable graphene as a promising *Q*-switcher to generate large-energy pulses via various laser systems. The very first investigations mainly focused on fibre lasers at the 1.5 μm telecommunication band [121], since this is the most important technical wavelength region. Ref. [122] reported an erbium-doped fibre (EDF) laser *Q*-switched by a GSA, producing shortest pulses of 2 μs at a 1.5 μm band with the highest repetition rate and a pulse energy of 103 kHz and 40 nJ, respectively. In particular, the laser wavelength can be continuously tuned from 1522 nm to 1555 nm, which is only limited by the bandwidth of the tunable filter, rather than the GSA, as shown in Fig. 4.4. Later, GSAs (e.g., graphene polymer composite, GO [123, 124], etc.) were employed to *Q*-switched lasers oscillating at a number of other laser wavelengths. To get high-power operation in fibre lasers, an effective method is using a double-clad active fibre as the gain medium. In Ref. [125], an average output power up to 1.8 W at 1044 nm was obtained from a ytterbium-doped double-clad fibre laser, accompanying a maximum pulse energy of 8.37 μJ. Using similar laser configuration in a thulium-doped double-clad fibre laser, Tang et al. subsequently reported a higher average output power of 5.2 W and a pulse energy of 18 μJ at 2005 nm [126].

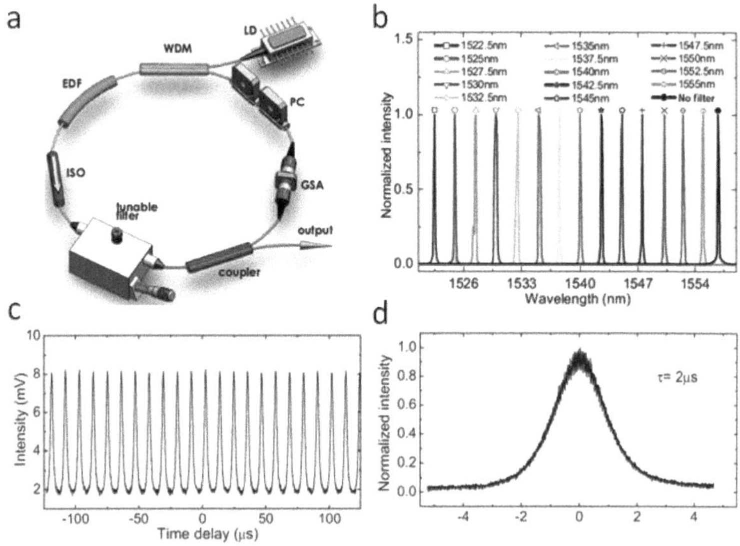

Figure 4.4 (a) Schematic setup of graphene *Q*-switched laser: laser diode (LD), wavelength division multiplexer (WDM), erbium-doped fibre (EDF), isolator (ISO), graphene-based saturable absorber (GSA) and polarisation controller (PC). (b) Output spectra for 14 tuning wavelengths. (c) Typical pulse train for 2.8 mW output power. (d) Single pulse envelope. Adapted from Ref. [122], with the permission of AIP Publishing.

For other laser formats (such as solid-state, waveguide), GSAs were also favoured for investigating the performance of the *Q*-switching operation. The typically dominant optical component (i.e., SAM) coated with large graphene films on transparent or reflective substrates plays a key role in pulse formation in laser cavities gained by various rare earth ion-doped crystals, including Nd:YAG [54, 55, 127], Nd:$GdVO_4$ [128], Er:YAG [129], Tm:YAG [130], Ho:YAG [131], etc. The wavelength-independent optical absorption of graphene enables effective modulation at all the laser frequencies emitted by the corresponding gain crystal with a specific cavity design. Compared to typical fibre lasers employing single-mode formats, solid-state lasers feature larger average power and higher pulse energy [54, 55].

4.3.2 Graphene Mode-Locked Lasers

The formation of mode-locked pulses in lasers is usually dependent on the modulation of cavity loss and the phase of the longitudinal modes [5]. A continuous wave can be converted to a mode-locked pulse train if a fixed phase relationship between the longitudinal modes in the gain bandwidth is induced. The circulating light in the laser cavity typically releases a single pulse per round trip through the output coupler; therefore, the pulse train usually has a constant repetition rate corresponding to the optical cavity length (e.g., a few tens of megahertz for a typical fibre laser with a few metres cavity length) [25]. The pulse duration of mode-locked lasers is generally in a short timescale ranging from a few hundreds of femtoseconds to a few tens of picoseconds [37]. In this section, we will discuss the current state of the art of graphene mode-locked lasers, including fibre (Section 4.2.1), solid-state (Section 4.2.2), waveguide and semiconductor (Section 4.2.3) lasers.

4.3.2.1 Graphene Mode-Locked Fibre Lasers

Ultra-fast fibre lasers are one of the most promising pulsed sources because of their mechanical flexibility, compactness, effective thermal dissipation and ease of cavity management [3]. Compared to solid-state lasers, fibre lasers could provide large gain [37]. Consequently, SAs (e.g., graphene) with inevitable non-saturable loss are tolerant for fibre lasers, owing to the gain compensation [10]. Beneficial from the broadband optical response and ultra-fast carrier relaxation time of graphene, various fibre lasers mode-locked with GSAs have been demonstrated since 2009 [24–26, 38, 40, 45, 47, 50, 52, 132–150], as listed in Table 4.1. To form an all-fibre laser cavity, GSAs have been typically integrated with fibre-compatible devices. One of the most popular techniques is to sandwich SA films between two fibre connectors (Fig. 4.5a), in which the intra-cavity light interacts with a graphene sheet perpendicularly [24–26]. In addition, an evanescent field interaction regime through graphene-coated fibres (D-shaped [50], tapered fibre [141]) and graphene-filled hollow-core fibres [52] enables optical interplay over a long distance, which is an effective

Table 4.1

Parametres of the SAs		Performance of the mode-locked lasers					
Graphene	**SAs types**	**Gain media**	**Wavelength (nm)**	**Repetition rate (MHz)**	**Output power (mW)**	**Pulse duration (ps)**	**References**
GO	Sandwich	YDF	1051.8-1068.2	14.2	2.1	340	[132]
GO-PVA			1029	40.9	6	191	[133]
CVD graphene			1069.8	0.9	0.37	580	[134]
			1035	16.29	/	65000	[38]
		EDF	1564	19.3	/	0.87	[38]
Graphene polymer			1557	/	/	-0.8	[24]
LPE graphene			1559	19.9	/	0.464	[25]
CVD graphene			1565	1.79	2	0.756	[26]
LPE graphene			1559	27.4	1.2	0.174	[135]
Mechanical cleavage			1566	6.22	/	0.88	[45]
LPE graphene			1525-1559	8	1	1	[40]
RGO	Hollow fibre		1555	15.36	80	0.51	[52]
CVD graphene	Tapered fibre		1539.4-1550	27	/	14-34	[136]
	Sandwich		1559.12	25.51	/	0.432	[137]
Mechanical cleavage			1560.5	106	9.6	0.9	[138]
			1557	114.1	2.8	0.57	[139]
Graphene-chitosan			1554	63.37	3.5	0.168	[140]

Parametres of the SAs		Performance of the mode-locked lasers					
Graphene	SAs types	Gain media	Wavelength (nm)	Repetition rate (MHz)	Output power (mW)	Pulse duration (ps)	References
RGO	Tapered fibre		1557.56	3.33	4.2	15.7	[141]
GO	GOSAM		1531	19.5	23.3	0.542	[142]
LPE graphene	Sandwich		1589.68	6.95	20.4	0.694	[143]
CVD graphene			1576.3	6.84	/	0.59	[47]
GO	D-shaped fibre	THDF	1950	33.25	1.6	0.578	[50]
CVD graphene	D-shaped fibre	THDF	1879.4	7.8	450 (Amplified)	4.7	[144]
CVD graphene	Sandwich	TDF	1884	20.5	1.35	1.2	[145]
LPE graphene			1940	6.46	2	3.6	[146]
			1953.3	16.937	1.41	2.1	[147]
CVD graphene		EDF, TDF	1565, 1944	20.19/18.43	0.5/1.3	0.933 , 1.03	[148]
			1585.5, 1938	20.5	0.5, 1.5	0.915 , 1.57	[149]
		THDF	1912-1918	1.82	/	6500	[38]
	GSAM	Er:ZBLAN	2784.5	25.4	18	42	[150]

approach for high-power operation. Ref. [52] reported a graphene composite (in solution)-filled hollow-core fibre component as an SA for mode-locked pulse output with an average power up to 80 mW, much larger than those typically investigated pulsed sources emitting power in the level of a few milli-watts.

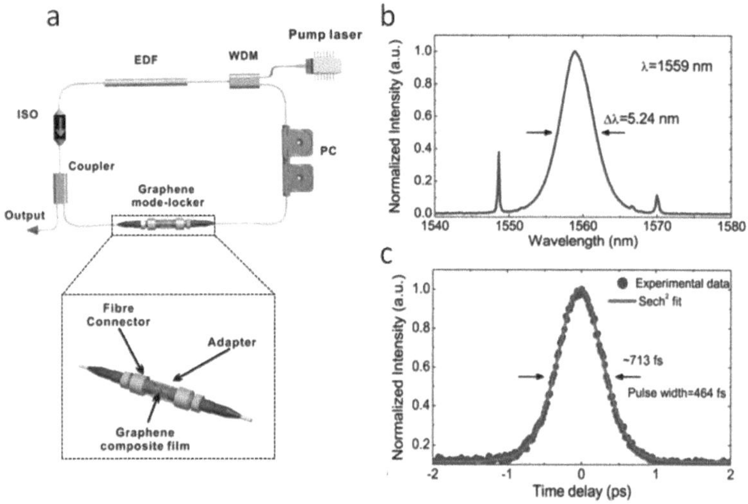

Figure 4.5 Graphene composite mode-locked fibre laser. (a) Ring fibre laser cavity and a close-up image of a graphene mode locker. (b) Mode-locked laser spectrum. (c) Auto-correlation trace of output pulse. Adapted with permission from Ref. [25]. Copyright (2010) American Chemical Society.

To realise multi-wavelength mode-locking with GSAs, gain fibres such as ytterbium-doped fibre (YDF) [132], EDF [139], thulium-doped fibre (TDF) [145] and thulium-holmium-co-doped fibre (THDF) [50] were used to emit light at several technical wavelengths. Ref. [38] reported mode-locked fibre lasers at 1, 1.5 and 2 μm by utilising a single GSA, demonstrating the practical feasibility of a graphene device with ultra-broadband optical absorption. Amongst these wavelengths, 1.5 μm lasers based on EDFs were the most intensively studied because of the low-cost fibre components in the telecommunication band and huge commercial requirements for ultra-fast sources at this waveband. Figure 4.5 illustrates a mode-locked fibre laser configured in a ring resonator and the characterised laser performance [25]. The narrow Kelly sidebands located at the spectral wings (in Fig. 4.5b) identify a typical anomalous soliton oscillation of the generated pulses. The strong interplay of optical non-linearity and group velocity dispersion gives rise to a pulse duration of 464 fs. It is noteworthy that the mode-locked pulse duration can be suppressed to a shortest value if the intra-cavity dispersion can be compensated to a minimum [135].

Using a dispersion map management cavity, graphene mode-locked fibre lasers have enabled short pulses with a duration down to sub-200 fs, as reported in Refs. [135, 140]. Both normal and anomalous dispersion fibres were used in this fibre cavity to periodically stretch and supress the pulse duration in one round-trip propagation, which significantly reduced the optical non-linearity by a directly generated short pulse. Dissipative solitons formed in a large normal dispersive cavity consisting of all-normal or net-normal dispersive fibres have also been investigated because of their feasibility for large-energy pulse generation. Ref. [151] reported a typical dissipative soliton cavity by employing a long dispersion compensation fibre, acquiring a high pulse energy up to 2.3 nJ. The formation of solitons in a dissipative regime is a multiple interaction of the normal dispersion, fibre Kerr non-linearity and effective laser gain bandwidth filtering [151]. The dynamics of the dissipative solitons is illustrated by the complex Ginzburg-Landau equation [152]; the repetition rate of mode-locked lasers is usually a constant only determined by the optical cavity length, as described by the formula $f = c/nL$ for a ring fibre laser, where c is the speed of light in vacuum, n is the refractive index of the fibre and L is the cavity length. High-repetition-rate mode-locking requires reduction of the cavity length so as to output the circulating pulse faster. Ref. [153] reported a graphene mode-locked laser with a repetition rate up to ~10 GHz by employing a 10 mm Fabry–Perot cavity. It is noteworthy that a similar cavity with a higher repetition rate is challenged because of the low optical gain in a short active fibre and precise alignment of a miniature cavity mirror and the SA. High-order harmonic mode-locking is a substituted approach to obtain high-repetition-rate pulses from a common fibre cavity by the multipulse regime, which generally operates at a relatively high power. GSAs have approached a repetition rate up to 5.882 GHz (corresponding to the 683rd harmonic wave) at 1.61 μm [154].

4.3.2.2 Graphene mode-locked solid-state lasers

Solid-state lasers, in which intra-cavity photons are amplified by bulk optical gain media, were the most extensively studied laser format. Indeed, the first experimentally realised mode-locked laser was obtained in a solid-state laser gained by Nd:glass and passively modulated by a dye-based SA [155]. After utilisation of SESAM and Kerr lens mode-locking technologies, ultra-fast solid-state lasers were rapidly developed [156, 157].

The ultra-broadband optical response of GSAs has enabled mode-locking operations at various infrared wavelengths. Laser media, including Ti:sapphire [158], Cr:LiSAF [159, 160], Nd:YVO_4 [161], Nd:YAG [56], Yb:GAGG [162], Yb:$CaYAlO_4$ [163], Nd:GdVO4 [164], Yb:KGW [165], Yb:KluW [53], Cr:forsterite [166], Cr:YAG [167], Er:Yb:glass [168], Tm:Lu_2O_3 [169], Tm:CLNGG [170], Tm:YAP [171], Cr:ZnS [172] and Cr:ZnSe [173], have been employed and modulated by GSAs to generate ultra-fast pulses at 0.8, 1, 1.2, 1.3, 1.5, 2, 2.4 and 2.5 μm [53, 56, 158–179], as listed in Table 4.2. To ensure effective optical modulation, graphene was usually deposited on a high-reflection mirror or a high-transmittive glass substrate and functioned as end feedback or an intra-cavity component in the lasers. Graphene films grown by CVD or graphene solution fabricated by LPE were directly transferred or spin-coated on the target substrates to form a large-area SAM [175, 178]. Ref. [174] reported a graphene oxide-based saturable absorber mirror (GOSAM) for mode-locking with an average output power up to 3.06 W, a repetition rate of 84 MHz and a pulse duration of 8.8 ps in an Nd:YVO_4 laser at 1065.5 nm, in which the power is larger than most of the graphene mode-locked lasers by other bulk media. Ref. [54] demonstrated the first radially polarised 1064 nm Nd:YAG laser mode-locked by a graphene-based SAM. The laser generates a 15.6 ps pulse with output power up to 2 W and a pulse repetition rate of 112 MHz. Compared to fibre lasers with large optical non-linearity and dispersion, solid-state lasers with light oscillating in free space can release a pulse in a timescale down to sub-100 fs. Figure 4.6 shows a graphene mode-locked Ti:sapphire laser setup and the characterised laser pulse and output spectrum [158]. The obtained pulse duration (63 fs) is one of the shortest laser pulses with graphene as the SA. As can be seen from the inset of Fig. 4.6b, the output spectrum has

Table 4.2

Parametres of the SAs		Performance of the mode-locked lasers					
Graphene	**SA type**	**Gain media**	**Wavelength (nm)**	**Repetition rate (MHz)**	**Output power (W)**	**Pulse duration (fs)**	**References**
CVD graphene	GSAM	Ti:sapphire	800	99.4	0.48	63	[158]
		Cr:LiSAF	850	132	0.01	68	[159]
			836-897	107	0.008	19	[160]
GO	GOSAM	Nd:YVO_4	1064	88	1.2	/	[161]
			1065.5	84	3.06	8800	[174]
LPE graphene	GSAM	Nd:YAG	1064	88	0.1	4000	[56]
RGO	RGOSAM	Yb:GAGG	1041.1	45	0.8	643	[162]
CVD graphene	GSAM	Yb:$CaYAlO_4$	1068	113.5	0.015	30	[163]
LPE graphene		Nd:$GdVO_4$	1065	43	0.36	16000	[164]
			1341.3	500	1.29	11000	[175]
GO	GOSAM		1064	70	1.1	4500	[176]
GO-PVA film			1063	120	0.68	12000	[177]

(Continued)

Table 4.2 *(Continued)*

Parametres of the SAs		Performance of the mode-locked lasers					
Graphene	**SA type**	**Gain media**	**Wavelength (nm)**	**Repetition rate (MHz)**	**Output power (W)**	**Pulse duration (fs)**	**References**
CVD graphene	GSAM	Nd:YVO_4	1064	140	1.6	<330000	[178]
LPE graphene		Yb:KGW	1031.1	86	0.504	428	[165]
CVD graphene		Yb:KLuW	1036-1050	92.5	0.35	160	[53]
		Cr:forsterite	1220-1270	75	0.23	94	[166]
		Cr:YAG	1516	85	0.5	91	[167]
		Er:Yb:glass	1550	88	0.0045	260	[168]
		Tm:Lu_2Os	2067	110	0.27	410	[169]
		Tm:CLNGG	2014.4	95	0.06	882	[170]
			2018	98.7	0.0602	729	[179]
GO	GOSAM	Tm:YAP	2023	71.8	0.268	<10000	[171]
CVD graphene	GSAM	Cr:ZnS	2400	108	0.25	41	[172]
			2500	77	0.08	226	[173]

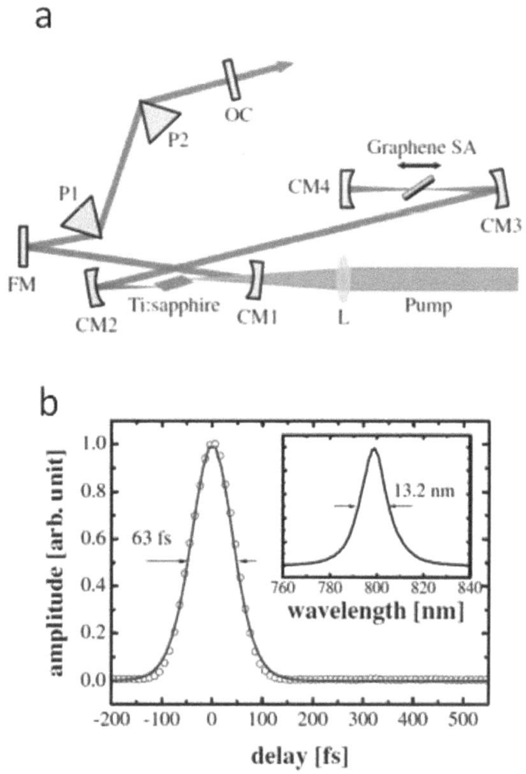

Figure 4.6 (a) Setup of the graphene SA mode-locked Ti:sapphire laser. CM1, CM2, CM3 and CM4: concave mirrors with radius of curvature (ROC) = –100 mm; FM: folding mirror; P1, P2: SF10 prisms; OC: output coupler with 10% transmission. (b) Auto-correlation trace of the monolayer graphene SA mode-locked Ti:sapphire laser and the corresponding optical spectrum (inset). Reproduced from Ref. [158]. Copyright (2012) The Japan Society of Applied Physics.

a smooth profile, different with the typical soliton spectra in fibre lasers with Kelly sidebands. In addition, large-area graphene (grown by the CVD method) SA mirrors were employed at two individual wavelengths (1.3 and 1.5 μm) to generate an ultra-fast pulse in the sub-100 fs scale [166, 167]. The high-quality monolayer graphene provides ~1% modulation depth and less than 2% non-saturable loss. For wavelengths longer than 2 μm, a few chromium-doped crystals gain lasers were investigated [172, 173]. Ref. [172] reported a 41 fs mode-locked laser with a super-wide spectral bandwidth up to 190 nm, which significantly compresses the shortest pulse duration realised by graphene. A very recent study reported in Ref. [160] demonstrated a 19 fs Cr^{3+}:LiSAF laser after optimising cavity dispersion with a prism pair, which is the shortest pulse experimentally observed to date by using graphene as the SA.

4.3.2.3 Graphene mode-locked waveguide and semiconductor lasers

The intrinsic compactness and small volume of lasers emitting from waveguide and semiconductor gain media provide alternative platforms for ultra-fast pulse generation [57]. Waveguide lasers are usually fabricated in rare earth-doped substrates (including glass, crystal and polymer [6]) by femtosecond laser writing technology. Semiconductor lasers (e.g., vertical external cavity surface emitting lasers [VECSELs] [180]) utilise direct-bandgap materials to achieve optical gain when they are pumped by laser diodes. Ultra-fast laser sources based on these substrates can radiate light with good beam quality, large average power, high repetition rate and wide spectral bandwidth [57, 181, 182], which are thus powerful tools for various applications in industrial materials processing, high-speed telecommunications, bio-imaging and non-linear conversion [57, 181, 182].

The flexible transplantation and easy integration of GSAs provide a possibility to mode-lock waveguide and semiconductor lasers. Waveguide lasers usually consist of a simple Fabry–Perot cavity to form optical oscillation and the intra-cavity light is fed back by a pair of end face-coated reflection mirrors [57]. A GSA can be directly deposited on one of the cavity mirrors to form a compact configuration, while the start of mode-locking requires precise alignment of each element. Up to date, the output wavelengths of graphene mode-locked waveguide lasers have covered the spectral range between 1 and 2 μm, by using Nd [58], Yb [57, 181], Er [181] or Tm [182] ion-doped gain media. Figure 4.7a shows an experimental setup of a Yb-doped bismuthate glass waveguide laser using an LPE graphene film as an SA [57]. It is noteworthy that the laser operates in a *Q*-switched mode-locking regime rather than a continuous-wave mode-locking state [58]. The short waveguide length is related to a high repetition rate of ~1.5 GHz at a ~ 1039 nm central wavelength [57], as depicted in Figs. 4.7b and 4.7c, respectively. Refs. [58, 181,

182] reported mode-locking or Q-switched mode-locking lasers at 1064, 1535 and 1943.5 nm, gained by Nd:YAG, Yb:Er-doped phosphate glass and Tm:YAG-based waveguides, corresponding to mode-locked repetition rates of 11.3, 6.8 and 7.8 GHz, respectively.

Semiconductor disk lasers mode-locked with graphene were also investigated [59, 183]. The large surface area and thin gain structure are beneficial for heat dissipation and thus can support high-power operation. Ref. [59] introduces a VECSEL mode-locked by monolayer graphene on a distributed Bragg reflector (DBR) structure, as schematically illustrated in Fig. 4.8a. Adjusting the thickness of SiO_2-filled space between graphene and the reflection substrate enables to change the graphene absorption via electric field enhancement; as a result, the modulation depth can be modified. A mode-locked laser with a pulse duration of 466 fs at ~950 nm was obtained, associated with a repetition rate of 2.48 GHz. Ref. [183] reported a graphene mode-locked semiconductor disk laser with average output power up to 10.2 W and a pulse energy of 2.8 nJ at a repetition rate of 1.76 GHz.

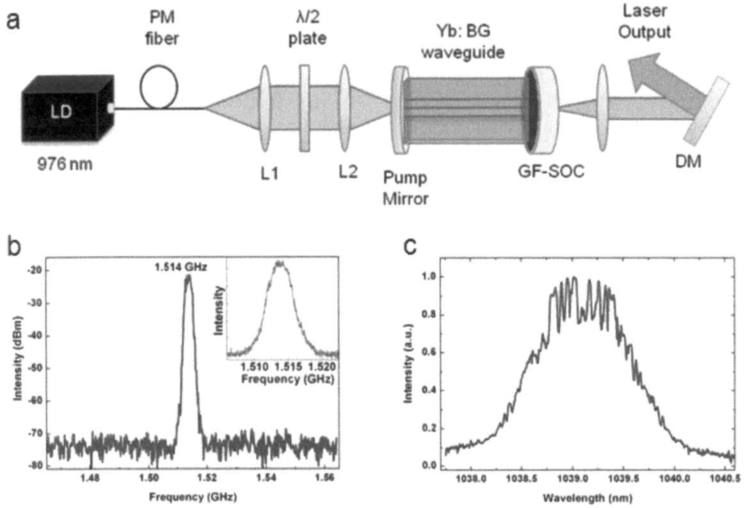

Figure 4.7 Graphene-modulated waveguide lasers. (a) Laser schematic. L1 and L2: coupling lenses; PM: polarisation-maintaining fibre; GFSOC: graphene film saturable output coupler; DM: dichroic mirror. (b) Radio-frequency spectrum measured at the maximum pump (zoom-out in the inset). (c) Corresponding output optical spectrum. Adapted with permission from Ref. [57], The Optical Society.

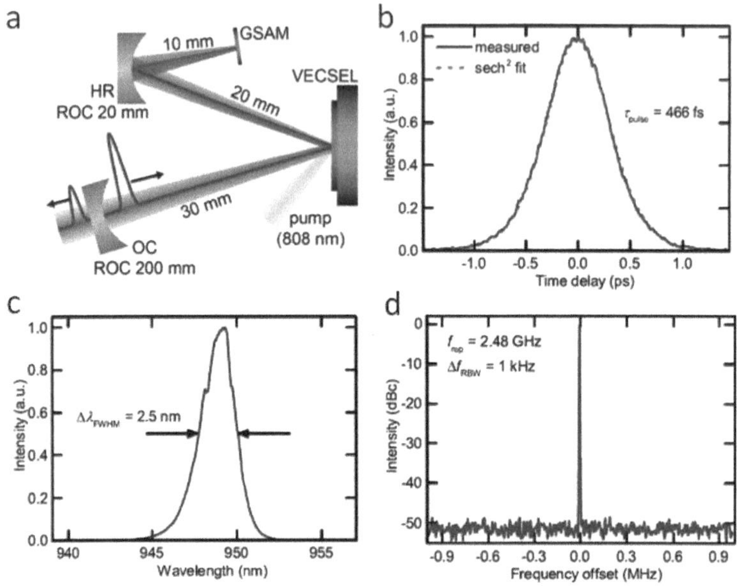

Figure 4.8 Graphene mode-locked semiconductor laser. (a) Schematic of the disk laser setup. OC: output coupler mirror; HR: high-reflective folding mirror; GSAM: graphene saturable absorber mirror. The total cavity length is 6 cm. (b) Auto-correlation signal and (dashed line) fit with the auto-correlation of an ideal sech^2-shaped pulse. (c) Optical spectrum. (d) Microwave spectrum centred around a repetition rate of 2.5 GHz, measured with a 1 kHz resolution bandwidth (RBW). Adapted with permission from Ref. [59], The Optical Society.

4.4 Transition Metal Dichalcogenide-Based Pulsed Lasers

Various TMDs have been demonstrated for *Q*-switching and mode locking [27, 29, 79–92, 94-99, 184–197]. Thus far, there are three most popular methods to fabricate SAs based on TMD materials. The first method is depositing the nanomaterials on a side-polished fibre (SPF), tapered fibre or other waveguides to realise non-linear interaction between the evanescent field and the TMD materials [27, 99, 184–186]. The parametres of such SAs depend on the deposition depth and evanescent field of light and thus are difficult to be precisely controlled. Attributing to the evanescent field and long interaction length, the TMDs nanosheets are not exposed to large optical intensity, which permits such SAs to work in a high-power regime [90]. The second method is directly coating the nanomaterials on the optical mirror, quartz glass or fibre facet [79, 82, 187–189]. This type of SA can be used in fibre laser or solid-state laser, while it is quite vulnerable because the nanomaterial is directly exposed in air. The third method is embedding the nanomaterials into a polymer (such as PVA [84, 91, 94] and fluorine mica [190]) to fabricate filmy SAs. By cutting the film composite into tiny pieces and sandwiching it between two fibre connectors, a large number of fibre-integrated SAs with similar performance can be obtained in a single experiment. However, the unwanted insert loss is inevitable and heat accumulation may destroy the polymer-TMD-based SAs. Here, in this section, we will discuss the current progress in TMD-based pulsed lasers.

4.4.1 Transition Metal Dichalcogenides *Q*-switched Lasers

Up till now, different types of TMD-based SAs have been successfully implemented in various types of *Q*-switched lasers: fibre lasers [80, 81, 84, 91, 92, 191–195], solid-state lasers [79, 88, 187] and waveguide lasers [186, 196]. A typical *Q*-switched solid-state laser based on a MoS_2 SA is illustrated in Fig. 4.9. By utilising a single MoS_2 SA, *Q*-switched operations were established in Nd:$GdVO_4$, Nd:YGG and Tm:Ho:YGG lasers at a wavelength of 1.06, 1.42 and 2.1 μm, respectively [79].

Table 4.3 summarises the characteristics of representative TMD SAs and the performance of *Q*-switched lasers [79, 80, 96, 186, 187, 189, 191–193, 197, 198]. The *Q*-switched lasers based on TMD SAs cover the wavelength range of 0.522 [187], 0.6351 [191], 1.0236 [186], 1.06 [79], 1.42 [79], 1.56 [80, 81, 87], 1.924 [96] and 2.1 [79]. The repetition rate of the *Q*-switched pulse is in the kilohertz regime and the pulse duration spans from hundreds of nanoseconds to tens of micro-seconds. The modulation depths of these TMD-based SAs spans from 0.4% to 29.4%, and the saturation intensities range from 837 kW/cm^2 to 2.45 GW/cm^2. Three factors may influence the modulation depths and saturation intensities of SAs. First, depending on the bandgap of the TMD materials, the absorption coefficient changes with the optical wavelength and will thus influence the properties of the SAs. Second, the thickness of the TMD materials

Table 4.3

Materials	Substrates type	Modulation depth	Saturation intensity (MW/cm^2)	Laser type	Wavelength (nm)	Repetition rate (kHz)	Output power (mW)	Pulse duration	References
MoS_2	PVA	2.15%	129.4	Fibre	1560	7.758-41.452	0.05-1.3	9.92-13.534	[80]
		/	/		635.5	240.4-438.6	7.1	0.227	[191]
		2%	10		1519.6-1567.7	8.77-43.47	0.5-5.9	26.7-3.3	[192]
	Quartz glass	7.7%	0.837	Solid-state	522	125	10	0.579	[187]
		27%	2450		1060	732	227	0.97	[79]
		27%	2450		1420	77	52	0.729	[79]
		27%	2450		2100	149	206	0.41	[79]
	Mirror	9%	-27	Fibre	1549.83	116-131	<20	0.66-0.76	[189]
	PVA	2.53%	148.2		1560	47.026-77.925	5.5-6.4	3.966-6.707	[80]
		>7%	/		635.1	232.7-512.8	8.7	0.207	[191]

Materials	Substrates type	Modulation depth	Saturation intensity (MW/cm^2)	Laser type	Wavelength (nm)	Repetition rate (kHz)	Output power (mW)	Pulse duration	References
WS_2	Waveguide	/	/	Wave-guide	1023.6	360	7.8	0.125-0.175	[186]
	D-shaped fibre	-0.4%	/	Fibre	1567.8	82-134	-0.3-2.5	0.71-2.82	[193]
	Au mirror	29.4%	>6		2865.7	25.6-131.6	-5-48.4	3.8-1.67	[197]
$MoSe_2$	PVA	4.7%	3.4		1060	60.0-74.9	6.26-8.72	4.6-2.8	[96]
		4.7%	3.4		1566	26.5-35.4	18.9-29.2	7.9-4.8	[96]
		4.7%	3.4		1924	14.0-21.8	0.13-1	16-5.5	[96]
		/	/		635.4	357.1555.1	6.2	0.24	[191]
		6.73%	132.5		1560	60.724-66.847	2.24-2.45	4.04-6.506	[80]
WSe_2		3.02%	270.4		1560	60.724-66.847	2.24-3.2	4.063-9.182	[80]
		4.31%	7.97		1560	92.64-138	<3.9	1.478-0.754	[198]

and the incorporation method also affect the performance of the SA. Third, the obtained modulation depths or saturation intensities are related to the duration of the illumination pulse used in the measurement system.

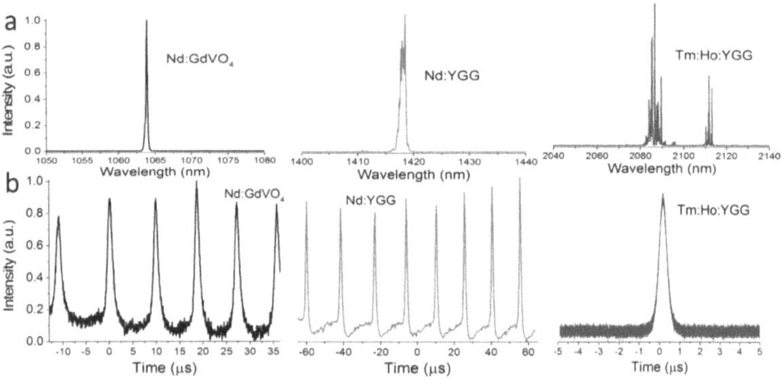

Figure 4.9 Passively *Q*-switched solid-state lasers based on a MoS_2 SA. (a) Nd:GdVO4, Nd:YGG and Tm:Ho:YGG laser spectra at centre wavelengths of 1.06, 1.42 and 2.1 μm, respectively. (b) Passively *Q*-switched Nd:GdVO4, Nd:YGG and Tm:Ho:YGG lasers with a pulse duration of 970, 729 and 410 ns, respectively. Adapted from Ref. [79]. Copyright © 2014, John Wiley and Sons.

4.4.2 Transition Metal Dichalcogenide Mode-Locked Lasers

4.4.2.1 Transition metal dichalcogenide mode-locked fibre lasers

Compared to *Q*-switched lasers or continuous-wave lasers, mode-locked lasers have a shorter pulse duration and a higher peak power. Three typical methods were used to fabricate fibre-integrated SAs for TMD mode-locked fibre lasers: depositing the nanomaterials on a tapered fibre or an SPF (such as SA_1 in Fig. 4.10a), coating a TMD film on the fibre facet [82] or embedding the nanomaterials into a polymer to fabricate filmy SAs (such as SA2 in Fig. 4.10a) [185]. The passively mode-locked operation based on a TMD SA is observed in the fibre laser [27, 29, 82–84, 90, 92, 95, 97–99, 184, 185, 199–203]. Depending on the setting of the cavity dispersion, dissipative soliton [27, 184, 185, 190] and standard soliton [29, 82, 83, 97-99,

199-203] can be delivered from the fibre laser. A typical soliton fibre laser mode-locked with a WS_2 SA is illustrated in Fig. 4.10. The fibre laser has a ring cavity with a net dispersion of −0.38 ps^2. The output pulses exhibit typical characteristics of a soliton with a bandwidth of 2.3 nm and a pulse duration of 1.32 ps [90].

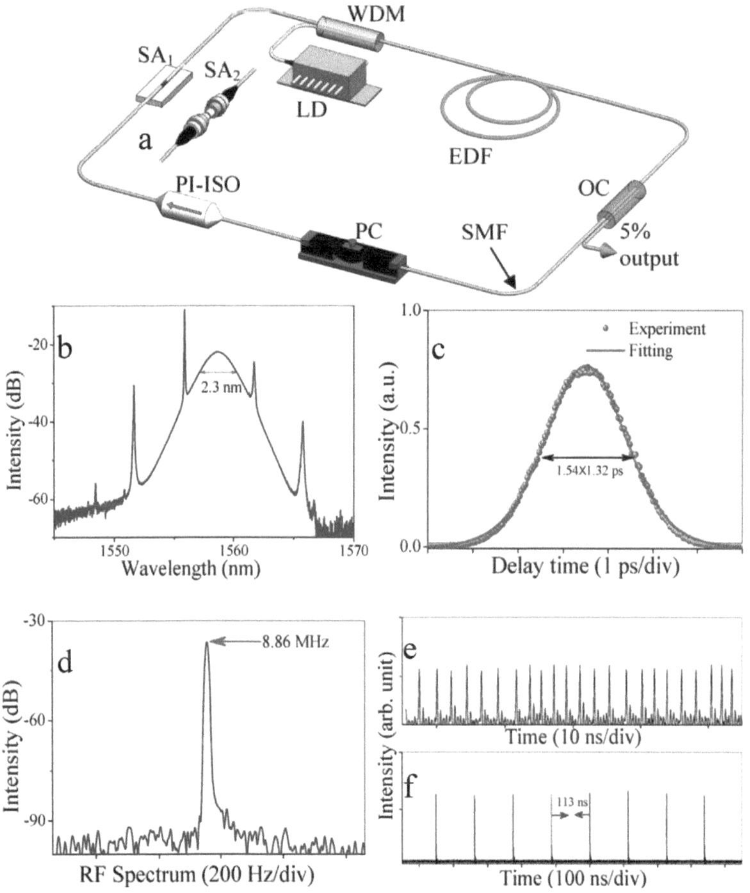

Figure 4.10 WS_2 mode-locked fibre laser. (a) Experimental setup. LD: laser diode; WDM: wavelength division multiplexer; EDF: erbium-doped fibre; OC: optical coupler; SMF: single-mode fibre; PC: polarisation controller; PI-ISO: polarisation-independent isolator; SAi: D-shaped fibre-based SA; SA_2: PVA-film based SA. (b) Output spectrum, (c) auto-correlation trace, (d) radio-frequency spectrum, (e) multiple pulses and (f) single pulse of the mode-locked laser. Reprinted by permission from Springer Nature Customer Service Centre GmbH: Springer Nature, *Scientific Reports*, Ref. [90], Copyright (2015).

Table 4.4

Materials	Substrates type	Modulation depth	Saturation power (MW/cm^2)	Laser type	Wavelength (nm)	Repetition rate (MHz)	Output power (mW)	Pulse duration (ps)	References
MoS_2	D-shaped fibre	2.5%	/	Dissipative soliton	1568	26.0	1-2.5	4.98	[27]
		2.5%	/	Soliton	1568	33.48	1-2.5	0.637	[27]
	Tapered fibre	10.47%	/	Dissipative soliton	1042.6	6.74	0.9-2.37	656	[184]
		4.3%	34	Soliton	1569.5	5.78-12.09	17.8	0.71-1.64	[201]
		4.4%	8000	Dissipative soliton	1595	3.63	1.8	121	[202]
	Fibre facet	35.4%	0.35	Soliton	1568.9	8.288	5.1	1.28	[82]
	PVA	10.69%	2.02		1535-1565	12.99	/	0.96-7.1	[84]
	Mirror	7%	/	Nd:YVO_4	1064.2	88.3	89	12.7	[88]
WS_2	D-shaped fibre	10.9%	3.8		1941	34.8	0.6	1.3	[199]
	D-shaped fibre	0.95%	600	Soliton	1557	8.86	/	1.32	[90]

Materials	Substrates type	Modulation depth	Saturation power (MW/cm^2)	Laser type	Wavelength (nm)	Repetition rate (MHz)	Output power (mW)	Pulse duration (ps)	References
	PVA	1.8%	750		1557	8.96	/	1.2	[90]
	D-shaped fibre	5.1	/	Dissipative soliton	1565	8.05	1.3	21.1	[185]
		2.9%	/		1064	5.57	7.6	630	[185]
	Tapered fibre	0.5%	/		1561	24.93	1.93	0.369	[203]
		0.6%	/	Soliton	1563	20.39	2.77	0.563	[203]
	PCF	3.53%	159		1563.8	19.57	2.64	0.808	[204]
	Mirror	4.3%	87 /	Yb:YAG	1057.5	86.7	<270	0.736	[205]
$MoSe_2$	D-shaped fibre	1.4%	200		1557	5.03	/	1.09	[95]
WSe_2		0.3%	50	Soliton	1556.7	5.31-3252.65	/	1.31	[95]
$MoTe_2$	PVA	0.9%	80		1561	5.26	/	1.22	[97]
WTe_2	D-shaped fibre	2.85%	/		1556.2	13.98	0.04	0.77	[98]

Table 4.4 summarises the characteristics of TMD-based SAs and the performance of mode-locked fibre lasers [27, 82, 84, 88, 90, 95, 97, 98, 184, 185, 199, 201–205]. The modulation depth ranges from 0.3% to 35.4%, and the saturable absorption intensity ranges from 0.35 MW/cm^2 to 750 MW/cm^2. The characteristics of these TMD SAs are similar to those of typical *Q*-switched fibre lasers. For mode-locking operations, the laser wavelengths are mainly focused at 1, 1.5 and 2 μm. The fundamental repetition rate of the mode-locked fibre laser depends on the cavity length, from several megahertz to tens of megahertz. Due to the high non-linearity of the fibres, multiple pulses tend to be formed in the fibre laser at the high-pump regime. With an SPF-based WSe_2 SA, 3252.65 MHz harmonically mode-locked pulses have been generated in an EDF laser [95]. The duration of the mode-locked pulse depends on both the cavity design and the saturable absorption property. For TMD SA mode-locked fibre lasers, the pulse duration in EDF lasers ranges from 0.369 ps to 21.1 ps, while in YDF lasers it is several hundreds of picoseconds.

4.4.2.2 Transition metal dichalcogenide mode-locked solid-state lasers

Solid-state lasers, usually exploiting crystal or doped glass as gain media, are the most commonly used lasers in industry, research and military. Compared to fibre lasers, solid-state lasers mode-locked with TMD-based SAs are less reported, partially attributed to the small modulation depth of such TMD materials. Ref. [88] demonstrated a passively mode-locked Nd:YVO_4 laser using a MoS_2 SA. The central wavelength, repetition rate and pulse duration are 1064.2 nm, 88.3 MHz and 12.7 ns, respectively. The characteristics of MoS_2/WS_2 SAs and the performance of mode-locked solid-state lasers are summarised in Table 4.4.

4.5 Other 2D Material-Based Pulsed Lasers

4.5.1 Other 2D Material *Q*-switched Lasers

Beside graphene and TMDs, there are also other 2D layered materials being investigated for pulsed lasers. BP, a recently rediscovered direct-bandgap semiconductor, has exhibited comparative properties to graphene. Non-linear optical absorption of BP identifies a new SA for large pulse energy exploration through passive *Q*-switching in various optical resonators, including generally studied fibres, solid-state lasers and waveguide lasers. BP also has been employed in lasers to generate *Q*-switched pulses in a wide spectral bandwidth, corresponding a series of typical wavelengths at near-infrared lasers of 1 μm [196, 206], 1.5 μm [30, 31, 207] and mid-infrared lasers of 2 μm [208–210], 2.4 μm [211], 2.7 μm [212], 2.8 μm [213] and 2.9 μm [214], respectively. In these investigations, the fabrication of BP-based SAs is mainly dominated by mechanical exfoliation and LPE approaches.

Refs. [30, 31] first reported experimental observations of mechanically exfoliated BP-based SAs for *Q*-switched laser generation at 1.55 μm fibre lasers, obtaining maximum pulse energies of 94.3 and 18 nJ, respectively. Figure 4.11 shows the typical *Q*-switched laser performance corresponding to the shortest pulse duration of 3.16 μs and the highest repetition rate of 40 kHz at 1532.5 nm. Further investigation demonstrated strong anisotropic linear and non-linear absorption in thin-layer BP films, which leads to linearly polarised pulse output at the same wavelength [30]. Other work in fibre lasers *Q*-switched by BP SAs was then reported at 2 THDF [210] and 2.8 Er:ZBLAN mid-infrared wavelengths [213]. In the latter one, pulsed energy and average output power up to 7.7 μJ and 485 mW were the highest values realised to date in BP-based fibre lasers, which are comparable with the values from GSAs.

For solid-state lasers, the interaction between light and SAs is usually in free space. In this case, a transmittive or reflective BP-coated SAM is more suitable for light modulation. To enable a large piece of device (e.g., BP-based SA film [108, 111]), the LPE technique is more feasible as a large amount of BP flakes in micro-or nanometre size can be exfoliated in specific solutions [28, 48]. BP-based SAMs have been extensively used for *Q*-switched pulse emission in bulk lasers gained by Yb:LuYAG [212], Er:Y_2O_3 [212], Tm:YAP [208, 209], Tm:CaYAlO4 [212] and Cr:ZnSe [211] media, covering the wavelengths of 1, 1.5, 2 and 2.4 μm. The results in Ref. [209] represent the current state-of-the-art performance of *Q*-switched laser with a BP-based SA, performing at the maximum average power of 3.1 W, maximum pulse energy of 39.5 μJ and minimum pulse duration of 181 ns. In addition, waveguide laser Q-switched by a BP SA was also carried out by using Nd:YAG ceramic as a gain medium [196]. Similar to the treatment in graphene-based waveguide lasers, a BP film was coated on one end face of the waveguide cavity to modulate the Q-factor [196]. Benefitting from the short cavity length, this waveguide laser was able to release Q-switched pulses with a pulse duration down to 55 ns in a high repetition rate up to 5.6 MHz, which is much different from other laser platforms using long cavities (usually a few metres for fibre and solid-state lasers). These investigations have sufficiently demonstrated the huge potential of BP for large-energy pulse generation. BP-based novel optical devices with enhanced non-linear optical properties are being developed for pulsed lasers, which would bring new findings in the near future.

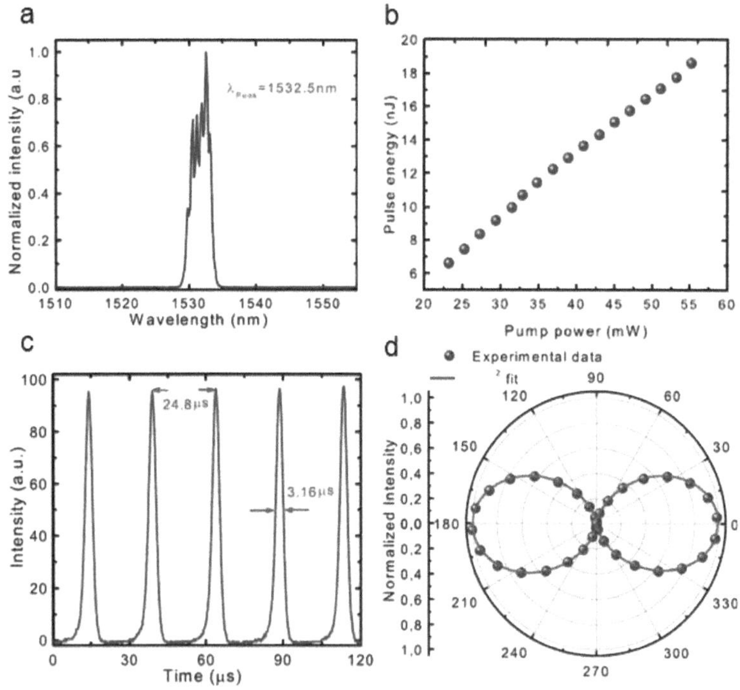

Figure 4.11 Typical performance of BP Q-switched fibre laser. (a) Output spectrum. (b) Pulse energy as a function of pump power. (c) Pulse train. (d) Output polarisation property. Adapted from Ref. [30] under a Creative Commons License Deed Attribution 4.0 International (CC BY 4.0), https://creativecommons.org/licenses/by/4.0/.

4.5.2 Other 2D Material Mode-Locked Lasers

Transient carrier transition and high electron mobility enable BPs competitive properties and potential for electronic and photonic applications [103, 104]. Considering the broad optical response and large non-linear optical absorption, BP-based ultra-fast pulse generation has been carried out recently [30-32, 207, 214–226]. Similarly with the aforementioned GSA components, BP films sandwiched by two fibre connectors were the most convenient approach for device integration in fibre lasers [30, 32, 207, 215, 216, 221]. In addition, evanescent field interaction with a D-shaped or tapered micro-fibre through surface depositing or coating was also an effective approach for laser cavity configuration [219, 220, 226]. For those pulsed lasers where solid-state components were used, BP-based saturable absorption mirrors were constructed and utilised. [214, 218, 222].

Amongst a large amount of BP SA studies via various laser platforms, the most intensively explored wavelength is still a 1.5 μm fibre laser [30–32, 207, 219–221], even for the first reports [30–32], because of the ease of soliton pulse formation with low-cost telecommunication fibre components. Figure 4.12 shows the laser setup and performance of a BP mode-locked laser with a ring cavity [30]. The Kelly sideband in the output spectrum infers a typical soliton wave output, as can be seen from Fig. 4.12b. Figure 4.12c illustrates an auto-correlation trace in which a pulse duration of 786 fs is revealed. Particularly, the mode-locked laser is identified with a linear output polarisation state (Fig. 4.12d) due to the anisotropic absorption of the mechanically exfoliated BP flakes. Ref. [32] reported a 272 fs mode-locked laser in an EDF laser, which is the shortest value enabled in all-anomalous cavity to date. At another technically important wavelength of 1 μm, mode-locked pulses with BP were investigated in both fibre and solid-state lasers [216, 218]. In a double-clad YDF laser, the mode-locked pulses present a high average output power up to 80 mW and 7.54 ps pulse duration, although the mechanically exfoliated BP flake has a non-saturable loss of 57% [216]. The solid-state laser adopted Nd:YVO_4 as a gain medium, in which an LPE-based BP SA mirror was used [218]. The mode-locked laser investigated here also generated a high average power up to 460 mW. To investigate the ultra-fast optical properties of BP in the mid-infrared region, several fibre lasers were subsequently employed, such as Ho^{3+}/Pr^{3+} co-doped fluoride fibre laser [214], Er:ZBLAN [213]. The average output powers of the mode-locked lasers were 87.8 mW at 2866.7 nm and 613 mW at 2783 nm.

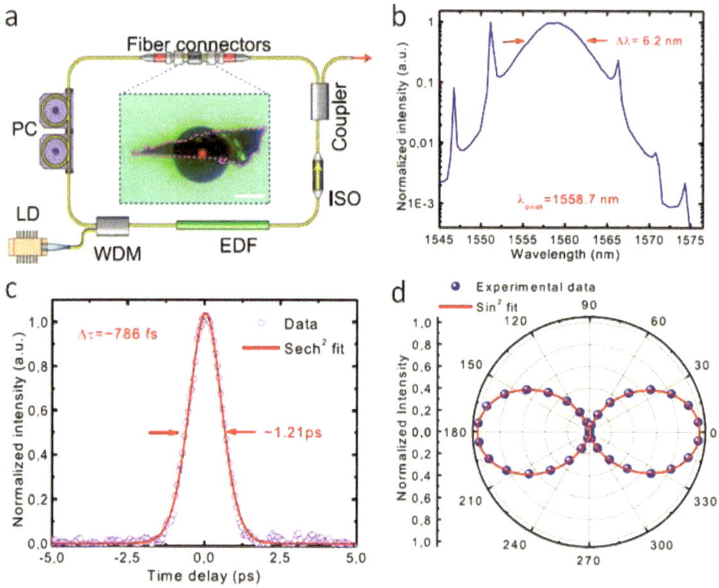

Figure 4.12 (a) BP mode-locked fibre laser setup. PC: polarisation controller; LD: laser diode; WDM: wavelength division multiplexer; EDF: erbium-doped fibre; ISO: isolator. The inset shows the transferred BP on the optical fibre end. Scale bar = 60 μm. The red point in the centre indicates the ~10-μm-diameter optical fibre core, with the outside concentric grey circle of the 125 μm fibre cladding. The area marked by the pink dotted line specifies the transferred BP. (b) Output spectrum. (c) Auto-correlation trace. (d) Output polarisation property. Adapted from Ref. [30] under a Creative Commons License Deed Attribution 4.0 International (CC BY 4.0), https://creativecommons.org/licenses/by/4.0/.

Similarly, TI-based SAs were mostly used in fibre and solid-state lasers for pulse generation at the typical wavelengths of 1, 1.5, 2 and 3 μm [33, 116, 119, 227–229]. The bulk structured SAs can be exfoliated and integrated evanescently with side-polished fibre, micro-fibre, fibre tip and quartz substrate to fit the laser system. For example, Ref. [228] reported a Sb_2Te_3 on side-polished fibre SA incorporating with EDF laser, which enables the output of 270 fs soliton pulses at 1560 nm.

4.6 Perspectives

Graphene has been playing a pioneer role in 2D materials from basic research to practical application, which also boosts the rapid development of other 2D layered materials (as introduced in the main text of TMDs, BP and TIs). The intensive photonics studies of these newly emerging semiconductors in the past decade have built up a novel SA family [2]. The zero-gap graphene and wide-gap distribution in bulk or few-layer forms of other 2D semiconductors have covered the band from visible to mid-infrared, in which various wavelengths emitted by fibre, solid-state, waveguide and semiconductor lasers are involved [24, 31, 79, 133, 146, 187, 191, 228]. Particularly, it is now possible to use one SA (such as graphene) as a broadband optical modulator to cover a wide bandwidth (up to >1000 nm) [38]; this is challenging with the previously used SAs. The understanding of ultra-fast relaxation dynamics of these semiconductors offers natural platforms for pulse formation in an ultra-short timescale, which deserves further studies to reach the technological limitation of developed ultra-fast sources.

Fibre and solid-state lasers are at the research frontier of ultra-fast sources by passive mode-locking or *Q*-switching technologies. The ultimate performance of a specific laser is an interactive result of the SA and laser dynamics, in which power, pulse duration, repetition rate and spectral bandwidth are important factors for practical application [1-6]. Despite the fact that fibre laser can support high-power operation up to kilowatts in continuous waves [230], the output power is still quite low for a pulsed operation regime (typically in milli-watts) because of the limitation of large optical non-linearity and damage threshold of the SAs [26, 30, 191]. An effective way to obtain higher average power and peak power is to use large-mode-area fibres [124]. In addition, evanescent-mode-coupled SAs (such as coating on D-shaped fibres, tapered fibres) also enable high-power operation by increasing the light-material interaction strength via the evanescent field. The thermal effect becomes a serious issue in solid-state lasers when a large gain is used to compensate the optical loss in cavity [37]. Reducing non-saturable loss of the SAs and thermal effect is an effective solution to enhance the power and pulse energy.

In fibre waveguides, optical non-linearity not only restricts the power but also limits the shortest pulse duration. Dispersion management can be used to enable large-energy dissipative solitons in normal dispersive map and reduce pulse duration by compensating the total dispersion close to zero [135, 151]. It should be noted that further performance improvement of the 2D material-based SAs and laser dynamics will break the shortest pulse duration record reported to date with 2D materials. Waveguide and semiconductor-based ultra-fast lasers were significantly developed because the versatile 2D material properties can fulfill the requirements for severe optical parametres and integration. The short cavity footprint of these lasers is beneficial for a high repetition rate (>10 GHz [58]), which is intrinsically challenging in fibre and solid-state lasers. Although short cavity design [153] and harmonic wave operation [138] have been investigated for the same purpose, the precise cavity alignment and multi-mode noise are still practical issues to be solved.

The broad operation bandwidth of 2D material-based SAs has drastically expanded the wavelength range of ultra-fast lasers, typically from the visible to the mid-infrared region [24, 31, 79, 133, 146, 187, 191, 228]. Graphene, TMDs, BP and Tis have all been demonstrated as effective SAs at near and mid-infrared wavelengths when layered flakes or bulk samples (TIs) are incorporated. Noteworthy that TMDs with a large bandgap are especially suitable SAs for the visible waveband [187, 191], which was a research blank due to the lack of available non-linear materials. TMDs with a bandgap larger than the photon energy were also successfully identified for ultra-fast pulse generation, as explained by theories of defect-induced bandgap decreasing [79, 90, 185], edge state of the materials [81], co-existence of semiconducting and metallic states [29] and two-photon absorption saturation [74].

Two-dimensional materials can combine various performances for specific utilisation. However, it still takes time to fully understand the physical and chemical properties of these semiconductors, as well as the performance change of a single material in different forms (number of layers, doping, defects). In practice, raw materials can be functionalised with polymer composites, solution or bio-compatible host [231] to optimise the SA parametres and stability. On the other hand, the development of fabrication technologies for large-scale yield is also a virtual challenge to push forward the successful commercialisation of 2D materials, which needs more research forces and economic investments in the future. Further, heterostructures formed by stacking different 2D materials (such as graphene and TMDs) and their hybrids (e.g., plasmonic structures [232–235]) can offer superior optical properties for passive optical modulation in various lasers, which is an effective way to modify the SA devices consisting of only one material [60].

References

1. Leahy-Hoppa, M. R., Miragliotta, J., Osiander, R., Burnett, J., Dikmelik, Y., McEnnis, C. and Spicer, J. B. (2010). Ultrafast laser-based spectroscopy and sensing: applications in LIBS, CARS, and THz spectroscopy, *Sensors*, **10**, pp. 4342–4372.
2. Keller, U. (2003). Recent developments in compact ultrafast lasers, *Nature*, **424**, pp. 831–838.
3. Fermann, M. E., Galvanauskas, A. and Sucha, G. (2002). *Ultrafast Lasers: Technology and Applications* (CRC Press), p. 80.
4. Peng, J. L., Ahn, H., Shu, R. H., Chui, H. C. and Nicholson, J. W. (2007). Highly stable, frequency-controlled mode-locked erbium fiber laser comb, *Appl. Phys. B*, **86**, pp. 49–53.
5. Rulliere, C. (2005). *Femtosecond Laser Pulses* (Springer Science).
6. Paschotta, R. (2008). *Encyclopedia of Laser Physics and Technology* (Wiley-VCH Berlin).
7. Guo, L., Hou, W., Zhang, H., Sun, Z., Dafu, C., Xu, Z., Wang, Y. and Ma, X. (2005). Diode-end-pumped passively mode-locked ceramic Nd:YAG laser with a semiconductor saturable mirror, *Opt. Express*, **13**, pp. 4085–4089.
8. Set, S. Y., Yaguchi, H., Tanaka, Y., Jablonski, M., Sakakibara, Y., Rozhin, A., Tokumoto, M., Kataura, H., Achiba, Y. and Kikuchi, K. (2003). Mode-locked fiber lasers based on a saturable absorber incorporating carbon nanotubes, *Optical Fiber Communication Conference*, PD44.
9. Wang, F., Rozhin, A. G., Scardaci, V., Sun, Z., Hennrich, F., White, I. H., Milne, W. I. and Ferrari, A. C. (2008). Wideband-tuneable, nanotube mode-locked, fibre laser, *Nat. Nanotechnol.*, **3**, pp. 738–742.
10. Martinez, A. and Sun, Z. (2013). Nanotube and graphene saturable absorbers for fibre lasers, *Nat. Photonics*, **7**, pp. 842–845.

11. Liu, X., Han, D., Sun, Z., Zeng, C., Lu, H., Mao, D., Cui, Y. and Wang, F. (2013). Versatile multi-wavelength ultrafast fiber laser mode-locked by carbon nanotubes, *Sci. Rep.*, **3**, p. 2718.
12. Kelleher, E. J. R., Travers, J. C., Sun, Z., Rozhin, A. G., Ferrari, A. C., Popov, S. V. and Taylor, J. R. (2009). Nanosecond-pulse fiber lasers mode-locked with nanotubes, *Appl. Phys. Lett.*, **95**, p. 111108.
13. Scardaci, V., Sun, Z., Wang, F., Rozhin, A. G., Hasan, T., Hennrich, F., White, I. H., Milne, W. I. and Ferrari, A. C. (2008). Carbon nanotube polycarbonate composites for ultrafast lasers, *Adv. Mater.*, **20**, pp. 4040–4043.
14. Sun, Z., Rozhin, A. G., Wang, F., Hasan, T., Popa, D., O'Neill, W. and Ferrari, A. C. (2009). A compact, high power, ultrafast laser mode-locked by carbon nanotubes, *Appl. Phys. Lett.*, **95**, p. 253102.
15. Sun, Z., Rozhin, A. G., Wang, F., Scardaci, V., Milne, W. I., White, I. H., Hennrich, F. and Ferrari, A. C. (2008). L-band ultrafast fiber laser mode locked by carbon nanotubes, *Appl. Phys. Lett.*, **93**, p. 061114.
16. Liu, X., Cui, Y., Han, D., Yao, X. and Sun, Z. (2015). Distributed ultrafast fibre laser, *Sci. Rep.*, **5**, p. 9101.
17. Kelleher, E. J. R., Travers, J. C., Ippen, E. P., Sun, Z., Ferrari, A. C., Popov, S. V. and Taylor, J. R. (2009). Generation and direct measurement of giant chirp in a passively mode-locked laser, *Opt. Lett.*, **34**, pp. 3526–3528.
18. Sun, Z., Hasan, T., Wang, F., Rozhin, A. G., White, I. H. and Ferrari, A. C. (2010). Ultrafast stretched-pulse fiber laser mode-locked by carbon nanotubes, *Nano Res.*, **3**, pp. 404–411.
19. Mao, D., Liu, X., Sun, Z., Lu, H., Han, D., Wang, G. and Wang, F. (2013). Flexible high-repetition-rate ultrafast fiber laser, *Sci. Rep.*, **3**, p. 3223.
20. Kelleher, E. J. R., Travers, J. C., Sun, Z., Ferrari, A. C., Golant, K. M., Popov, S. V. and Taylor, J. R. (2010). Bismuth fiber integrated laser-mode locked by carbon nanotubes, *Laser Phys. Lett.*, **7**, pp. 790–794.
21. Castellani, C. E. S., Kelleher, E. J. R., Travers, J. C., Popa, D., Hasan, T., Sun, Z., Flahaut, E., Ferrari, A. C., Popov, S. V. and Taylor, J. R. (2011). Ultrafast Raman laser mode-locked by nanotubes, *Opt. Lett.*, **36**, pp. 3996–3998.
22. Hasan, T., Sun, Z., Tan, P. H., Popa, D., Flahaut, E., Kelleher, E. J. R., Bonaccorso, F., Wang, F., Jiang, Z., Torrisi, F., Privitera, G., Nicolosi, V. and Ferrari, A. C. (2014). Double-wall carbon nanotubes for wide-band, ultrafast pulse generation, *ACS Nano*, **8**, pp. 3996–3998.
23. Mao, D., Feng, T., Zhang, W., Lu, H., Jiang, Y., Li, P., Jiang, B., Sun, Z. and Zhao, J. (2017). Ultrafast all-fiber based cylindrical-vector beam laser, *Appl. Phys. Lett.*, **110**, p. 021107.

24. Hasan, T., Sun, Z., Wang, F., Bonaccorso, F., Tan, P. H., Rozhin, A. G. and Ferrari, A. C. (2009). Nanotube-polymer composites for ultrafast photonics, *Adv. Mater.*, **21**, pp. 3874–3899.

25. Sun, Z., Hasan, T., Torrisi, F., Popa, D., Privitera, G., Wang, F., Bonaccorso, F., Basko, D. M. and Ferrari, A. C. (2010). Graphene mode-locked ultrafast laser, *ACS Nano*, **4**, pp. 803–810.

26. Bao, Q., Zhang, H., Wang, Y., Ni, Z., Yan, Y., Shen, Z. X., Loh, K. P. and Tang, D. Y. (2009). Atomic-layer graphene as a saturable absorber for ultrafast pulsed lasers, *Adv. Funct. Mater.*, **19**, pp. 3077–3083.

27. Khazaeizhad, R., Kassani, S. H., Jeong, H., Yeom, D. I. and Oh, K. (2014). Mode-locking of Er-doped fiber laser using a multilayer MoS_2 thin film as a saturable absorber in both anomalous and normal dispersion regimes, *Opt. Express*, **22**, pp. 23732–23742.

28. Bonaccorso, F. and Sun, Z. (2014). Solution processing of graphene, topological insulators and other 2d crystals for ultrafast photonics, *Opt. Mater. Express*, **4**, pp. 63–78.

29. Zhang, H., Lu, S. B., Zheng, J., Du, J., Wen, S. C., Tang, D. Y. and Loh, K. P. (2014). Molybdenum disulfide (MoS_2) as a broadband saturable absorber for ultra-fast photonics, *Opt. Express*, **22**, pp. 7249–7260.

30. Li, D., Jussila, H., Karvonen, L., Ye, G., Lipsanen, H., Chen, X. and Sun, Z. (2015). Polarization and thickness dependent absorption properties of black phosphorus: new saturable absorber for ultrafast pulse generation, *Sci. Rep.*, **5**, p. 15899.

31. Chen, Y., Jiang, G., Chen, S., Guo, Z., Yu, X., Zhao, C., Zhang, H., Bao, Q., Wen, S., Tang, D. and Fan, D. (2015). Mechanically exfoliated black phosphorus as a new saturable absorber for both Q-switching and mode-locking laser operation, *Opt. Express*, **23**, pp. 12823–12833.

32. Sotor, J., Sobon, G., Macherzynski, W., Paletko, P. and Abramski, K. M. (2015). Black phosphorus saturable absorber for ultrashort pulse generation, *Appl. Phys. Lett.*, **107**, p. 051108.

33. Yan, P., Lin, R., Ruan, S., Liu, A., Chen, H., Zheng, Y., Chen, S., Guo, C. and Hu, J. (2015). A practical topological insulator saturable absorber for mode-locked fiber laser, *Sci. Rep.*, **5**, p. 8690.

34. Geim, A. K. and Novoselov, K. S. (2007). The rise of graphene, *Nat. Mater.*, **6**, p. 183.

35. Bonaccorso, F., Sun, Z., Hasan, T. and Ferrari, A. C. (2010). Graphene photonics and optoelectronics, *Nat. Photonics*, **4**, pp. 611–622.

36. Ferrari, A. C., Bonaccorso, F., Fal'ko, V., Novoselov, K. S., Roche, S., Boggild, P., Borini, S., Koppens, F. H. L., Palermo, V., Pugno, N., Garrido,

J. A., Sordan, R., Bianco, A., Ballerini, L., Prato, M., Lidorikis, E., Kivioja, J., Marinelli, C., Ryhanen, T., Morpurgo, A., Coleman, J. N., Nicolosi, V., Colombo, L., Fert, A., Garcia-Hernandez, M., Bachtold, A., Schneider, G. F., Guinea, F., Dekker, C., Barbone, M., Sun, Z., Galiotis, C., Grigorenko, A. N., Konstantatos, G., Kis, A., Katsnelson, M., Vandersypen, L., Loiseau, A., Morandi, V., Neumaier, D., Treossi, E., Pellegrini, V., Polini, M., Tredicucci, A., Williams, G. M., Hong, B. H., Ahn, J. H., Kim, J. M., Zirath, H., van, Wees, B. J., van der Zant, H., Occhipinti, L., Matteo, A. D., Kinloch, I. A., Seyller, T., Quesnel, E., Feng, X., Teo, K., Rupesinghe, N., Hakonen, P., Neil, S. R. T., Tannock, Q., Lofwander, T. and Kinaret, J. (2015). Science and technology roadmap for graphene, related two-dimensional crystals, and hybrid systems, *Nanoscale*, **7**, pp. 4598–4810.

37. Sun, Z., Hasan, T. and Ferrari, A. C. (2012). Ultrafast lasers mode-locked by nanotubes and graphene, *Physica E*, **44**, pp. 1082–1091.

38. Fu, B., Hua, Y., Xiao, X., Zhu, H., Sun, Z. and Yang, C. (2014). Broadband graphene saturable absorber for pulsed fiber lasers at 1, 1.5, and 2 μm, *IEEE J. Sel. Top. Quantum Electron.*, **20**, p. 1100705.

39. Novoselov, K. S., Geim, A. K., Morozov, S. V., Jiang, D., Katsnelson, M. I., Grigorieva, I. V., Dubonos, S. V. and Firsov, A. A. (2005). Two-dimensional gas of massless Dirac fermions in graphene, *Nature*, **438**, pp. 197–200.

40. Sun, Z., Popa, D., Hasan, T., Torrisi, F., Wang, F., Kelleher, E. J. R., Travers, J. C., Nicolosi, V. and Ferrari, A. C. (2010). A stable, wideband tunable, near transform-limited, graphene mode-locked, ultrafast laser, *Nano Res.*, **3**, pp. 653–660.

41. Wang, J., Hernandez, Y., Lotya, M., Coleman, J. N. and Blau, W. J. (2009). Broadband nonlinear optical response of graphene dispersions, *Adv. Mater.*, **21**, pp. 2430–2435.

42. Lu, S., Chen, S., Zheng, Z., Zhang, H., Zhao, C. and Wen, S. (2012). Saturable absorption in graphene at 800-nm band, *Proc. SPIE*, **8555**, p. 12.

43. Breusing, M., Ropers, C. and Elsaesser, T. (2009). Ultrafast carrier dynamics in graphite, *Phys. Rev. Lett.*, **102**, p. 086809.

44. Ruzicka, B. A., Wang, S., Liu, J., Loh, K. P., Wu, J. Z. and Zhao, H. (2009). Spatially resolved pump-probe study of single-layer graphene produced by chemical vapor deposition, *Opt. Mater. Express*, **2**, pp. 708–716.

45. Martinez, A., Fuse, K. and Yamashita, S. (2011). Mechanical exfoliation of graphene for the passive mode-locking of fiber lasers, *Appl. Phys. Lett.*, **99**, p. 121107.

46. Hasan, T., Torrisi, F., Sun, Z. and Ferrari, A. C. (2010). Solution-phase exfoliation of graphite for ultrafast photonics, *Phys. Status Solidi B*, **247**, pp. 2953–2957.

47. Zhang, H., Tang, D. Y., Zhao, L. M., Bao, Q. L. and Loh, K. P. (2010). Large energy mode locking of an erbium-doped fiber laser with atomic layer graphene, *Opt. Express*, **17**, pp. 17630–17635.

48. Bonaccorso, F., Lombardo, A., Hasan, T., Sun, Z., Colombo, L. and Ferrari, A. C. (2012). Production and processing of graphene and 2d crystals, *Mater. Today*, **15**, pp. 564–589.

49. Bao, Q., Zhang, H., Yang, J. X., Wang, S., Tang, D. Y., Jose, R., Ramakrishna, S., Lim, C. T. and Loh, K. P. (2010). Graphene-polymer nanofiber membrane for ultrafast photonics, *Adv. Funct. Mater.*, **20**, pp. 782–791.

50. Jung, M., Koo, J., Park, J., Song, Y. W., Jhon, Y. M., Lee, K., Lee, S. and Lee, J. H. (2013). Mode-locked pulse generation from an all fiberized, Tm-Ho-codoped fiber laser incorporating a graphene oxide-deposited side-polished fiber, *Opt. Express*, **21**, pp. 20062–20072.

51. Li, X., Wu, K., Sun, Z., Meng, B., Wang, Y., Wang, Y., Yu, X., Yu, X., Zhang, Y., Shum, P. P. and Wang, Q. J. (2016). Single-wall carbon nanotubes and graphene oxide-based saturable absorbers for low phase noise mode-locked fiber lasers, *Sci. Rep.*, **6**, p. 25266.

52. Choi, S. Y., Cho, D. K., Song, Y. W., Oh, K., Kim, K., Rotermund, F. and Yeom, D. I. (2012). Graphene-filled hollow optical fiber saturable absorber for efficient soliton fiber laser mode-locking, *Opt. Express*, **20**, pp. 5652–5657.

53. Ugolotti, E., Schmidt, A., Petrov, V., Kim, J. W., Yeom, D. I., et al. (2012). Graphene mode-locked femtosecond Yb:KLuW laser, *Appl. Phys. Lett.*, **101**, p. 161112.

54. Li, L., Ren, Z., Chen, X., Qi, M., Zheng, X., Bai, J. and Sun, Z. (2013). Passively mode-locked radially polarized Nd-doped yttrium aluminum garnet laser based on graphene-based saturable absorber, *Appl. Phys. Express*, **6**, p. 082701.

55. Li, L., Zheng, X., Jin, C., Qi, M., Chen, X., Ren, Z., Bai, J. and Sun, Z. (2014). High repetition rate Q-switched radially polarized laser with a graphene-based output coupler, *Appl. Phys. Lett.*, **105**, p. 221103.

56. Tan, W. D., Su, C. Y., Knize, R. J., Xie, G. Q., Li, L. J. and Tang, D. Y. (2010). Mode locking of ceramic Nd:yttrium aluminum garnet with graphene as a saturable absorber, *Appl. Phys. Lett.*, **96**, p. 031106.

57. Mary, R., Brown, G., Beecher, S. J., Torrisi, F., Milana, S., Popa, D., Hasan, T., Sun, Z., Lidorikis, E., Ohara, S., Ferrari, A. C. and Kar, A. K. (2013). 1.5

GHz picosecond pulse generation from a monolithic waveguide laser with a graphene-film saturable output coupler, *Opt. Express*, **21**, pp. 7943–7950.

58. Okhrimchuk, A. G. and Obraztsov, P. A. (2015). 11-GHz waveguide Nd:YAG laser CW mode-locked with single-layer graphene, *Sci. Rep.*, **5**, p. 11172.

59. Zaugg, C. A., Sun, Z., Wittwer, V. J., Popa, D., Milana, S., Kulmala, T. S., Sundaram, R. S., Mangold, M, Sieber, O. D., Golling, M., Lee, Y., Ahn, J. H., Ferrari, A. C. and Keller, U. (2013). Ultrafast and widely tuneable vertical-external-cavity surface-emitting laser, mode-locked by a graphene-integrated distributed Bragg reector, *Opt. Express*, **21**, pp. 31548–31559.

60. Sun, Z., Martinez, A. and Wang, F. (2016). Optical modulators with two-dimensional layered materials, *Nat. Photonics*, **10**, pp. 227–238.

61. Li, D., Xue, H., Qi, M., Aksimsek, S., Chekurov, N., Kim, W., Li, C., Riikonen, J., Ye, F., Dai, Q., Ren, Z., Bai, J., Hasan, T., Lipsanen, H. and Sun, Z. (2017). Graphene actively Q-switched lasers, *2D Mater.*, **4**, p. 025095.

62. Mak, K. F., Lee, C., Hone, J., Shan, J. and Heinz, T. F. (2010). Atomically thin MoS_2: a new direct-gap semiconductor, *Phys. Rev. Lett.*, **105**, p. 136805.

63. Kang, K., Xie, S., Huang, L., Han, Y., Huang, P. Y., Mak, K. F., Kim, C. J., Muller, D. and Park, J. (2015). High-mobility three-atom-thick semiconducting films with wafer-scale homogeneity, *Nature*, **520**, pp. 656–660.

64. Wang, Q. H., Kalantar-Zadeh, K., Kis, A., Coleman, J. N. and Strano, M. S. (2012). Electronics and optoelectronics of two-dimensional transition metal dichalcogenides, *Nat. Nanotechnol.*, **7**, pp. 699–712.

65. Mohanraj, J., Velmurugan, V. and Sivabalan, S. (2016). Transition metal dichalcogenides based saturable absorbers for pulsed laser technology, *Opt. Mater.*, **60**, pp. 601–617.

66. Saynatjoki, A., Karvonen, L., Rostami, H., Autere, A., Mehravar, S., Lombardo, A., Norwood, R. A., Hasan, T., Peyghambarian, N., Lipsanen, H., Kieu, K., Ferrari, A. C., Polini, M. and Sun, Z. (2016). Ultra-strong nonlinear optical processes and trigonal warping in MoS_2 layers, *Nat. Commun.*, **8**, p. 893.

67. Coleman, J. N., Lotya, M., O'Neill, A., Bergin, S. D., King, P. J., Khan, U., Young, K., Gaucher, A., De, S., Smith, R. J., Shvets, I. V., Arora, S. K., Stanton, G., Kim, H. Y., Lee, K., Kim, G. T., Duesberg, G. S., Hallam, T., Boland, J. J., Wang, J. J., Donegan, J. F., Grunlan, J. C., Moriarty, G., Shmeliov, A., Nicholls, R. J., Perkins, J. M., Grieveson, E. M., Theuwissen, K., McComb,

D. W., Nellist, P. D. and Nicolosi, V. (2011). Two-dimensional nanosheets produced by liquid exfoliation of layered materials, *Science*, **331**, pp. 568–571.

68. Ramakrishna Matte, H. S. S., Gomathi, A., Manna, A. K., Late, D. J., Datta, R., Pati, S. K. and Rao, C. N. R. (2010). MoS_2 and WS_2 analogues of graphene, *Angew. Chem.*, **122**, pp. 4153–4156.

69. Huang, X., Zeng, Z. and Zhang, H. (2013). Metal dichalcogenide nanosheets: preparation, properties and applications, *Chem. Soc. Rev.*, **42**, pp. 1934–1946.

70. Nicolosi, V., Chhowalla, M., Kanatzidis, M. G., Strano, M. S. and Coleman, J. N. (2013). Liquid exfoliation of layered materials, *Science*, **340**, pp. 1–18.

71. Bonaccorso, F. and Sun, Z. (2013). Solution processing of graphene, topological insulators and other 2d crystals for ultrafast photonics, *Opt. Mater. Express*, **4**, pp. 63–78.

72. Radisavljevic, B., Radenovic, A., Brivio, J., Giacometti, V. and Kis, A. (2011). Single-layer MoS_2 transistors, *Nat. Nanotechnol.*, **6**, pp. 147–150.

73. Kumar, A. and Ahluwalia, P. K. (2012). Electronic structure of transition metal dichalcogenides monolayers 1H-MX2 (M = Mo, W; X = S, Se, Te) from ab-initio theory: new direct band gap semiconductors, *Eur. Phys. J. B*, **85**, p. 186.

74. Wang, K., Feng, Y., Chang, C., Zhan, J., Wang, C., Zhao, Q., Coleman, J. N., Zhang, L., Blau, W. J. and Wang, J. (2014). Broadband ultrafast nonlinear absorption and nonlinear refraction of layered molybdenum dichalcogenide semiconductors, *Nanoscale*, **6**, pp. 10530–10535.

75. Mak, K. F., McGill, K. L., Park, J. and McEuen, P. L. (2014). Valleytronics. The valley Hall effect in MoS_2 transistors, *Science*, **344**, pp. 1489–1492.

76. Lopez-Sanchez, O., Lembke, D., Kayci, M., Radenovic, A. and Kis, A. (2013). Ultrasensitive photodetectors based on monolayer MoS_2, *Nat. Nanotechnol.*, **8**, pp. 497–501.

77. Tsai, M. L., Su, S. H., Chang, J. K., Tsai, D. S., Chen, C. H., Wu, C. I., Li, L. J., Chen, L. J. and He, J. H. (2014). Monolayer MoS_2 heterojunction solar cells, *ACS Nano*, **8**, pp. 8317–8322.

78. Wang, K., Wang, J., Fan, J., Lotya, M., O'Neill, A., Fox, D., Feng, Y., Zhang, X., Jiang, B., Zhao, Q., Zhang, H., Coleman, J. N., Zhang, L. and Blau, W. J. (2013). Ultrafast saturable absorption of two-dimensional MoS_2 nanosheets, *ACS Nano*, **7**, pp. 9260–9267.

79. Wang, S., Yu, H., Zhang, H., Wang, A., Zhao, M., Chen, Y., Mei, L. and Wang, J. (2014). Broadband few-layer MoS_2 saturable absorbers, *Adv. Mater.*, **26**, pp. 3538–3544.

80. Chen, B., Zhang, X., Wu, K., Wang, H., Wang, J. and Chen, J. (2015). Q-switched fiber laser based on transition metal dichalcogenides MoS_2, $MoSe_2$, WS_2, and WSe_2, *Opt. Express*, **23**, pp. 26723–26737.

81. Woodward, R. I., Kelleher, E. J., Howe, R. C., Hu, G., Torrisi, F., Hasan, T., Popov, S. V. and Taylor, J. R. (2014). Tunable Q-switched fiber laser based on saturable edge-state absorption in few-layer molybdenum disulfide (MoS_2), *Opt. Express*, **22**, pp. 31113–31122.

82. Xia, H., Li, H., Lan, C., Li, C., Zhang, X., Zhang, S. and Liu, Y. (2014). Ultrafast erbium-doped fiber laser mode-locked by a CVD-grown molybdenum disulfide (MoS_2) saturable absorber, *Opt. Express*, **22**, pp. 17341–17348.

83. Wang, Y. D., Mao, D., Gan, X. T., Han, L., Ma, C. J., Xi, T. L., Zhang, Y., Shang, W. Y., Hua, S. J., Zhao, J. L. (2015). Harmonic mode locking of bound-state solitons fiber laser based on MoS_2 saturable absorber, *Opt. Express*, **23**, pp. 205–210.

84. Zhang, M., Howe, R. C. T., Woodward, R. I., Kelleher, E. J. R., Torrisi, F., Hu, G., Popov, S. V., Taylor, J. R. and Hasan, T. (2015). Solution processed MoS_2-PVA composite for sub-bandgap mode-locking of a wideband tunable ultrafast Er:fiber laser, *Nano Res.*, **8**, pp. 1522–1534.

85. Luo, Z., Huang, Y., Zhong, M., Li, Y., Wu, J., Xu, B., Xu, H., Cai, Z., Peng, J. and Weng, J. (2014). 1-, 1.5-, and 2-μm fiber lasers Q-switched by a broad-band few-layer MoS_2 saturable absorber, *J. Lightwave Technol.*, **32**, pp. 4077–4084.

86. Zhen, T., Kan, W., Lingchen, K., Nan, Y., Yao, W., Rong, C., Weisheng, H., Jianqiu, X. and Yulong, T. (2015). Mode-locked thulium fiber laser with MoS_2, *Laser Phys. Lett.*, **12**, p. 065104.

87. Woodward, R. I., Howe, R. C. T., Hu, G., Torrisi, F., Zhang, M., Hasan, T. and Kelleher, E. J. R. (2015). Few-layer MoS_2 saturable absorbers for short-pulse laser technology: current status and future perspectives [Invited], *Photonics Res.*, **3**, pp. A30–A42.

88. Feng, C., Zhang, X., Wang, J., Liu, Z., Cong, Z., Rao, H., Wang, Q. and Fang, J. (2016). Passively mode-locked Nd3+:YVO4 laser using a molybdenum disulfide as saturable absorber, *Opt. Mater. Express*, **6**, p. 1358.

89. Yan, P., Liu, A., Chen, Y., Chen, H., Ruan, S., Guo, C., Chen, S., Li, I. L., Yang, H., Hu, J. and Cao, G. (2015). Microfiber-based WS_2-film saturable absorber for ultra-fast photonics, *Opt. Mater. Express*, **5**, pp. 479–489.

90. Mao, D., Wang, Y. D., Ma, C. J., Han, L., Jiang, B. Q., Gan, X. T., Hua, S. J., Zhang, W. D., Mei, T. and Zhao, J. L. (2015). WS_2 mode-locked ultrafast fiber laser, *Sci. Rep.*, **5**, p. 7965.

91. Zhang, M., Hu, G., Hu, G., Howe, R. C., Chen, L., Zheng, Z. and Hasan, T. (2015). Yb- and Er-doped fiber laser Q-switched with an optically uniform, broadband WS_2 saturable absorber, *Sci. Rep.*, **5**, p. 17482.

92. Wu, K., Zhang, X., Wang, J., Li, X. and Chen, J. (2015). WS_2 as a saturable absorber for ultrafast photonic applications of mode-locked and Q-switched lasers, *Opt. Express*, **23**, pp. 11453–14161.

93. Yan, P., Chen, H., Yin, J., Xu, Z., Li, J., Jiang, Z., Zhang, W., Wang, J., Li, I. L., Sun, Z. and Ruan, S. (2017). Large-area tungsten disulfide for ultrafast photonics, *Nanoscale*, **9**, pp. 1871–1877.

94. Luo, Z., Li, Y., Zhong, M., Huang, Y., Wan, X., Peng, J. and Weng, J. (2015). Non-linear optical absorption of few-layer molybdenum diselenide ($MoSe_2$) for passively mode-locked soliton fiber laser [Invited], *Photonics Res.*, **3**, pp. A79–A86.

95. Mao, D., She, X., Du, B., Yang, D., Zhang, W., Song, K., Cui, X., Jiang, B., Peng, T. and Zhao, J. (2016). Erbium-doped fiber laser passively mode locked with few-layer $WSe_2/MoSe_2$ nanosheets, *Sci. Rep.*, **6**, p. 23583.

96. Woodward, R. I., Howe, R. C., Runcorn, T. H., Hu, G., Torrisi, F., Kelleher, E. J. and Hasan, T. (2015). Wideband saturable absorption in few-layer molybdenum diselenide ($MoSe_2$) for Q-switching Yb-, Er- and Tm-doped fiber lasers, *Opt. Express*, **23**, pp. 20051–20061.

97. Mao, D., Du, B., Yang, D., Zhang, S., Wang, Y., Zhang, W., She, X., Cheng, H., Zeng, H. and Zhao, J. (2016). Nonlinear saturable absorption of liquid-exfoliated molybdenum/tungsten ditelluride nanosheets, *Small*, **12**, pp. 1489–1497.

98. Koo, J., Jhon, Y. I., Park, J., Lee, J., Jhon, Y. M. and Lee, J. H. (2016). Near-infrared saturable absorption of defective bulk-structured WTe_2 for femtosecond laser mode-locking, *Adv. Funct. Mater.*, **26**, pp. 7454–7461.

99. Cui, Y., Lu, F. and Liu, X. (2017). Nonlinear saturable and polarization-induced absorption of rhenium disulfide, *Sci. Rep.*, **7**, p. 40080.

100. Li, L., Yu, Y., Ye, G. J., Ge, Q., Ou, X., Wu, H., Feng, D., Chen, X. H. and Zhang, Y. (2014). Black phosphorus field-effect transistors, *Nat. Nanotechnol.*, **9**, pp. 372–377.

101. Liu, H., Neal, A. T., Zhu, Z., Luo, Z., Xu, X., Tomanek, D. and Ye, P. D. (2014). Phosphorene: an unexplored 2D semiconductor with a high hole mobility, *ACS Nano*, **8**, pp. 4033–4041.

102. Koenig, S. P., Doganov, R. A., Schmidt, H., Neto, A. H. C. and Ozyilmaz, B. (2014). Electric field effect in ultrathin black phosphorus, *Appl. Phys. Lett.*, **104**, p. 103106.

103. Xia, F., Wang, H. and Jia, Y. (2014) Rediscovering black phosphorus as an anisotropic layered material for optoelectronics and electronics, *Nat. Commun.*, **5**, p. 54458.

104. Xia, F., Wang, H., Xiao, D., Dubey, M. and Ramasubramaniam, A. (2014). Two-dimensional material nanophotonics, *Nat. Photonics*, **8**, pp. 899–907.

105. Autere, A., Ryder, C., Saynatjoki, A., Karvonen, L., Amirsolaimani, B., Norwood, R. A., Peyghambarian, N., Kieu, K., Lipsanen, H., Hersam, M. C. and Sun, Z. (2017). Rapid and large-area characterization of exfoliated black phosphorus using third-harmonic generation microscopy, *J. Phys. Chem. Lett.*, **8**(7), pp. 1343–1350.

106. Rodin, A. S., Carvalho, A. and Neto Castro, A. H. (2014). Strain-induced gap modification in black phosphorus, *Phys. Rev. Lett.*, **112**, p. 176801.

107. Lu, S., Miao, L. L., Guo, Z. N., Qi, X., Zhao, C. J., Zhang, H., Wen, S. C., Tang, D. Y. and Fan, D. Y. (2015). Broadband nonlinear optical response in multilayer black phosphorus: an emerging infrared and mid-infrared optical material, *Opt. Express*, **23**, pp. 11183–11194.

108. Li, D., Castillo, S. E. D. R., Jussila, H., Ye, G., Ren, Z., Bai, J., Chen, X., Lipsanen, H., Sun, Z. and Bonaccorso, F. (2016). Black phosphorus polycarbonate polymer composite for pulsed fibre lasers, *Appl. Mater. Today*, **4**, pp. 17–23.

109. Ge, S., Li, C., Zhang, Z., Zhang, C., Zhang, Y., Qiu, J., Wang, Q., Liu, J., Jia, S., Feng, J. and Sun, D. (2015). Dynamical evolution of anisotropic response in black phosphorus under ultrafast photoexcitation, *Nano Lett.*, **15**, pp. 4650–4656.

110. Castellanos-Gomez, A., Vicarelli, L., Prada, E., Island, J. O., Narasimha-Acharya, K. L., Blanter, S. I., Groenendijk, D. J., Buscema, M., Steele, G. A., Alvarez, J. V., Zandbergen, H. W., Palacios, J. J. and van der Zant, H. S. J. (2014). Isolation and characterization of few-layer black phosphorus, *2D Mater.*, **1**, p. 025001.

111. Mao, D., Li, M., Cui, X., Zhang, W., Lu, H., Song, K. and Zhao, J. (2016). Stable high-power saturable absorber based on polymer-black-phosphorus films, *Opt. Commun.*, **406**, pp. 254–259.

112. Luo, Z., Rahbarihagh, Y., Hwang, J., Liu, H., Du, Y. and Ye, P. D. (2014). Temporal and thermal stability of Al_2O_3-passivated phosphorene MOSFETs, *IEEE Electron Device Lett.*, **35**, pp. 1314–1316.

113. Wood, J. D., Wells, S. A., Jariwala, D., Chen, K. S., Cho, E. K., Sangwan, V. K., Liu, X., Lauhon, L. J., Marks, T. J. and Hersam, M. C. (2014). Effective passivation of exfoliated black phosphorus transistors against ambient degradation, *Nano Lett.*, **14**, pp. 6964–6970.

114. Chen, X., Wu, Y., Wu, Z., Han, Y., Xu, S., Wang, L., Ye, W., Han, T., He, Y., Cai, Y. and Wang, N. (2015). High-quality sandwiched black phosphorus heterostructure and its quantum oscillations, *Nat. Commun.*, **6**, p. 7315.

115. Boguslawski, J., Sotor, J., Sobon, G., Tarka, J., Jagiello, J., Macherzynski, W., Lipinska, L. and Abramski, K. M. (2014). Mode-locked Er-doped fiber laser based on liquid phase exfoliated Sb_2Te_3 topological insulator, *Laser Phys.*, **24**, p. 105111.

116. Chi, C., Lee, J., Koo, J. and Lee, J. H. (2014). All-normal-dispersion dissipative-soliton fiber laser at 1.06 μm using a bulk-structured Bi_2Te_3 topological insulator-deposited side-polished fiber, *Laser Phys.*, **24**, p. 105106.

117. Luo, Z., Huang, Y., Weng, J., Cheng, H., Lin, Z., Xu, B., Cai, Z. and Xu, H. (2013). 1.06 μm Q-switched ytterbium-doped fiber laser using few-layer topological insulator Bi_2Se_3 as a saturable absorber, *Opt. Express*, **21**, pp. 29516–29522.

118. Tang, P., Zhang, X., Zhao, C., Wang, Y., Zhang, H., Shen, D., Wen, S., Tang, D. and Fan, D. (2013). Topological insulator: Bi_2Te_3 saturable absorber for the passive Q-switching operation of an in-band pumped 1645-nm Er:YAG ceramic laser, *IEEE Photonics J.*, **5**, p. 1500707.

119. Jung, M., Lee, J., Koo, J., Park, J., Song, Y. W., Lee, K., Lee, S. and Lee, J. H. (2013). A femtosecond pulse fiber laser at 1935 nm using a bulk-structured Bi_2Te_3 topological insulator, *Opt. Express*, **22**, pp. 7865–7874.

120. Svelto, O. (1998). *Principles of Lasers* (Plenum, New York: 1976).

121. Luo, Z., Zhou, M., Weng, J., Huang, G., Xu, H., Ye, C. and Cai, Z. (2010). Graphene-based passively Q-switched dual-wavelength erbium-doped fiber laser, *Opt. Lett.*, **35**, pp. 3709–3711.

122. Popa, D., Sun, Z., Hasan, T., Torrisi, F., Wang, F. and Ferrari, A. C. (2011). Graphene Q-switched, tunable fiber laser, *Appl. Phys. Lett.*, **98**, p. 073106.

123. Liu, J., Wu, S., Yang, Q. H. and Wang, Pu. (2011). Stable nanosecond pulse generation from a graphene-based passively Q-switched Yb-doped fiber laser, *Opt. Lett.*, **36**, pp. 4008–4010.

124. Liu, C., Ye, C., Luo, Z., Cheng, H., Wu, D., Zheng, Y., Liu, Z. and Qu, B. (2013). High-energy passively Q-switched 2 μm Tm3+-doped double-clad fiber laser using graphene oxide-deposited fiber taper, *Opt. Express*, **21**, p. 1.

125. Yu, Z., Song, Y., Dong, X., Li, Y., Tian, J. and Wang, Y. (2013). Watt-level passively Q-switched double-cladding fiber laser based on graphene oxide saturable absorber, *Appl. Opt.*, **52**, p. 29.

126. Tang, Y., Yu, X., Li, X., Yan, Z. and Wang, Q. J. (2014). High-power thulium fiber laser Q switched with single-layer graphene, *Opt. Lett.*, **39**, pp. 614–617.

127. Yu, H., Chen, X., Zhang, H., Xu, X., Hu, X., Wang, Z., Wang, J., Zhuang, S. and Jiang, M. (2010). Large energy pulse generation modulated by graphene epitaxially grown on silicon carbide, *ACS Nano*, **4**, pp. 7582–7586.

128. Li, X., Xu, J., Wu, Y., He, J. and Hao, X. (2011). Large energy laser pulses with high repetition rate by graphene Q-switched solid-state laser, *Opt. Express*, **19**, pp. 9950–9955.

129. Gao, C., Wang, R., Zhu, L., Gao, M., Wang, Q., Zhang, Z., Wei, Z., Lin, J. and Guo, L. (2012). Resonantly pumped 1.645 μm high repetition rate Er:YAG laser Q-switched by a graphene as a saturable absorber, *Opt. Lett.*, **37**, pp. 632–634.

130. Wang, Q., Teng, H., Zou, Y., Zhang, Z., Li, D., Wang, R., Gao, C., Lin, J., Guo, L. and Wei, Z. (2012). Graphene on SiC as a Q-switcher for a 2 μm laser, *Opt. Lett.*, **37**, pp. 395–397.

131. Zhao, T., Wang, Y., Chen, H. and Shen, D. (2014). Graphene passively Q-switched Ho:YAG ceramic laser, *Appl. Phys. B*, **116**, pp. 947–950.

132. Huang, S., Wang, Y., Yan, P., Zhao, J., Li, H. and Lin, R. (2014). Tunable and switchable multi-wavelength dissipative soliton generation in a graphene oxide mode-locked Yb-doped fiber laser, *Opt. Express*, **22**, pp. 11417–11426.

133. Li, X. H., Wang, Y. G., Wang, Y. S., Zhang, Y. Z., Wu, K., Shum, P. P., Yu, X., Zhang, Y. and Wang, Q. J. (2013). All-normal-dispersion passively mode-locked Yb-doped fiber ring laser based on a graphene oxide saturable absorber, *Laser Phys. Lett.*, **10**, p. 075108.

134. Zhao, L. M., Tang, D. Y., Zhang, H., Wu, X., Bao, Q. and Loh, K. P. (2010). Dissipative soliton operation of an ytterbium-doped fiber laser mode locked with atomic multilayer graphene, *Opt. Lett.*, **35**, pp. 3622–3624.

135. Popa, D., Sun, Z., Torrisi, F., Hasan, T., Wang, F. and Ferrari, A. C. (2010). Sub 200 fs pulse generation from a graphene mode-locked fiber laser, *Appl. Phys. Lett.*, **97**, p. 203106.

136. He, X., Liu, Z. and Wang, D. N. (2012). Wavelength-tunable, passively mode-locked fiber laser based on graphene and chirped fiber Bragg grating, *Opt. Lett.*, pp. 2394–2396.

137. Huang, P. L., Lin, S. C., Yeh, C. Y., Kuo, H. H., Huang, S. H., Lin, G. R., Li, L. J., Su, C. Y. and Cheng, W. H. (2012). Stable mode-locked fiber laser based on CVD fabricated graphene saturable absorber, *Opt. Express*, **20**, pp. 2460–2465.

138. Sobon, G., Sotor, J. and Abramski, K. M. (2012). Passive harmonic mode-locking in Er-doped fiber laser based on graphene saturable absorber with repetition rates scalable to 2.22 GHz, *Appl. Phys. Lett.*, **100**, p. 161109.

139. Sotor, J., Sobon, G. and Abramski, K. M. (2012). Scalar soliton generation in all-polarization-maintaining, graphene mode-locked fiber laser, *Opt. Lett.*, **37**, pp. 2166–2168.

140. Tarka, J., Sobon, G., Boguslawski, J., Sotor, J., Jagiello, J., Aksienionek, M., Lipinska, L., Zdrojek, M., Judek, J. and Abramski, K. M. (2014). 168 fs pulse generation from graphene-chitosan mode-locked fiber laser, *Opt. Mater. Express*, **4**, pp. 1981–1986.

141. Wang, J., Luo, Z., Zhou, M., Ye, C., Fu, H., Cai, Z., Cheng, H., Xu, H. and Qi, W. (2012). Evanescent-light deposition of graphene onto tapered fibers for passive Q-switch and mode-locker, *IEEE Photonics J.*, **4**, pp. 1295–1305.

142. Xu, J., Wu, S., Li, H., Liu, J., Sun, R., Tan, F., Yang, Q. H. and Wang, P. (2012). Dissipative soliton generation from a graphene oxide mode-locked Er-doped fiber laser, *Opt. Express*, **20**, pp. 23653–23658.

143. Zhang, H., Bao, Q., Tang, D., Zhao, L. and Loh, K. (2009). Large energy soli-ton erbium-doped fiber laser with a graphene-polymer composite mode locker, *Appl. Phys. Lett.*, **95**, p. 141103.

144. Li, X., Yu, X., Sun, Z., Yan, Z., Sun, B., Cheng, Y., Yu, X., Zhang, Y. and Wang, Q. J. (2015). High-power graphene mode-locked Tm/Ho co-doped fiber laser with evanescent field interaction, *Sci. Rep.*, **5**, p. 16624.

145. Sobon, G., Sotor, J., Pasternak, I., Krajewska, A., Strupinski, W. and Abramski, K. M. (2013). Thulium-doped all-fiber laser mode-locked by CVD-graphene/PMMA saturable absorber, *Opt. Express*, **21**, pp. 127971–127976.

146. Zhang, M., Kelleher, E. J. R., Torrisi, F., Sun, Z., Hasan, T., Popa, D., Wang, F., Ferrari, A. C., Popov, S. V. and Taylor, J. R. (2012). Tm-doped fiber laser mode-locked by graphene-polymer composite, *Opt. Express*, **20**, pp. 25077–25084.

147. Wang, Q. Q., Chen, T., Zhang, B., Li, M., Lu, Y. and Chen, K. P. (2013). All-fiber passively mode-locked thulium-doped fiber ring laser using optically deposited graphene saturable absorbers, *Appl. Phys. Lett.*, **102**, p. 131117.

148. Sotor, J., Sobon, G., Pasternak, I., Krajewska, A., Strupinski, W. and Abramski, K. M. (2013). Simultaneous mode-locking at 1565 nm and 1944 nm in fiber laser based on common graphene saturable absorber, *Opt. Express*, **21**, pp. 18994–19002.

149. Sotor, J., Sobon, G., Tarka, J., Pasternak, I., Krajewska, A., Strupinski, W. and Abramski, K. M. (2014). Passive synchronization of erbium and thulium doped fiber mode-locked lasers enhanced by common graphene saturable absorber, *Opt. Express*, **22**, pp. 5536–5543.

150. Zhu, G., Zhu, X., Wang, F., Xu, S., Li, Y., Guo, X., Balakrishnan, K., Norwood, R. A. and Peyghambarian, N. (2016). Graphene mode-locked fiber laser at 2.8 μm, *IEEE Photonics Technol. Lett.*, **28**, pp. 7–10.

151. Zhang, H., Tang, D., Knize, R. J., Zhao, L., Bao, Q. and Loh, K. P. (2010). Graphene mode locked, wavelength-tunable, dissipative soliton fiber laser, *Appl. Phys. Lett.*, **96**, p. 111112.

152. Zhao, L. M., Tang, D. Y. and Wu, J. (2006). Gain-guided soliton in a positive group-dispersion fiber laser, *Opt. Lett.*, **31**, pp. 1788–1790.

153. Martinez, A. and Yamashita, S. (2012). 10 GHz fundamental mode fiber laser using a graphene saturable absorber, *Appl. Phys. Lett.*, **101**, p. 041118.

154. Meng, Y., Niang, A., Guesmi, K., Salhi, M. and Sanchez, F. (2014). 1.61 μm high-order passive harmonic mode locking in a fiber laser based on graphene saturable absorber, *Opt. Express*, **22**, pp. 29921–29926.

155. Maria, A. J. D., Stetser, D. A. and Heynau, H. (1966). Self mode-locking of lasers with saturable absorbers, *Appl. Phys. Lett.*, **8**, pp. 174–176.

156. Keller, U., Weingarten, K. J., Kartner, F. X., Kopf, D., Braun, B., Jung, I. D., Fluck, R., Honninger, C., Matuschek, N. and Aus der Au, J. (1996). Semiconductor saturable absorber mirrors (SESAMs) for femtosecond to nanosecond pulse generation in solid-state lasers, *IEEE J. Sel. Top. Quantum Electron.*, **2**, pp. 435–453.

157. Stenger, J., Binnewies, T., Wilpers, G., Riehle, F., Telle, H. R. (2001). Phase-coherent frequency measurement of the Ca intercombination line at 657 nm with a Kerr-lens mode-locked femtosecond laser, *Phys. Rev. A*, **63**, p. 021802.

158. Baek, I. H., Lee, H. W., Bae, S., Hong, B. H., Ahn, Y. H., Yeom, D. I. and Rotermund, F. (2012). Efficient mode-locking of sub-70-fs Ti:Sapphire laser by graphene saturable absorber, *Appl. Phys. Express*, **5**, p. 032701.

159. Canbaz, F., Kakenov, N., Kocabas, C., Demirbas, U. and Sennaroglu, A. (2015). Graphene mode-locked Cr:LiSAF laser at 850 nm, *Opt. Lett.*, **40**, pp. 4110–4113.

160. Canbaz, F., Kakenov, N., Kocabas, C., Demrbas, U. and Sennaroglu, A. (2017). Generation of sub-20-fs pulses from a graphene mode-locked laser, *Opt. Express*, **25**, pp. 2834–2839.

161. Zhang, L., Wang, Y. G., Yu, H. J., Zhang, S. B., Hou, W., Lin, X. C. and Li, J. M. (2011). High power passively mode-locked Nd:YVO4 laser using graphene oxide as a saturable absorber, *Laser Phys.*, **21**, pp. 2072–2075.

162. Lou, F., Cui, L., Li, Y. B., Hou, J., He, J. L., Jia, Z. T., Liu, J. Q., Zhang, B. T., Yang, K. J., Wang, Z. W. and Tao, X. T. (2013). High-efficiency femtosecond Yb:Gd3Al0:5Ga4:5O12 mode-locked laser based on reduced graphene oxide, *Opt. Lett.*, **38**, pp. 4189–4192.

163. Ma, J., Huang, H., Ning, K., Xu, X., Xie, G., Qian, L., Loh, K. P. and Tang, D. (2016). Generation of 30 fs pulses from a diode-pumped graphene mode-locked Yb:CaYAlO4 laser, *Opt. Express*, **41**, pp. 890–893.

164. Xu, J. L., Li, X. L., Wu, Y. Z., Hao, X. P., He, J. L. and Yang, K. J. (2011). Graphene saturable absorber mirror for ultra-fast-pulse solid-state laser, *Opt. Lett.*, **36**, pp. 1948–1950.

165. Xu, J. L., Li, X. L., He, J. L., Hao, X. P., Wu, Y. Z., Yang, Y. and Yang, K. J. (2011). Performance of large-area few-layer graphene saturable absorber in femtosecond bulk laser, *Appl. Phys. Lett.*, **99**, p. 261107.

166. Cho, W. B., Kim, J. W., Lee, H. W., Bae, S., Hong, B. H., Choi, S. Y., Baek, I. H., Kim, K., Yeom, D. I. and Rotermund, F. (2011). High-quality, large-area monolayer graphene for efficient bulk laser mode-locking near 1.25 μm, *Opt. Lett.*, **36**, pp. 4089–4091.

167. Davide Di Dio Cafiso, S., Ugolotti, E., Schmidt, A., Petrov, V., Griebner, U., Agnesi, A., Cho, W. B., Jung, B. H., Rotermund, F., Bae, S., Hong, B. H., Reali, G. and Pirzio, F. (2013). Sub-100-fs Cr:YAG laser mode-locked by monolayer graphene saturable absorber, *Opt. Lett.*, **38**, pp. 1745–1747.

168. Lee, C. C. and Schibli, T. R. (2010). Ultra-short optical pulse generation with single-layer graphene, *J. Nonlinear Opt. Phys. Mater.*, **19**, pp. 767–771.

169. Lagatsky, A. A., Sun, Z., Kulmala, T. S., Sundaram, R. S., Milana, S., Torrisi, F., Antipov, O. L., Lee, Y., Ahn, J. H., Brown, C. T. A., Sibbett, W. and Ferrari, A. C. (2013). 2 μm solid-state laser mode-locked by single-layer graphene, *Appl. Phys. Lett.*, **102**, p. 013113.

170. Xie, G. Q., Ma. J., Lv, P., Gao, W. L., Yuan, P., Qian, L. J., Yu, H. H., Zhang, H. J., Wang, J. Y. and Tang, D. Y. (2012). Graphene saturable absorber

for Q-switching and mode locking at 2 μm wavelength [Invited], *Opt. Mater. Express*, **2**, pp. 878–883.

171. Liu, J., Wang, Y. G., Qu, Z. S., Zheng, L. H., Su, L. B. and Xu, J. (2012). Graphene oxide absorber for 2 μm passive mode-locking Tm:YAlO3 laser, *Laser Phys. Lett.*, **9**, p. 1519.

172. Tolstik, N., Sorokin, E. and Sorokina, I. T. (2014). Graphene mode-locked Cr:ZnS laser with 41 fs pulse duration, *Opt. Express*, **22**, pp. 5564–5571.

173. Cizmeciyan, M. N., Kim, J. W., Bae, S., Hong, B. H., Rotermund, F. and Sennaroglu, A. (2013). Graphene mode-locked femtosecond Cr:ZnSe laser at 2500 nm, *Opt. Lett.*, **38**, pp. 341–343.

174. Feng, C., Wang, Y., Liu, J., Tsang, Y. H., Song, Y. and Yu, Z. (2013). 3W high-power laser passively mode-locked by graphene oxide saturable absorber, *Opt. Commun.*, **298–299**, pp. 168–170.

175. Xu, J. L., Li, X. L., He, J. L., Hao, X. P., Yang, Y., Wu, Y. Z., Liu, S. D. and Zhang, B. T. (2012). Efficient graphene Q switching and mode locking of 1.34 μm neodymium lasers, *Opt. Lett.*, **37**, pp. 2652–2654.

176. Wang, Y., Qu, Z. S., Liu, J. and Tsang, Y. H. (2012). Graphene oxide absorbers for watt-level high-power passive mode-locked Nd:GdVO4 laser operating at 1 μm, *J. Lightwave Technol.*, **30**, pp. 3259–3262.

177. Wang, Y. G., Chen, H., Hsieh, W. and Tsang, Y. H. (2013). Mode-locked Nd:GdVO4 laser with graphene oxide/polyvinyl alcohol composite material absorber as well as an output coupler, *Opt. Commun.*, **289**, pp. 119–122.

178. Xu, S. C., Mana, B. Y., Jiang, S. Z., Chen, C. S., Yang, C., Liu, M., Huang, Q. J., Zhang, C., Bi, D., Meng, X. and Liu, F. Y. (2014). Watt-level passively Q-switched mode-locked YVO4/Nd:YVO4 laser operating at 1.06 μm using graphene as a saturable absorber, *Opt. Laser Technol.*, **56**, pp. 393–397.

179. Ma, J., Xie, G. Q., Lv, P., Gao, W. L., Yuan, P., Qian, L. J., Yu, H. H., Zhang, H. J., Wang, J. Y. and Tang, D. Y. (2012). Graphene mode-locked femtosecond laser at 2 μm wavelength, *Opt. Lett.*, **37**, pp. 2085–2087.

180. Gaafar, M., Nakdali, D. A., Möller, C., Fedorova, K. A., Wichmann, M., Shakfa, M. K., Zhang, F., Rahimi-Iman, A., Rafailov, E. U. and Koch, M. (2014). Self-mode-locked quantum-dot vertical-external-cavity surface-emitting laser, *Opt. Lett.*, **39**, pp. 4623–4626.

181. Choudhary, A., Dhingra, S., D'Urso, B., Kannan, P. and Shepherd, D. P. (2015). Graphene Q-switched mode-locked and Q-switched ion-exchanged waveguide lasers, *IEEE Photonics Technol. Lett.*, **27**, pp. 646–649.

182. Ren, Y., Brown, G., Mary, R., Demetriou, G., Popa, D., Torrisi, F., Ferrari, A. C., Chen, F. and Kar, A. K. (2015). 7.8-GHz graphene-based 2-μm monolithic waveguide laser, *IEEE J. Sel. Top. Quantum Electron.*, **21**, p. 1602106.

183. Husaini, S. and Bedford, R. G. (2014). Graphene saturable absorber for high power semiconductor disk laser mode-locking, *Appl. Phys. Lett.*, **104**, p. 161107.

184. Du, J., Wang, Q., Jiang, G., Xu, C., Zhao, C., Xiang, Y., Chen, Y., Wen, S. and Zhang, H. (2014). Ytterbium-doped fiber laser passively mode locked by few-layer molybdenum disulfide (MoS_2) saturable absorber functioned with evanescent field interaction, *Sci. Rep.*, **4**, p. 6346.

185. Mao, D., Zhang, S., Wang, Y., Gan, X., Zhang, W., Mei, T., Wang, Y., Wang, Y., Zeng, H. and Zhao, J. (2015). WS_2 saturable absorber for dissipative soliton mode locking at 1.06 and 1.55 μm, *Opt. Express*, **23**, pp. 27509–27519.

186. Ma, L., Tan, Y., Wang, S., Akhmadaliev, S., Zhou, S., Yu, H., Zhang, H. and Chen, F. (2016). Continuous-wave and Q-wwitched YbSGG waveguide laser, *J. Lightwave Technol.*, **1**, pp. 0733–8724.

187. Zhang, Y., Wang, S., Wang, D., Yu, H., Zhang, H., Chen, Y., Mei, L., Di Lieto, A., Tonelli, M. and Wang, J. (2016). Atomic-layer molybdenum sulfide passively modulated green laser pulses, *Photonics Technol. Lett.*, **28**, pp. 197–200.

188. Yan, P., Chen, H., Liu, A., Li, K., Ruan, S., Ding, J., Qiu, X. and Guo, T. (2017). Self-starting mode-locking by fiber-integrated WS_2 saturable absorber mirror, *IEEE J. Sel. Top. Quantum Electron.*, **23**, p. 1100106.

189. Ren, J., Wang, S., Cheng, Z., Yu, H., Zhang, H., Chen, Y., Mei, L. and Wang, P. (2015). Passively Q-switched nanosecond erbium-doped fiber laser with MoS_2 saturable absorber, *Opt. Express*, **23**, pp. 5607–5613.

190. Li, L., Jiang, S., Wang, Y., Wang, X., Duan, L., Mao, D., Li, Z., Man, B. and Si, J. (2015). WS_2/fluorine mica (FM) saturable absorbers for all-normal-dispersion mode-locked fiber laser, *Opt. Express*, **23**, pp. 28698–28706.

191. Luo, Z., Wu, D., Xu, B., Xu, H., Cai, Z., Peng, J., Weng, J., Xu, S., Zhu, C., Wang, F., Sun, Z. and Zhang, H. (2015). Two-dimensional material-based saturable absorbers: towards compact visible-wavelength all-fiber pulsed lasers, *Nanoscale*, **8**, pp. 1066–1072.

192. Huang, Y., Luo, Z., Li, Y., Zhong, M., Xu, B., Che, K., Xu, H., Cai, Z., Peng, J. and Weng, J. (2014). Widely-tunable, passively Q-switched erbium-doped fiber laser with few-layer MoS_2 saturable absorber, *Opt. Express*, **22**, pp. 25258–25266.

193. Kassani, S. H., Khazaeizhad, R., Jeong, H., Nazari, T., Yeom, D. I. and Oh, K. (2015). All-fiber Er-doped Q-switched laser based on tungsten disulfide saturable absorber, *Opt. Mater. Express*, **5**, pp. 373–379.

194. Li, W., Peng, J., Zhong, Y., Wu, D., Lin, H., Cheng, Y., Luo, Z., Weng, J., Xu, H. and Cai, Z. (2016). Orange-light passively Q-switched Pr3+-doped all-fiber lasers with transition-metal dichalcogenide saturable absorbers, *Opt. Mater. Express*, **6**, p. 2031.

195. Luan, C., Yang, K., Zhao, J., Zhao, S., Song, L., Li, T., Chu, H., Qiao, J., Wang, C., Li, Z., Jiang, S., Man, B. and Zheng, L. (2016). WS_2 as a saturable absorber for Q-switched 2 micron lasers, *Opt. Lett.*, **41**, pp. 3783–3786.

196. Tan, Y., Guo, Z., Ma, L., Zhang, H., Akhmadaliev, S., Zhou, S. and Chen, F. (2016). Q-switched waveguide laser based on two-dimensional semiconducting materials: tungsten disulfide and black phosphorous, *Opt. Express*, **24**, pp. 2858–2866.

197. Chen, W., Hongyu, L., Han, Z., Chun, L., Jitao, X., Jianfeng, L. and Yong, L. (2016). Passively Q-switched mid-infrared fluoride fiber laser around 3 μm using a tungsten disulfide (WS_2) saturable absorber, *Laser Phys. Lett.*, **13**, p. 105108.

198. Guo, C., Chen, B., Wang, H., Zhang, X., Wang, J., Wu, K. and Chen, J. (2016). Investigation on the stability of WSe_2-PVA saturable absorber in an all PM Q-Switched fiber laser, *IEEE Photonics J.*, **8**, p. 1503612.

199. Jung, M., Lee, J., Park, J., Koo, J., Jhon, Y. M. and Lee, J. H. (2015). Mode-locked, 1.94-μm, all-fiberized laser using WS_2-based evanescent field interaction, *Opt. Express*, **23**, pp. 19996–20006.

200. Liu, M., Zheng, X. W., Qi, Y. L., Liu, H., Luo, A. P., Luo, Z. C., Xu, W. C., Zhao, C. J. and Zhang, H. (2014). Microfiber-based few-layer MoS_2 saturable absorber for 2.5 GHz passively harmonic mode-locked fiber laser, *Opt. Express*, **22**, pp. 22841–22846.

201. Liu, H., Luo, A. P., Wang, F. Z, Tang, R., Liu, M., Luo, Z. C., Xu, W. C., Zhao, C. J. and Zhang, H. (2014). Femtosecond pulse erbium-doped fiber laser by a few-layer MoS_2 saturable absorber, *Opt. Lett.*, **39**, pp. 4591–4594.

202. Cui, Y., Lu, F. and Liu, X. (2016). MoS_2-clad microfibre laser delivering conventional, dispersion-managed and dissipative solitons, *Sci. Rep.*, **6**, p. 30524.

203. Khazaeinezhad, R., Kassani, S. H., Jeong, H., Park, K. J., Kim, B. Y., Yeom, D. I. and Oh, K. (2015). Ultrafast pulsed all-fiber laser based on tapered fiber enclosed by few-layer WS_2 nanosheets, *Photonics Technol. Lett.*, **27**, pp. 1581–1584.

204. Yan, P., Liu, A., Chen, Y., Wang, J., Ruan, S., Chen, H. and Ding, J. (2015). Passively mode-locked fiber laser by a cell-type WS_2 nanosheets saturable absorber, *Sci. Rep.*, **5**, p. 12587.

205. Hou, J., Zhao, G., Wu, Y., He, J. and Hao, X. (2015). Femtosecond solid-state laser based on tungsten disulfide saturable absorber, *Opt. Express*. **23**, pp. 27292–27298.

206. Ma, J., Lu, S., Guo, Z., Xu, X., Zhang, H., Tang, D. Y. and Fan, D. Y. (2015). Few-layer black phosphorus based saturable absorber mirror for pulsed solid-state lasers, *Opt. Express*, **23**, pp. 22643–22648.

207. Ismail, E. I., Kadir, N. A., Latiff, A. A., Ahmad, H. and Harun, S. W. (2016). Black phosphorus crystal as a saturable absorber for both a Q-switched and mode-locked erbium doped fiber laser, *RSC Adv.*, **6**, p. 72692.

208. Chu, Z. Z., Liu, J., Guo, Z. N. and Zhang, H. (2016). 2 μm passively Q-switched laser based on black phosphorus, *Opt. Mater. Express*, **6**, pp. 2374–2379.

209. Zhang, H., He, J., Wang, Z., Hou, J., Zhang, B., Zhao, R., Han, K., Yang, K., Nie, H. and Sun, X. (2016). Dual-wavelength, passively Q-switched Tm:YAP laser with black phosphorus saturable absorber, *Opt. Mater. Express*, **6**, pp. 2328–2335.

210. Yu, H., Zheng, X., Yin, K., Cheng, X. and Jiang, T. (2016). Nanosecond passively Q-switched thulium/holmium-doped fiber laser based on black phosphorus nanoplatelets, *Opt. Mater. Express*, **6**, pp. 603–609.

211. Wang, Z., Zhao, R., He, J., Zhang, B., Ning, J., Wang, Y., Su, X., Hou, J., Lou, F., Yang, K., Fan, Y., Bian, J. and Nie, J. (2016). Multi-layered black phosphorus as saturable absorber for pulsed Cr:ZnSe laser at 2.4 μm, *Opt. Express*, **24**, pp. 1598–1603.

212. Kong, L., Qin, Z., Xie, G., Guo, Z., Zhang, H., Yuan, P. and Qian, L. (2016). Black phosphorus as broadband saturable absorber for pulsed lasers from 1 μm to 2.7 μm wavelength, *Laser Phys. Lett.*, **13**, p. 045801.

213. Qin, Z., Xie, G., Zhang, H., Zhao, C., Yuan, P., Wen, S. and Qian, L. (2015). Black phosphorus as saturable absorber for the Q-switched Er:ZBLAN fiber laser at 2.8 μm, *Opt. Express*, **23**, pp. 24713–24718.

214. Li, J., Luo, H., Zhai, B., Lu, R., Guo, Z., Zhang, H. and Liu, Y. (2016). Black phosphorus: a two-dimension saturable absorption material for mid-infrared Q-switched and mode-locked fiber lasers, *Sci. Rep.*, **6**, p. 30361.

215. Sotor, J., Sobon, G., Kowalczyk, M., Macherzynski, W., Paletko, P. and Abramski, K. M. (2015). Ultrafast thulium-doped fiber laser mode locked with black phosphorus, *Opt. Lett.*, **40**, pp. 3885–3888.

216. Hisyam, M. B., Rusdi, M. F. M., Latiff, A. A. and Harun, S. W. (2015). Generation of mode-locked Ytterbium doped fiber ring laser using few-layer black phosphorus as a saturable absorber, *IEEE J. Sel. Top. Quantum Electron.*, **23**, pp. 39–43.

217. Yu, H., Zheng, X., Yin, K., Cheng, X. and Jiang, T. (2015). Thulium/holmium-doped fiber laser passively mode locked by black phosphorus nanoplatelets-based saturable absorber, *Appl. Opt.*, **54**, pp. 10290–10294.

218. Zhang, B., Lou, F., Zhao, R., He, J., Li, J., Su, X., Ning, J. and Yang, K. (2015). Exfoliated layers of black phosphorus as saturable absorber for ultrafast solid-state laser, *Opt. Lett.*, **40**, pp. 3691–3694.

219. Luo, Z. C., Liu, M., Guo, Z. N., Jiang, X. F., Luo, A. P., Zhao, C. J., Yu, X. F., Xu, W. C. and Zhang, H. (2015). Microfiber-based few-layer black phosphorus saturable absorber for ultra-fast fiber laser, *Opt. Express*, **23**, pp. 20030–20039.

220. Park, K., Lee, J., Lee, Y. T., Choi, W. K., Lee, J. H. and Song, Y. W. (2015). Black phosphorus saturable absorber for ultrafast mode-locked pulse laser via evanescent field interaction, *Ann. Phys.*, **527**, pp. 770–776.

221. Wang, Z., Xu, Y., Dhanabalan, S. C., Sophia, J., Zhao, C., Xu, C., Xiang, Y., Li, J. and Zhang, H. (2016). Black phosphorus quantum dots as an efficient saturable absorber for bound soliton operation in an Erbium doped fiber laser, *IEEE Photonics J.*, **7**, p. 1503310.

222. Qin, Z., Xie, G., Zhao, C., Wen, S., Yuan, P. and Qian, L. (2016). Mid-infrared mode-locked pulse generation with multilayer black phosphorus as saturable absorber, *Opt. Lett.*, **41**, pp. 56–59.

223. Lee, D., Park, K., Debnath, P. C., Kim, I. and Song, Y. W. (2016). Thermal damage suppression of a black phosphorus saturable absorber for high-power operation of pulsed fiber lasers, *Nanotechnology*, **27**, p. 365203.

224. Song, Y., Chen, S., Zhang, Q., Li, L., Zhao, L., Zhang, H. and Tang, D. (2016). Vector soliton fiber laser passively mode locked by few layer black phosphorus-based optical saturable absorber, *Opt. Express*, **24**, pp. 25933–25942.

225. Chen, Y., Mu, H., Li, P., Lin, S., Nanjunda, S. B. and Bao, Q. (2016). Optically driven black phosphorus as a saturable absorber for mode-locked laser pulse generation, *Opt. Eng.*, **55**, p. 081317.

226. Chen, Y., Chen, S., Liu, J., Gao, Y. and Zhang, W. (2016). Sub-300 femtosecond soliton tunable fiber laser with all-anomalous dispersion passively mode locked by black phosphorus, *Opt. Express*, **24**, pp. 13316–13324.

227. Yu, H., Zhang, H., Wang, Y., Zhao, C., Wang, B., Wen, S., Zhang, H. and Wang, J. (2013). Topological insulator as an optical modulator for pulsed solid-state lasers, *Laser Photonics Rev.*, **7**, pp. L77–L83.

228. Sotor, J., Sobon, G., Grodecki, K. and Abramski, K. M. (2014). Mode-locked erbium-doped fiber laser based on evanescent field interaction with Sb_2Te_3 topological insulator, *Appl. Phys. Lett.*, **104**, p. 251112.

229. Li, J., Luo, H., Wang, L., Zhao, C., Zhang, H., Li, H. and Liu, Y. (2015). 3-μm mid-infrared pulse generation using topological insulator as the saturable absorber, *Opt. Lett.*, **40**, pp. 3659–3662.

230. Jeong, Y., Sahu, J., Payne, D. and Nilsson, J. (2005). Ytterbium-doped large-core fiber laser with 1.36 kW continuous-wave output power, *Opt. Express*, **12**, pp. 6088–6092.

231. Kujala, S., Mannila, A., Karvonen, L., Kieu, K. and Sun, Z. (2016). Natural silk as a photonics component: a study on its light guiding and nonlinear optical properties, *Sci. Rep.*, **6**, p. 22358.

232. Grigorenko, A. N., Polini, M. and Novoselov, K. S. (2012). Graphene plasmonics, *Nat. Photonics*, **6**, pp. 749–758.

233. Yang, X., Zhai, F., Hu, H., Hu, d., Liu, R., Zhang, S., Sun, M., Sun, Z., Chen, J. and Dai, Q. (2016). Far-field spectroscopy and near-field optical imaging of coupled plasmon-phonon polaritons in 2D van der Waals heterostructures, *Adv. Mater.*, **28**, pp. 2931–2938.

234. Jussila, H., Yang, H., Granqvist, N. and Sun, Z. (2016). Surface plasmon resonance for characterization of large-area atomic-layer graphene film, *Optica*, **3**, pp. 151–158.

235. Hu, H., Yang, X., Zhai, F., Hu, D., Liu, R., Liu, K., Sun, Z. and Dai, Q. (2016). Far-field nanoscale infrared spectroscopy of vibrational fingerprints of molecules with graphene plasmons, *Nat. Commun.*, **7**, p. 12334.

Chapter 5

Plasmonic-Nanostructure-Enhanced Optical Properties of 2D Materials

Zhuo Wang[a,b] **and Andrew T. S. Wee**[a]
[a]*Physics Department, National University of Singapore, 2 Science Drive 3, Singapore 117542, Singapore*
[b]*SZU- NUS Collaborative Innovation Centre for Optoelectronic Science & Technology, College of Optoelectronic Engineering, Shenzhen University, Shenzhen 518060, China*
wzhuo@szu.edu.cn, phyweets@nus.edu.sg

As a typical 2D materials system, transition metal dichalcogenides (TMDs) such as molybdenum disulfide (MoS_2), molybdenum diselenide ($MoSe_2$), tungsten disulfide (WS_2) and tungsten diselenide (WSe_2) have impressive properties arising from their atomically thin nature and have been demonstrated to be potentially useful building blocks for optoelectronic integrated circuits. An important research focus on 2D TMDs is their applications in ultra-thin photo-emitters or photo-detectors. However, being atomically thin limits light–matter interaction in 2D TMDs, resulting in low light-to-light conversion efficiency. Fortunately, integration of photonic nanostructures such as nanocavities and plasmonic nanostructures provides a promising solution to this challenge.

Two-Dimensional Materials in Nanophotonics: Developments, Devices, and Applications
Edited by Yuerui Lu

ISBN 978-981-4800-22-8 (Hardcover), 978-0-429-42877-7 (eBook)
www.jennystanford.com

This chapter will focus on plasmonic-nanostructure-enhanced optical properties of 2D TMDs and related devices. First we will present an introduction to 2D TMDs and a brief analysis of the physical process of light–matter interactions, followed by a detailed review of the resulting optical properties: absorption, photoluminescence (PL), Raman scattering and SHG. We conclude with a brief outlook of this field.

5.1 Introduction to 2D TMDs

When TMDs thin down to a monolayer, a unique crystal symmetry and electronic band structure give rise to unique optical and optoelectronic properties that are absent in their bulk counterparts. In this section, we present the crystal and electronic band structure of 2D TMDs and their related physical properties.

5.1.1 Crystal Structure

Monolayer TMDs with MX_2 stoichiometry, where M is defined as a transition metal element (M = Mo, W, etc.) and X is defined as a chalcogen (X = S, Se, etc.), are composed of a layer of M atoms sandwiched by two layers of X atoms, as shown in Fig. 5.1a [1]. The monolayer is only several Ångströms thick, resulting in quantum confinement in the out-of-plane direction. The quantum confinement leads to novel optical and electronic properties that are tunable by electric field, doping, strain and environment effects. Furthermore, the flexibility and lack of surface dangling bonds enable the 2D materials to be integrated with plasmonic structures or stacked with other 2D materials with different lattice constants. Bulk crystal polytypes 2H, 3R or 1T are formed due to differences in stacking order and metal atom coordination, as shown in Fig. 5.1b. The structural phase determines the electronic and optical properties of TMDs. For instance, 2H-phase TMDs are generally semiconducting, while 1T-phase TMDs are generally metallic. The 2H monolayer phase has broken in-plane inversion symmetry, allowing second-harmonic emission, while the SHG does not exist in the 1T phase with an inversion centre.

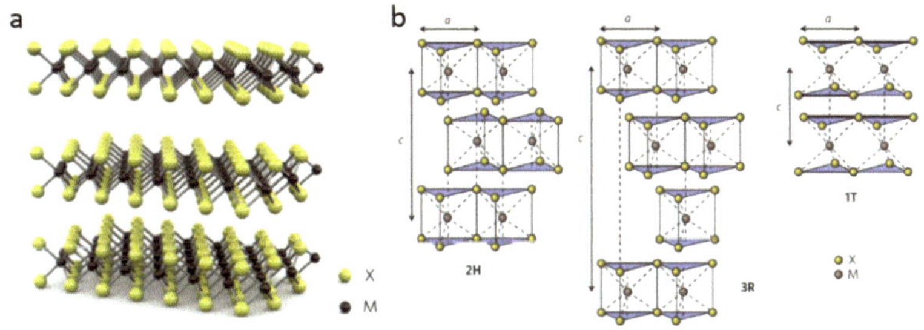

Figure 5.1 Structure of TMD materials [1]. (a) Three-dimensional schematic representation of a typical MX_2 structure, with the chalcogen atoms (X) in yellow and the metal atoms (M) in grey. (b) Schematics of the structural polytypes: 2H (hexagonal symmetry, two layers per repeat unit, trigonal prismatic coordination), 3R (rhombohedral symmetry, three layers per repeat unit, trigonal prismatic coordination) and 1T (tetragonal symmetry, one layer per repeat unit, octahedral coordination). The lattice constants *a* are in the range of 3.1 to 3.7 Å for different materials. The stacking index *c* indicates the number of layers in each stacking order, and the inter-layer spacing is ~6.5 Å. Reprinted by permission from Springer Nature Customer Service Centre GmbH: Springer Nature, *Nature Nanotechnology*, Ref. [1], Copyright (2012).

5.1.2 Electronic Band Structure

The applications of TMDs in optical and optoelectronic devices rely on their electronic properties, such as band structure and density of states. TMDs have sizable bandgaps that change from indirect in the bulk to direct in the monolayer, allowing applications such as light emitters, transistors, photo-detectors and electro-luminescent devices. The bulk TMDs such as WSe_2 and MoS_2 are indirect-bandgap semiconductors with a conduction band minimum (CBM) at the *Q* point and a valence band maximum (VBM) at the Γ point in reciprocal space (Fig. 5.2a,b). In the monolayer, they cross over to become direct-bandgap semiconductors with gaps located at the *K* and the degenerate *K′* points (Fig. 5.2c,d) [2]. All MX_2 compounds undergo a similar transition from indirect to direct bandgap with decreasing layer number, covering the bandgap energy range between 1.1 and 1.9 eV.

5.2 Light–Matter Interaction in 2D TMDs

The unique crystal symmetry and electronic band structure introduced before bring in novel optical and optoelectronic properties. These properties arise from light–mater interaction in TMDs, that is, when light is absorbed by 2D TMDs, it undergoes light-to-light, light-to-electron and light-to-heat conversion processes, resulting in PL, SHG, photo-carriers, thermal effects, etc. (Fig. 5.3). This section presents the physics involved in light absorption, PL, Raman scattering and SHG.

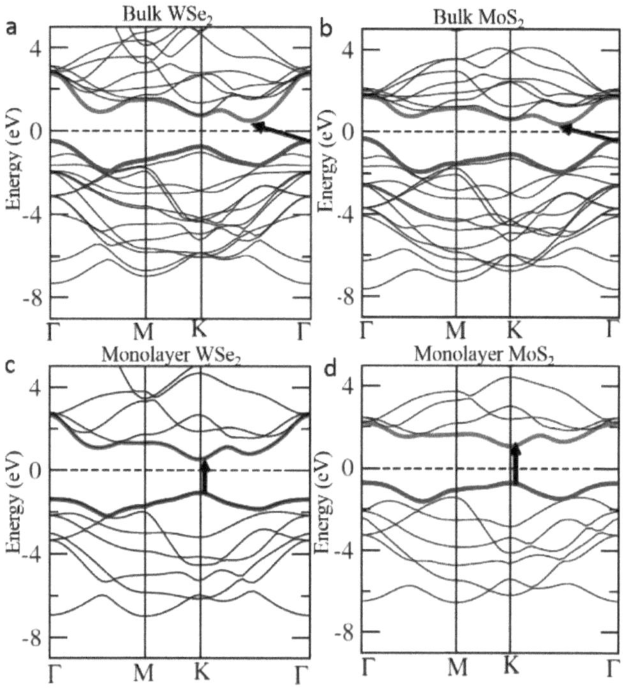

Figure 5.2 Electronic band structure of TMDs [2]. (Top) Calculated band structure of bulk WSe_2 (a) and bulk MoS_2 (b). (Bottom) Band structure of monolayer WSe_2 (c) and monolayer MoS_2 (d). The arrows indicate the lowest energy transitions. Reprinted by permission from Springer Nature Customer Service Centre GmbH: Springer Nature, *The European Physical Journal B: Condensed Matter and Complex Systems*, Ref. [2], Copyright (2012).

2D TMDs have absorption coefficients of more than 5×10^7 m^{-1} in the visible range after normalisation to the corresponding atomic thickness, which is an order of magnitude larger than the absorption coefficients of GaAs and Si [4]. The enhanced light–matter interactions arise from the phenomenon of 'band nesting' in TMDs, which leads to enhanced photon absorption, according to

Castro Neto's density functional theory (DFT) calculations [3]. As shown in the band structure of a TMD (left panel of Fig. 5.4), band nesting refers to regions with the conduction band and valence band parallel to each other. When the materials absorb a photon, the generated electrons and holes propagate with the same velocities but in opposite directions. As shown in the right panel of Fig. 5.4, these regions have van Hove singularities (VHS), leading to high electronic density of states, that is, the number of electrons per unit volume participating in the absorption process at a given energy is high.

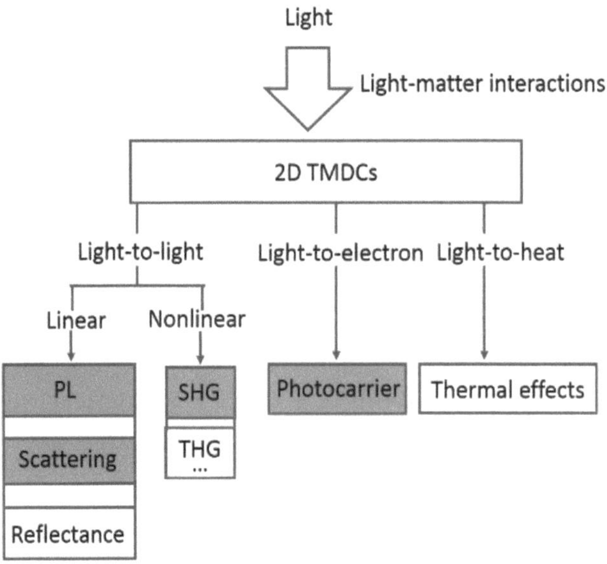

Figure 5.3 Effects resulting from light–matter interactions in 2D TMDs. Light–matter interaction leads to PL, SHG, photo-carriers, thermal effects, etc.

In the light absorption process, the band-to-exciton mechanism plays an important role in the absorption spectra of TMDs. Electrons in the valence band absorb sufficient energy to be excited to the conduction band, and the electrons and holes they leave behind remain electrostatically 'bound' by an exciton-binding energy. The exciton-binding energy accounts for the difference between electronic bandgap and optical bandgap. As shown in Fig. 5.5a [5], the optical absorption spectra of monolayer TMDs is dominated by direct transitions of electrons from the valence to the conduction band near the *K* and *K'* points. For instance, monolayer WSe_2 (MoS_2) can absorb nearly 10% of incident light at the exciton resonance wavelength of 750 nm (650 nm) and even more strongly at shorter wavelengths where there are band-nesting regions. Other TMDs have similar spectral trends except that the resonances lie in different spectral regions.

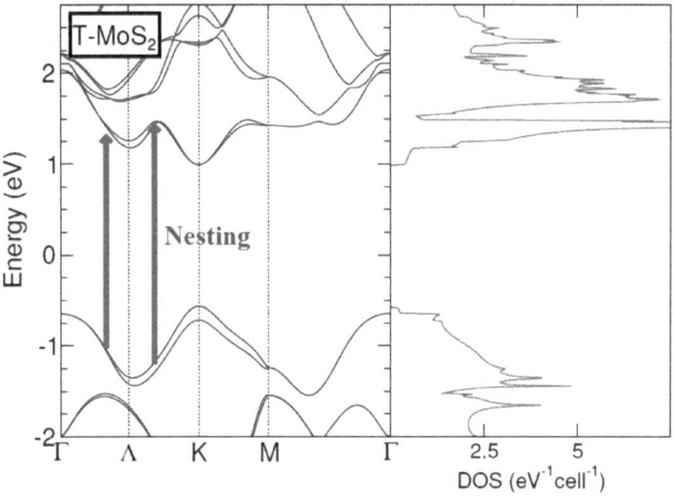

Figure 5.4 Band structures and density of states of TMDs [3]. The arrows indicate the transitions corresponding to the first prominent peaks in the optical conductivity. Reprinted (figure) with permission from Ref. [3]. Copyright (2013) by the American Physical Society.

The electron–hole pairs bonded by the Coulomb force are called excitons. In addition to excitons, high-order excitonic quasi-particles such as trions (bound states of two holes and one electron or two electrons and one hole) have also been reported in TMDs. Excitons and trions in 2D TMDs have binding energies roughly an order of magnitude larger than those in conventional quasi-2D semiconductors, thus allowing them to be observed in the absorption spectrum even at room temperature. The behaviour of excitons and trions has a strong influence on the optical properties in 2D TMDs, including light absorption and emission.

Efficient light absorption in 2D TMDs results in optical and optoelectronic properties that are superior or complementary to those observed in the first-discovered 2D material, graphene [7].

Graphene, a monolayer of carbon atoms arranged in a honeycomb structure, is a gapless semi-metal with excellent electron conductivity, making it a promising electrode material to collect electrons in electronic devices. However, the lack of bandgap in graphene and the existence of inversion symmetry limit its application in light-emitting devices in the visible light range based on linear and non-linear optical processes. In contrast, 2D TMDs are direct-bandgap semiconductors with inversion symmetry-breaking structures when thinned down to a monolayer, thus possessing intriguing PL [5] and SHG [6] properties in the visible wavelength range, as shown in Fig. 5.5a,b.

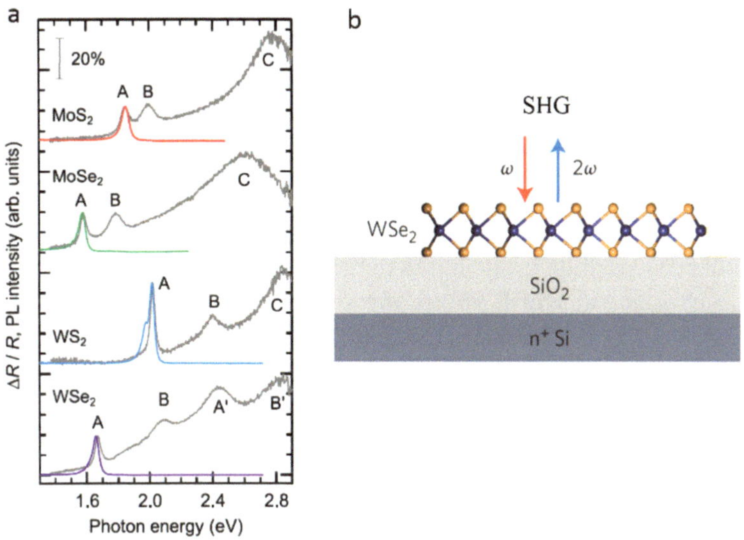

Figure 5.5 Optical response of 2D TMDs. (a) Absorption (grey lines) and PL (coloured lines) spectra of single-layer semiconducting TMD crystals (MoS_2, $MoSe_2$, WS_2, WSe_2). Reprinted with permission from Ref. [5]. Copyright (2013) by the American Chemical Society. (b) SHG from WSe_2. Reprinted by permission from Springer Nature Customer Service Centre GmbH: Springer Nature, *Nature Nanotechnology*, Ref. [6], Copyright (2015).

Although light–matter interactions in TMDs are strong as compared to graphene, it is inherently weak in absolute terms because the optical absorption depth is extremely short (<1 nm for monolayer). For practical optical and optoelectronic devices based on 2D materials, a key challenge is to maximise light–matter interactions in the materials and enhance the resulting PL, Raman scattering and SHG. To optimise these properties, the related physics and progress in this field are reviewed in the following sections.

5.3 PL of 2D TMDs

5.3.1 Physics of PL

Upon light illumination, electrons in the materials can be excited from the valence band to the conduction band. In the relaxation process, excited electrons will decay to the ground state by two pathways: radiative (in the form of photon emission) and non-radiative (in the form of phonon emission) processes. The radiative decay process involving absorption of light and subsequent emission of light with lower energy is defined as photo-luminescence (PL). As shown in Fig. 5.5a, the absorption spectra and PL are correlated. For WSe_2 ($MoSe_2$), the absorption peak at a wavelength of 750 nm (800 nm) corresponds well with the PL peak at this wavelength. In contrast, for MoS_2 (WS_2), the two peaks are slightly shifted. The Stokes shift of peak position is attributed to the inhomogeneity of the dielectric environment, which affects the surface interaction of monolayer MoS_2 (WS_2) and changes the binding energy of the excitons, that is, the electron–hole pairs bounded by the Coulomb force.

In TMDs, the PL intensity is influenced by many factors, including layer number, strain and interfacial interactions such as charge transfer, plasmonic nanostructures, etc. First, we discuss how the layer number of the material influences the electronic structure and optical properties. As WSe_2 has a higher quantum yield at room temperature than TMDs such as $MoSe_2$, MoS_2 and WS_2, we use it as the model system here. A systematic study of WSe_2 with different layer numbers was reported by Zeng et al. [8]. As WSe_2 thins to a few layers, the PL intensity from the direct inter-band transition drops and reaches a maximum at one monolayer, over 3 orders of magnitude higher than that from the bulk and nearly 1 order of magnitude stronger than the bilayer. This light emission behaviour is fully consistent with the calculated band structure. In addition, an interesting 'valley optical selection rule' for PL of monolayer TMDs is obeyed such that its circularly polarised component has the same helicity as the excitation laser. This can be used as a tool to identify the optical transition in the degenerate *K* and *K*′ valleys. Details about this valley selection rule are beyond the scope of this chapter and can be found elsewhere [9, 10].

5.3.2 PL Enhancement Methods

The direct bandgap in monolayer TMDs leads to much stronger PL as compared to their few-layer and bulk counterparts, opening up the possibility of developing ultra-thin light-emitting devices. However, the low light absorption of monolayer TMDs in absolute terms and the large non-radiative recombination rate of excitons which far exceeds the radiative recombination rate result in low PL efficiency (less than 10^{-2}), limiting their application in practical devices [11]. Fortunately, the integration of TMDs with external photonic structures such as optical cavities and plasmonic nanostructures offers possible solutions. In particular, integration with plasmonic nanostructures enables the significant manipulation of the near-field incident light and local optical density of states surrounding the 2D TMDs, thus resulting in modified absorption/emission properties. Furthermore, this integration procedure is easily achieved, either by transferring 2D TMDs on plasmonic or cavity nanostructures or by patterning nanostructures on TMDs. The transfer method, if done properly, should not sacrifice their physical and chemical properties as the 2D membrane is flexible and has no dangling bonds.

5.3.2.1 Tuning the emission process

The coupling of 2D TMDs with optical cavities has been demonstrated to significantly enhance the PL of TMDs. The first example was shown by Xuetao Gan and co-workers [12]. In their experiment, monolayer MoS_2 was transferred onto a nanocavity made of semiconducting a gallium phosphide (GaP) membrane with a cavity resonance that overlaps with the monolayer MoS_2 exciton resonance (Fig. 5.6a). As shown in Fig. 5.6b, a fivefold PL enhancement in MoS_2 at the cavity with resonance mode (green) is obtained due to the Purcell effect. The Purcell effect enhances the spontaneous emission rate of excitons in monolayer MoS_2. The enhancement originates from a larger optical density of states of the nanocavity than that of free space by a Purcell factor of $(3/4\pi^2)(Q/V_m)(\lambda_0/n)$ [3], where Q, V_m, λ_0 and n denote the cavity quality factor, mode volume, resonance wavelength and refractive index of the substrate, respectively [12]. These high Q and sub-wavelength V_m lead to enhancement in the radiative decay rate.

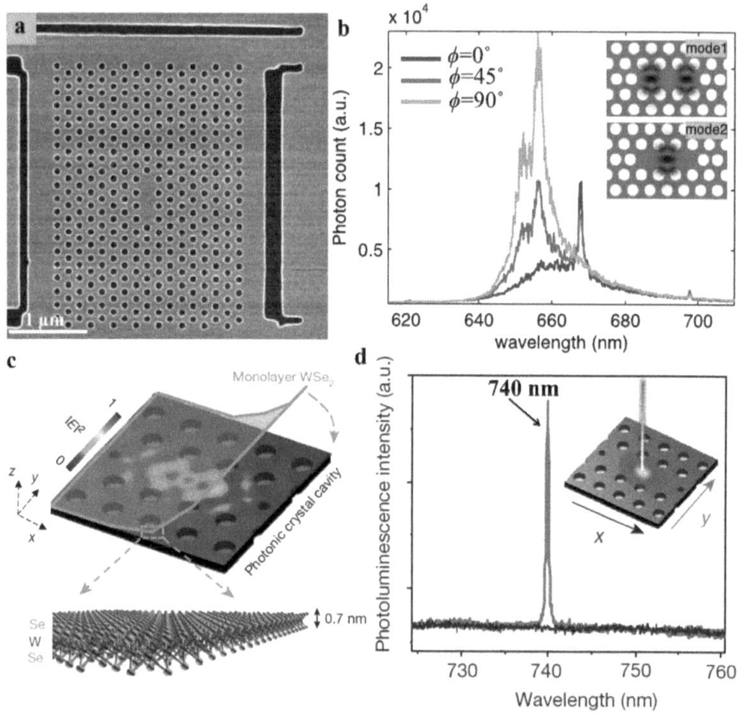

Figure 5.6 Coupled TMDs–cavity photonic devices. (a) SEM image of the GaP nanocavity before transfer of the MoS_2 monolayer [12]. (b) Enhanced PL of monolayer MoS_2 at two cavity modes due to the Purcell effect. ϕ: polarisation analyser angle with respect to the long edge of the cavity [12]. (c) Cartoon depiction of hybrid monolayer WSe_2–cavity nanolasers [13]. (d) Polarisation-resolved PL spectra of a WSe_2–cavity device [13]. (a, b) Reproduced from Ref. [12] under a Creative Commons License Attribution 3.0 Unported (CC BY 3.0), https://creativecommons.org/licenses/by/3.0/. (c, d) Reprinted by permission from Springer Nature Customer Service Centre GmbH: Springer Nature, *Nature*, Ref. [13], Copyright (2015).

A continuous-wave nanolaser operating in the visible regime by introducing monolayer WSe_2 as a gain media on a pre-fabricated photonic crystal on a GaP thin membrane has also been reported

(Fig. 5.6c) [13]. Lasing at a low threshold power of 27 nW is achieved at 130 K by enhancing spontaneous emission into the resonant cavity mode. As shown in Fig. 5.6d, the laser emission is the sharp feature located at ~ 740 nm with a line width of 0.3 nm at the half-maximum of this spectrum. The peak is polarised in the *y* direction, consistent with the fundamental mode of the cavity.

5.3.2.2 Combined excitation and emission enhancement

Further PL enhancement of 2D TMDs has been obtained by designing metallic plasmonic nanostructures with multiple resonances. These resonances simultaneously couple to the excitation and emission wavelengths, increasing the light absorption owing to the excitation field localisation, and strengthening the light emission due to increased scattering cross section of the plasmonic nanostructures. As a result, PL is enhanced. Butun et al. demonstrated enhanced PL with an enhancement factor of 12 from monolayer MoS_2 by plasmonic silver nanodisc arrays (Fig. 5.7a) [14]. Lee et al. also demonstrated PL enhancement (over 15-fold) in MoS_2 using silver bowtie nanoantenna arrays (Fig. 5.7b) [15]. The plasmon resonances of these nanostructures can be identified from the reflectance spectra (Fig. 5.7c,d).

In a report by Wang et al., giant PL enhancement was obtained in monolayer WSe_2 on a gold plasmonics nanostructure with sub-20-nm-wide trenches (Fig. 5.8a) [16]. The trenches in the gold nanostructure support strong field enhancements, as shown in the finite-difference time-domain (FDTD) simulation results in Fig. 5.8b, where a monolayer WSe_2 flake is placed over the trench. The resonances are in the form of lateral gap plasmons in the sub-20-nm trenches with electric fields predominantly parallel to the plane of WSe_2 to promote strong light absorption [17]. Figure 5.8c presents a scanning electron microscope (SEM) image of a triangular WSe_2 single-crystal monolayer transferred onto the gold substrate with a trench width of ~12 nm. The lateral gap plasmon resonance can be tuned to be matched with the pump laser wavelength (633 nm) by varying the pitch of the structures (Fig. 5.9a), which are critical for plasmon coupling with light to achieve optimal field confinement [18, 19].

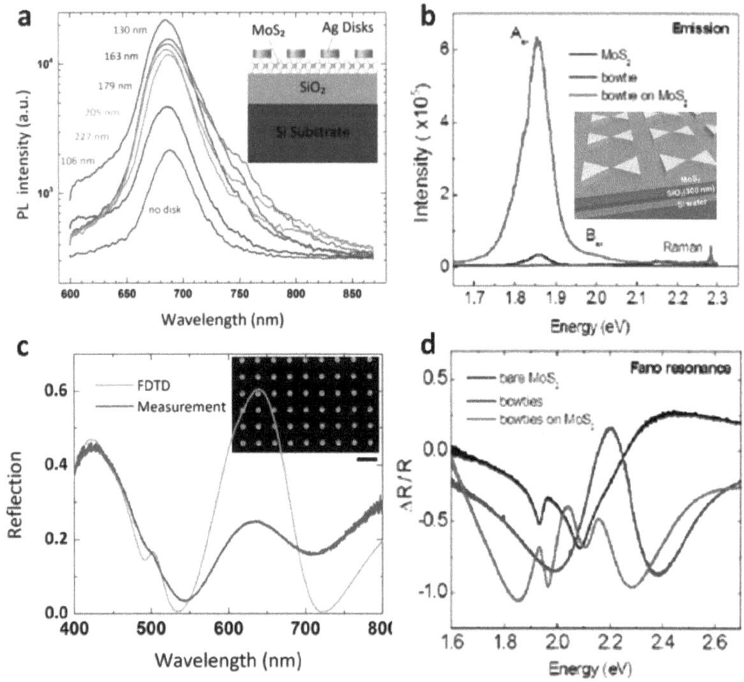

Figure 5.7 Coupled TMDs–plasmonic nanostructures. (a) PL intensity of heterostructure of MoS_2/silver discs plasmonic nanostructures. Inset: Cross section of the heterostructure [14]. (b) PL spectra of bare MoS_2, bowtie array and bowtie MoS_2 [15]. Inset: Device schematic. (c) Reflection curves of silver disk plasmonic nanostructure with a diameter of 130 nm and a periodicity of 400 nm which is used in (a). (d) Experimental $\Delta R/R$ spectra of bare MoS_2, bowtie and bowtie MoS_2 system for three different samples which were used for PL measurement, as shown in (b) [15]. (a, c) Reprinted with permission from Ref. [14]. Copyright (2015) American Chemical Society. (b, d) Reprinted with permission from Ref. [15]. Copyright (2015) American Chemical Society.

The dependence of the plasmon resonance on the pitch size of the plasmonic nanostructures is summarised in Fig. 5.9a, where the plasmon resonance was red-shifted from 507 to 633 nm as the pitch size is increased from 60 to 200 nm. A 633 nm pump laser was used to investigate the effect of plasmon-enhanced PL emission from monolayer WSe_2 (Fig. 5.9b), where the plasmonic nanostructure with a pitch size of 200 nm has the maximum PL intensity. The simulation in Fig. 5.9c,d shows that this nanostructure has the maximum product of near-field enhancement at the excitation wavelength and Purcell factor at the emission wavelength. The near-field enhancement increases the excitation rate of the electron–hole pairs, and the Purcell factor enhancement increases the optical density of state in the emission process, thus increasing the PL intensity. Given the occupation of the trenches over the laser spot, the PL enhancement of WSe_2 over the trenches is estimated to be 20,000-fold as compared to emission from WSe_2 on an un-patterned gold substrate.

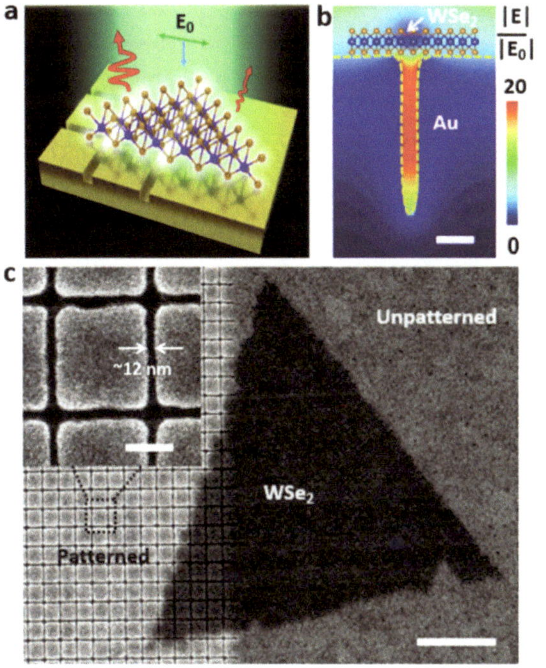

Figure 5.8 Schematic of WSe_2–gold plasmonic hybrid structure with strong optical absorption. (a) Schematic of PL emission from a single-crystal monolayer of a WSe_2 flake on a gold substrate. Part of the triangular flake rests on the patterned region of the substrate consisting of sub-20-nm-wide trenches. (b) Representative simulation of the electric field distribution of the lateral gap plasmons with a WSe_2 monolayer flake suspended over a single trench. The polarisation of the incident laser field is across the gap. The dashed yellow line denotes the boundary between air and gold. The scale bar is 20 nm. (c) Representative SEM image of WSe_2 on square arrays with a trench width of 12 nm (patterned) and un-patterned gold film (unpatterned). The scale bars in the main figure and the inset are 1 μm and 100 nm, respectively. Reproduced from Ref. [16] under a Creative Commons License Attribution 4.0 International (CC BY 4.0), https://creativecommons.org/licenses/by/4.0/.

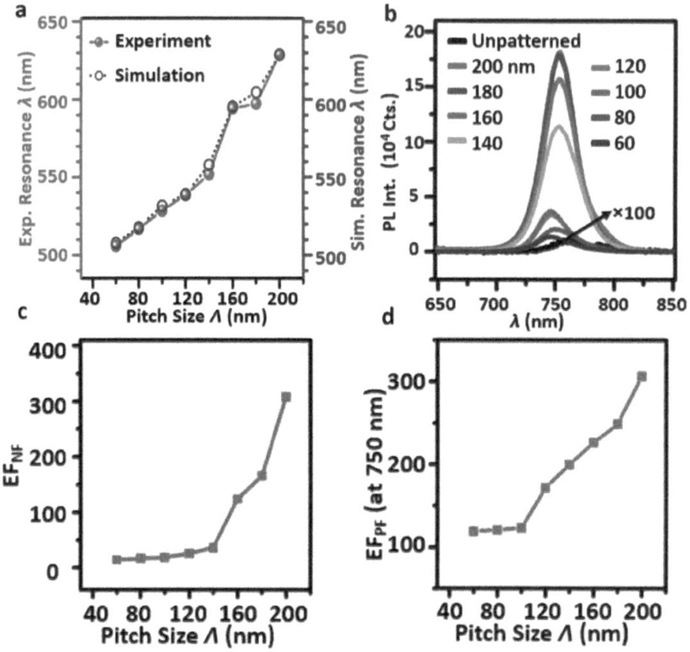

Figure 5.9 Plasmon resonance, PL spectra, simulated near-field enhancement and simulated Purcell factor enhancement. (a) Lateral gap plasmon resonance of a nanostructure with different pitches. (b) PL spectra of WSe_2 on a gold substrate measured by a 633 nm pump laser. (c) Simulated enhancement factor of the near-field intensity for the WSe_2 monolayer flake as placed on the trench region of the gold substrate compared to that on an un-patterned gold film at the pump laser wavelength of 633 nm. This enhancement factor for the near-field intensity is denoted as 'EF_{NF}', which accounts for the excitation process. (d) Simulated enhancement factor for the Purcell factor at the PL emission wavelength of WSe_2 (750 nm) denoted as 'EF_{PF}', accounting for the emission process. Reproduced from Ref. [16] under a Creative Commons License Attribution 4.0 International (CC BY 4.0), https://creativecommons.org/licenses/by/4.0/.

The polarisation-dependent characteristics of the PL emission from WSe_2 on patterned gold nanostructures is also investigated (Fig. 5.10). To resolve the PL-enhanced regions, a 920-nm-pitch substrate and a 532 nm pump laser were used to do the PL mapping. The 532 nm pump laser has a smaller spot size due to the shorter wavelength as compared to the 633 nm pump laser. As a result, the imaging resolution of the PL mapping as obtained by the 532 nm pump laser is much clearer and shows enhanced PL emissions from trench regions. Figure 5.10a shows an SEM image of the sample. Figure 5.10b,c presents the simulated electric field distributions of lateral gap plasmons at the intersection of two trenches with 0° and 45° excitation laser polarisations.

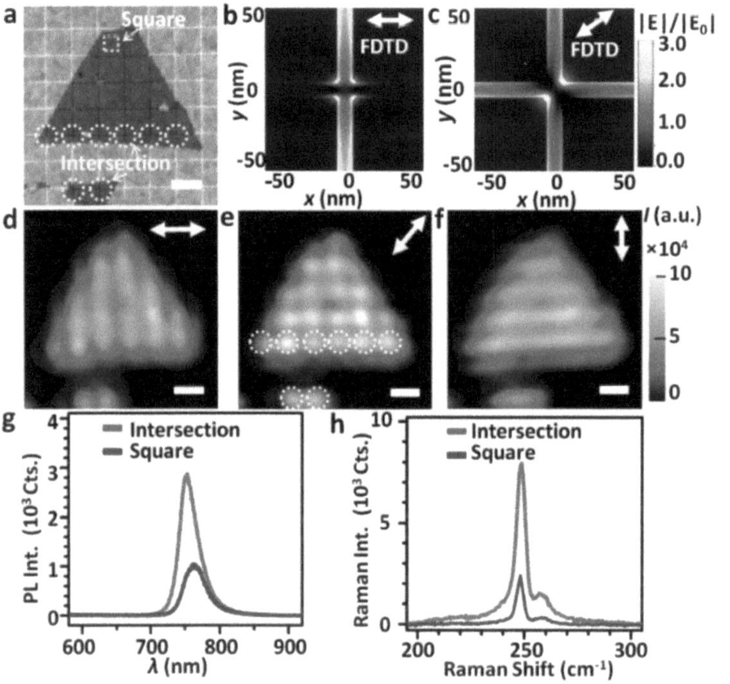

Figure 5.10 Polarisation-dependent characteristics of PL emission from a WSe_2 flake on gold nanostructures with a pitch size of 920 nm and a pump laser of 532 nm. (a) SEM image of a WSe_2 flake on gold nanostructures. The white dashed circles denote the region where the trenches in the gold film intersect, and the dashed square denotes the centre of the square region. (b, c) Simulated electric field distribution at a plane 1 nm above the surface of patterned gold nanostructures with 0° and 45° polarisations, respectively. (d–f) Corresponding PL intensity mappings with polarisation angles of 0°, 45° and 90°, respectively. The intensity value at each pixel was obtained by integrating the PL spectrum across the spectrum window of 700–820 nm. The white arrows show the polarisation directions of the pump laser. (g, h) PL and Raman spectra of WSe_2 taken at the intersection of gold trenches and the square centre, respectively. All the scale bars are 1 μm. The laser power and integration time for measuring PL (Raman) are 30 μW and 0.5 s (450 μW and 15 s), respectively. Reproduced from Ref. [16] under a Creative Commons License Attribution 4.0 International (CC BY 4.0), https://creativecommons.org/licenses/by/4.0/.

The simulated electric field distributions of this nanostructure agree with the experimental PL mappings in Fig. 5.10d–f. The maximum field enhancement occurs in trenches perpendicular to the polarisation direction (Fig. 5.10b) and at the intersection of the trenches under 45° polarisation (Fig. 5.10c). A plot of PL intensity intersecting several trenches indicates a laser beam waist of ~600 nm. These results provide further evidence that the highest PL enhancements are from WSe_2 directly above the trenches, where the optical fields are localised. It is noted that when the pitch size is larger than the pump laser wavelength, both lateral gap plasmons at the trenches and surface plasmon polariton (SPP) on the flat squares are excited. However, the intensity of the SPP is only 1/6 that of the lateral gap plasmon. Thus, the contribution to the PL enhancement of WSe_2 still predominantly stems from the lateral gap plasmons supported by the trenches. Figure 5.10g,h presents the PL and Raman spectra of WSe_2 at the intersection and the square centre, respectively. Both Raman and PL intensities of WSe_2 at the intersection are found to be larger than the ones at the square centre. The un-shifted Raman peak of WSe_2 over trenches and square centre suggests that no strain is induced by the trench, confirming that the observed PL enhancement was not induced by strain [20].

The incorporation of gold arrays with sub-20-nm-wide trenches onto WSe_2 monolayers opens up a new platform for investigating novel electrical/optical properties in 2D materials, such as electro-luminescence and SHG, enabling potential applications in novel optoelectronic devices.

5.3.2.3 Optimising excitation, emission and collection efficiencies

The local PL enhancement of ~2000-fold was reported by simultaneously optimising light absorption, emission and collection efficiencies [21]. This large enhancement was obtained by integrating MoS_2 into a plasmonic nanostructure consisting of placing a silver nanocube over a gold film separated by a nanoscale dielectric spacer (<10 nm), as demonstrated in Fig. 5.11a,b. The resonance mode in the nanostructure is aligned with the quasi-particle gap and excitons, respectively, enabling simultaneous enhancement of the absorption and the PL quantum yield. Furthermore, as the light emission from the nanostructures is directional with a single radiation lobe normal to the substrate, the light collection efficiency from MoS_2 with nanostructures is optimized, thus increasing the total detected local PL.

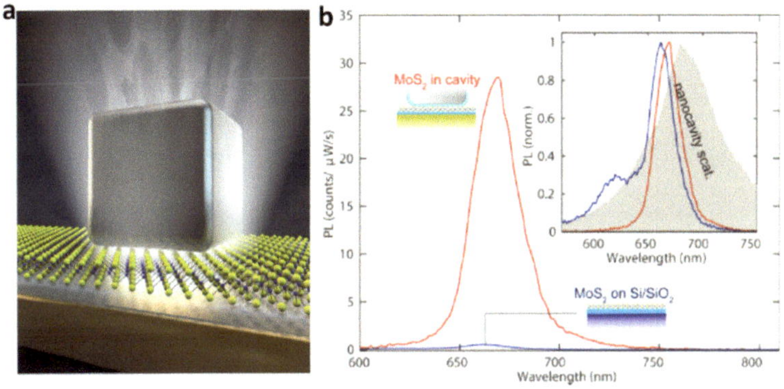

Figure 5.11 Coupled TMD–nanocavity consisting of placing a silver nanocube over a gold substrate [21]. (a) 3D illustration of the nanocavity, consisting of a silver nanocube on a gold substrate, separated by a monolayer of MoS_2, HfO_2 and polymer layers. (b) PL spectra from a MoS_2 monolayer on a SiO_2/Si substrate (blue) and in the nanocavity (red) obtained using a diffraction-limited excitation spot. Reprinted with permission from Ref. [21]. <https://pubs.acs.org/doi/abs/10.1021/acs.nanolett.5b01062>. Further permissions related to the material excerpted should be directed to the ACS.

Interestingly, metallic plasmonic nanostructures have also been reported to quench the PL of TMDs. To take full advantage of the metallic nanostructures in enhancing PL, we need to have an in-depth comprehension of the interaction between light, metallic plasmonic nanostructures and TMDs so as to avoid or reduce possible quenching effects.

5.3.3 Interaction between Light, Plasmonic Nanostructures and TMDs

5.3.3.1 Excitation process

Light coupling with metallic plasmonic nanostructures can generate localised surface plasmons, that is, collective oscillations of free electrons, resulting in confined and enhanced fields at the edges or gaps of the nanostructures as compared to the excitation field (Fig. 5.12) [22]. This enhanced field works on TMDs by enhancing the excitation rate of electron–hole pairs, so metallic plasmonic nanostructures always make a positive contribution on the final PL emission in the excitation process.

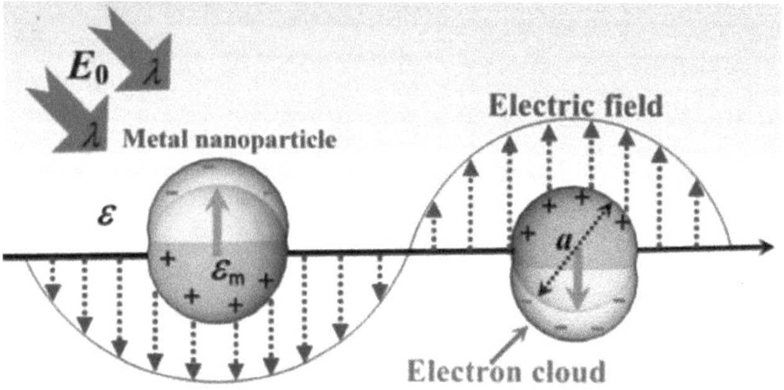

Figure 5.12 Illustration of the localised surface plasmon effect on a plasmonic nanostructure [22]. Gold nanoparticles are taken as an example.

5.3.3.2 Radiative emission

However, in the emission process, plasmonic nanostructures can cause not only PL enhancement but also quenching in TMDs. We can understand the enhancement mechanism from two different but equally valid perspectives. The existence of plasmonic nanostructures near TMDs increases the optical density of states, owing to the confinement of the emission field in the vicinity of the plasmonic nanostructures. This increases the radiative decay rate of the TMDs due to the decrease in the modal volume, namely the effective cavity where the mode is confined within, according to the Purcell effect. The increase in radiative decay rate leads to a net increase in the emission intensity, as shown previously. Another way to look at the enhanced emission is to treat plasmonic nanostructures as efficient antennas that radiate the emission in the far field in a more efficient way compared to the normal substrate, thus increasing the probability of far-field emission for the TMDs.

5.3.3.3 Non-radiative emission

Both PL enhancement and PL quenching are competing effects in plasmonic nanostructures. PL quenching occurs when non-radiative energy transfers from the excitons of TMDs to the 'dark' surface plasmon mode of the plasmonic nanostructures, which is an induced surface plasmon that dissipates this energy by absorption. This quenching process occurs via dipole–dipole resonant non-radiative energy transfer which is highly sensitive to the donor–acceptor separation, dipole–dipole orientation, dipole moment strength of both the donor and the acceptor and dipole–dipole resonance. For instance, if TMDs are placed extremely close to plasmonic nanostructures, the non-radiative relaxation rates can dominate, sacrificing the advantage of the radiative decay rate enhancement of the plasmonic nanostructures. In the weak pumping regime, these competing enhancement and quenching processes are quantified by the quantum yield η, $\eta = \Gamma_r/(\Gamma_{nr} + \Gamma_r)$, where Γ_r is the radiative and Γ_{nr} is the non-radiative decay rate.

5.3.4 Other Factors Influencing PL

In addition to plasmonic effects, there are many other factors such as charge transfer, strain, type of substrate, etc., influencing the PL intensity of TMDs. In the Au–MoS_2 hybrid system, Bhanu et al. reported the PL quenching phenomenon (Fig. 5.14a) and attributed it to charge transfer from monolayer MoS_2 to a gold nanoantenna via a Schottky barrier of 0.4 eV (Fig. 5.14b) [24]. Conley et al. reported a strain-induced decrease of PL intensity in a MoS_2 monolayer [25] (Fig. 5.14c), indicating the direct-to-indirect transition of the optical bandgap of MoS_2 with an applied strain of ~1%. Strain can also influence Raman spectra (Fig. 5.14d). As the Raman peak position and intensity are sensitive to strain, they can serve as a good indicator for the existence of strain at the interface of plasmonic nanostructures and TMDs. The PL intensity of TMDs can be influenced by the substrate as well. Mak et al. reported that suspended monolayer MoS_2 on etched holes on a silicon substrate has much larger PL than that on an un-etched Si substrate (Fig. 5.14e,f) [26]. Buscema et al. quantitatively studied PL emission from monolayer MoS_2 on dielectrics (SiO_2, hexagonal boron nitride, mica and the polymeric dielectric gel films) and conducting substrates (Au and few-layer graphene) [27]. They found that on all studied substrates, monolayer MoS_2 shows a ~fourfold PL enhancement, relative to the commonly used SiO_2 substrate [27]. This is attributed to the combined doping effects of the substrate in MoS_2 and on the radiative decay rates of neutral and charged excitons.

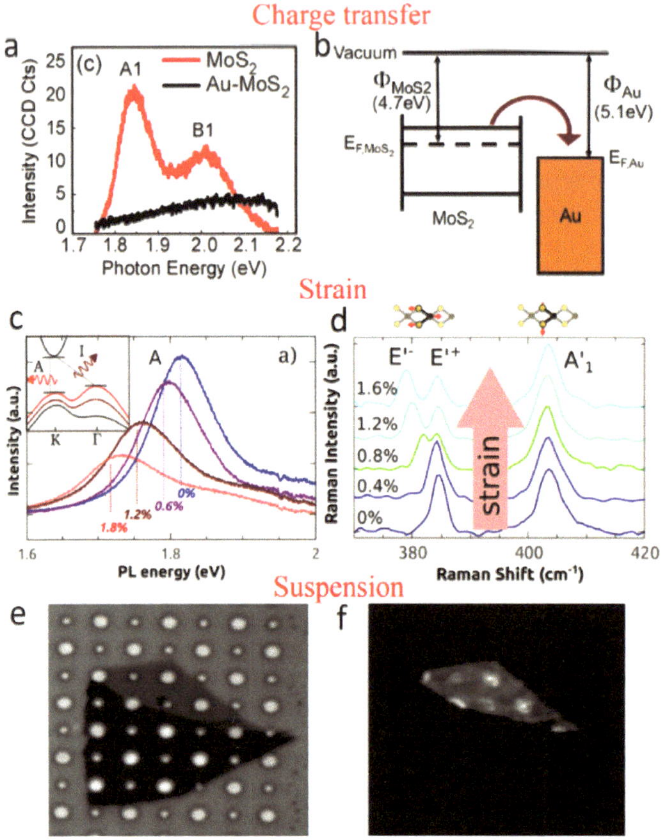

Figure 5.13 PL is influenced by charge transfer, strain and suspension. (a) PL spectra of a bilayer MoS_2 flake as exfoliated (red) and after 2.0 nm Au deposition (black) [24]. (b) The energy band diagram for MoS_2 and Au shows the relative positions of the Fermi level with respect to the vacuum level before establishing contact [24]. The direction of the arrow represents the transfer of electrons from MoS_2 to Au after the contact is established [25]. (c) Evolution of the PL spectrum as monolayer MoS_2 is strained from 0% to 1.6%. (d) Raman spectra of monolayer MoS_2 as it is strained from 0% to 1.8% [25]. (e) Representative optical image of mono- and few-layer MoS_2 crystals on a silicon substrate with etched holes of 1.0 and 1.5 μm diameter [26]. (f) PL image of the same samples in (e). The PL quantum yield is much enhanced for suspended regions of the monolayer samples, and the emission from the few-layer sample is too weak to be seen in this image [26]. (a, b) Reprinted by permission from Springer Nature Customer Service Centre GmbH: Springer Nature, *Scientific Reports*, Ref. [24], Copyright (2014). (c, d) Reprinted with permission from Ref. [25]. Copyright (2013) American Chemical Society. (e, f) Reprinted (figure) with permission from Ref. [26]. Copyright (2010) by the American Physical Society.

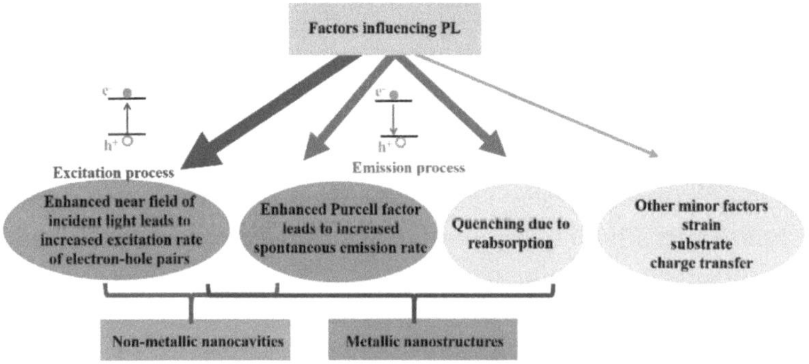

Figure 5.14 Summary of the factors influencing the PL of TMDs integrated with metallic nanostructures and non-metallic nanocavities.

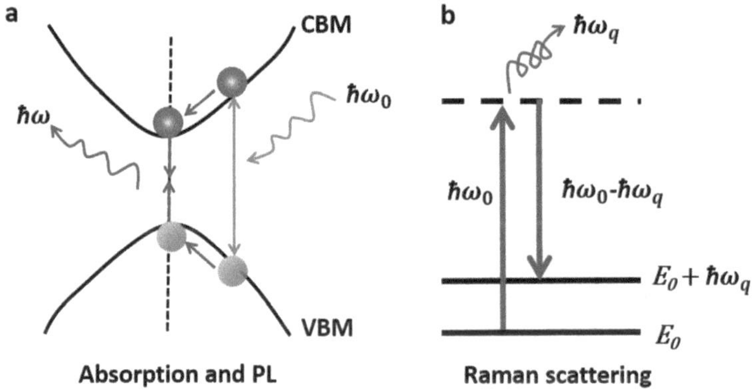

Figure 5.15 (a) Light absorption and emission process. $\hbar\omega_0$ is the energy of the incident light, and $\hbar\omega$ is the energy of the emitted light in the PL process. (b) Raman scattering process. $\hbar\omega_q$ is the energy of the emitted phonon.

The various factors that influence the PL of TMDs integrated with metallic nanostructures and non-metallic nanocavities are summarised in Fig. 5.15. The PL intensity is predominantly determined by the excitation and emission processes. In the excitation process, spectral overlap of the resonance mode of metallic nanostructures or non-metallic nanocavities with pump laser wavelength gives rise to enhanced absorption of incident light, leading to an increased excitation rate of electron–hole pairs in TMDs. Generally, the metallic nanostructure leads to a much larger near-field enhancement than non-metallic nanocavities. In the emission process, the spectral overlap of the resonance mode in metallic nanostructures and non-metallic cavities with the emission wavelength of the TMDs can enhance the Purcell factor, leading to an enhanced spontaneous emission rate of the electron–hole pairs. It is noted that in the emission process, the metallic nanostructure may induce a quench channel due to re-absorption of the emitted light from TMDs. These two major processes together with other minor factors such as strain, substrate, charge transfer, etc., need to be considered when we design metallic nanostructures or non-metallic nanocavities to enhance the PL of TMDs.

5.4 Raman Spectroscopy of 2D TMDs

5.4.1 Origin of Raman Processes

PL and Raman spectroscopies are important non-destructive techniques to characterise the quality and thickness of TMDs. In monolayer TMDs, electrons in the VBM can be excited to the CBM by incident photons, and the excess carriers can also recombine, releasing the energy by emitting light, as shown in Fig. 5.16a. The former process can be characterised by light absorption measurements, and the latter process can be characterised by PL measurements. If lattice vibrations or phonons participate in the optical transitions, the generated photon will have a lower energy than the incident photon (Stoke shift), and the energy difference is associated with a specific phonon mode of the material. This inelastic light scattering process is called Raman scattering, as shown in Fig. 5.16b.

The Raman spectrum of monolayer WSe_2 shows two peaks at around 250 cm^{-1} (Fig. 5.17a) [28]. Little change is observed on both modes as the thickness increases. In contrast, the Raman spectrum of MoS_2 has two prominent peaks: an in-plane (E^1_{2g}) mode located at around 383cm^{-1} and an out-of-plane (A_{1g}) mode at around 407cm^{-1}, as shown in Fig. 5.17b [29]. The in-plane mode is caused by sulphur atoms vibrating in one direction and the molybdenum atom in the other, while the out-of-plane mode results from just the sulphur atoms vibrating out of plane. These two modes evolve with the thickness of MoS_2. In the monolayer, the in-plane mode up-shifts to 386 cm^{-1} and the out-of-plane mode down-shifts to

404 cm^{-1} relative to the bulk. The peak wavenumber difference between the two modes (~18 cm^{-1}) is a reliable identification for monolayer MoS_2.

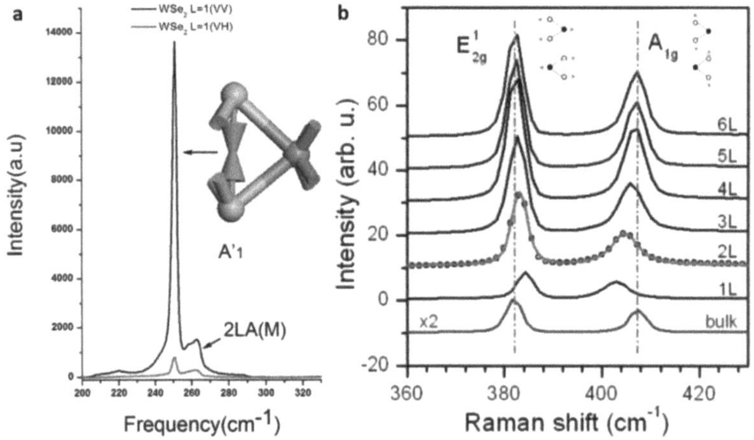

Figure 5.16 Raman spectra of TMDs. (a) Raman spectra of monolayer WSe_2 taken with a laser under conventional geometry (VV) and cross-polarisation (VH). Reprinted by permission from Springer Nature Customer Service Centre GmbH: Springer Nature, *Scientific Reports*, Ref. [28], Copyright (2014). (b) Raman spectra of MoS_2 with various thickness. Reprinted with permission from Ref. [29]. Copyright (2010) American Chemical Society.

5.4.2 Raman Effects in TMDs Integrated with Plasmonic Nanostructures

The hybridisation of TMDs with metal nanostructures can enhance the Raman intensity, induce phase transition and cause strain (Fig. 5.18). Kim et al. reported that the Raman intensities of monolayer WSe_2 with functionalised gold nanoparticles (Fig. 5.18a) is enhanced (Fig. 5.18b) by 1.76-fold owing to energy transfer from surface plasmons [30]. Kang et al. reported that hot electrons generated from plasmon decay of gold nanoparticles (Fig. 5.18c) can transfer into the MoS_2 layer, inducing a transient reversible 2H to 1T phase transition in MoS_2 (Fig. 5.18d). The evidence for this phase transition is the appearance of Raman modes characteristic of the octahedral phase and a red shift and intensity change of the PL spectrum. Sun et al. probed local strain at MoS_2–gold nanoparticle boundaries with surface plasmon–enhanced Raman scattering. The local mechanical strain splits both the in-plane vibration mode E^1_{2g} and the out-of-plane vibration mode A_{1g} in monolayer MoS_2, and activates the in-plane mode E_{1g} which is normally forbidden in the back-scattering Raman process.

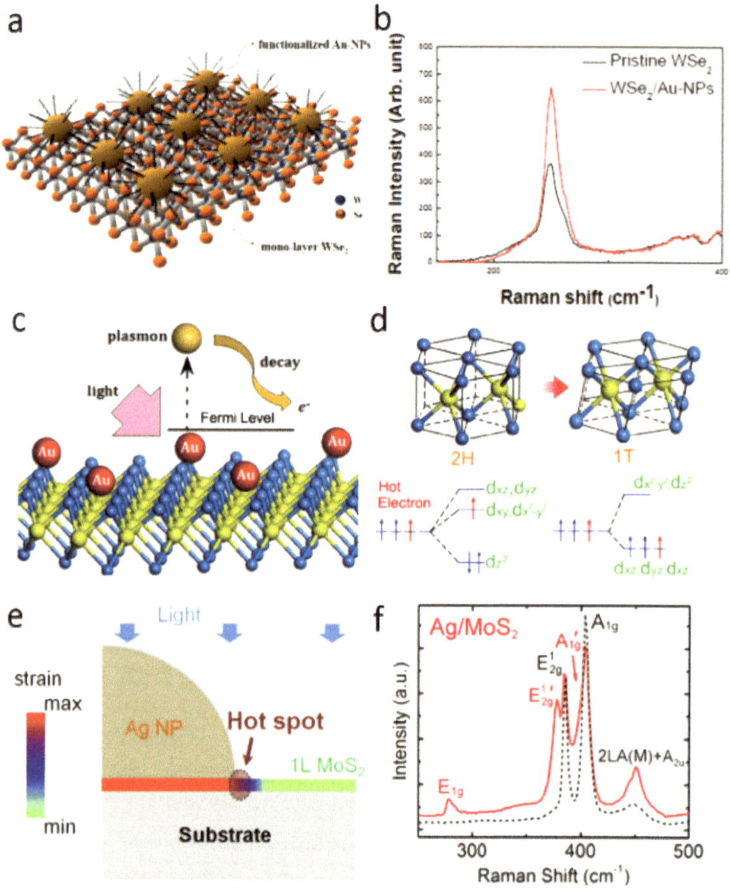

Figure 5.17 Schematic illustration (a) and Raman spectrum (b) of WSe_2/gold nanoparticles hybrid nanostructure [30]. (c) The principle of hot-electron generation: the gold plasmon decays into hot electron–hole pairs with the highest electron energies above the Fermi level. The hot electrons can transfer into the conduction band of the MoS_2 [31]. (d) The transition between 2H and 1T lattice structures [31]. (e) Schematic diagrams of the distribution of local strain in monolayer MoS_2 [32]. (f) Raman spectra of monolayer WS_2 before (black dashed lines) and after (red lines) the deposition of 1 nm silver [32]. (e, f) Reprinted with permission from Ref. [32]. Copyright (2014) American Chemical Society.

5.5 SHG of 2D TMDs

5.5.1 Basic Concepts of SHG

Another optical effect resulting from light–matter interaction in TMDs is SHG, namely converting the incident light with frequency ω into light with frequency 2ω. SHG, the lowest-order frequency-mixing non-linear optical process, plays an important role in frequency up-conversion in lasers and optoelectronic hybrid devices. Recent studies show that 2D TMDs are an interesting non-linear material for SHG. This is because 2D TMD_S have a thickness far below the optical coherence length, eliminating the requirement for phase matching, and a relative large second-order susceptibility in terms of thickness, thus enabling the design of smaller and thinner non-linear optical and optoelectronic devices [33]. This section will briefly discuss the basic concept of non-linear optics related to SHG, the motivation for SHG spectroscopy and the progress of nanostructure-enhanced SHG research on TMDs.

When an electric field $\mathbf{E}(\omega)$ is applied to a material, a macroscopic electric dipole moment called polarisation, $P(\omega)$, will be induced [34]. The induced polarisation exhibits a linear dependence on the electric field, as defined by the equation

$$\mathbf{P}(\omega) = \varepsilon_0 \chi(\omega) \times \mathbf{E}(\omega), \tag{5.1}$$

where $\chi(\omega)$ is the optical susceptibility of the material and ω is the angular frequency. As defined in Eq. 5.1, all the linear optical properties of a material with the corresponding phenomena such as absorption, refraction and reflection are associated with polarisation, $\mathbf{P}(\omega)$. In these processes, the polarisation oscillates at the same frequency as the incoming electric field, namely acting as a source of radiation with frequency ω. When the electric field of the light becomes intense, χ itself depends on the electric field, and thus the polarisation can be expressed as a power series of $\mathbf{E}$:

$$\mathbf{P} = \mathbf{P}^{L} + \mathbf{P}^{NL} = \varepsilon_0 [\chi^{(1)} \times \mathbf{E} + \chi^{(2)} \times \mathbf{E} \times \mathbf{E} + \chi^{(3)} \times \mathbf{E} \times \mathbf{E} \times \mathbf{E}], \tag{5.2}$$

where $\chi^{(1)}$ is the linear optical susceptibility, corresponding to the susceptibility at low electric field intensity; $\chi^{(2)}$ is the second-order non-linear optical susceptibility and is a third rank tensor with 3^3 elements, corresponding to the possible combinations of the three Cartesian components of the polarisation and of the two interacting electric fields; and $\chi^{(3)}$ is the third-order non-linear optical susceptibility and is a fourth rank tensor with 3^4 elements. The order of magnitude of the elements for $\chi^{(1)}$, $\chi^{(2)}$ and $\chi^{(3)}$ is 1, 10^{-12} m/V and 10^{-23} m^2/V^2, respectively. When the light intensity equals 10^9 W/cm^2, the related electric field $\mathbf{E}_0$ is 4.34×10^7 V/m, reaching a value that is in the same order as those existing between the charges, nuclei and electrons constituting matter. Therefore, the optical properties of the material are modified. Such a strong electric field can be easily realised by a focused pulsed laser with an energy of at least 10 nJ/pulse.

Second-order polarisation is the origin of well-known non-linear phenomena such as SHG, sum frequency generation (SFG) and difference frequency mixing (DFM). Given a material with a second-order non-linear susceptibility irradiated by an optical field oscillating at two frequencies, ω_1 and ω_2, the polarisation is given by

$$\mathbf{P}_i^{(2)} = \varepsilon_0 \chi_{ijk}^{(2)} \times \mathbf{E}_j(\omega_1,\omega_2) \times \mathbf{E}_k(\omega_1,\omega_2), \tag{5.3}$$

where the indices i, j and k represent the Cartesian components of the polarisation and of the incident fields, respectively, and $\chi_{ijk}^{(2)}$ denotes the relevant tensor element of the second-order non-linear susceptibility, $\chi^{(2)}$. This equation shows that the non-linear polarisation oscillates at new frequencies of ω_3, which equals $2\omega_1$, $2\omega_2$, $\omega_1+\omega_2$ or $|\omega_1-\omega_2|$, instead of ω_1 and ω_2, thus acting as the source of radiation for achieving SHG, SFG or DFM. The intensity, I, of the field generated by the polarisation at the new frequency, ω_3, is given by

$$I(\omega_3) \propto \left|\chi_{ijk}^{(2)}\right|^2 I_j I_k L^2 \cdot \mathrm{sinc}^2\left(\frac{\Delta k L}{2}\right), \tag{5.4}$$

where $\sin c(x) = \sin(x)/x$, L is the optical path length and Δk is the phase (wave vector) mismatch:

$$\Delta k = \mathbf{K}_{out} - \Sigma\, \mathbf{K}_{in}, \tag{5.5}$$

where $\mathbf{K}_{out}$ and $\mathbf{K}_{in}$ are the wave vectors of the new outgoing field and original incident field, respectively, ($|k|=k=n\omega/c$). The new field intensity I is the largest when $\Delta k = 0$, as it satisfies the momentum conservation condition (momentum $\mathbf{p} = \hbar\mathbf{k}$). Take SFG ($\omega_3 = \omega_1 + \omega_2$) as an example. The momentum conservation, namely phase-matching condition ($\Delta k = 0$), requires that $k_3 = k_1 + k_2$. If this condition is not satisfied, the fields generated at different locations along the optical path in the material will destructively interfere with each other. As a result, no efficient frequency conversion can be achieved. For the second-harmonic process, the severity of the phase mismatch can be easily quantified by the coherence length L_{coh}

$$L_{\text{coh}} = \frac{\pi}{|\Delta k|} = \frac{\lambda}{4|n_2 - n_1|}, \tag{5.6}$$

where n_1 and n_2 are refractive indices at the fundamental and second-harmonic waves, respectively, and λ is the wavelength of the fundamental wave in the vacuum. For typical optical materials, L_{coh} is a few tens of microns (μm).

To observe these second-order non-linear phenomena, the light has to fulfill phase-matching conditions and the material should have no central symmetric crystal structure. If the material has an inversion centre, the second-order susceptibility will be zero. This can be demonstrated in a simple way by considering even-order polarisations:

$$\mathbf{P}^{(2n)} = \varepsilon_0 \chi^{(2n)} \mathbf{E}^{2n} \ (n = 1, 2, 3, \ldots) \tag{5.7}$$

In a centro-symmetric material, the sign of $\mathbf{P}(2n)$ is reversed as the electric field is reversed:

$$-\mathbf{P}^{(2n)} = \varepsilon_0 \chi^{(2n)} (-\mathbf{E})^{2n} \tag{5.8}$$

$$-\mathbf{P}^{(2n)} = \varepsilon_0 \chi^{(2n)} - \mathbf{E}^{2n} \tag{5.9}$$

Therefore $-\mathbf{P}(2n) = \mathbf{P}(2n)$. This is only possible when $\chi^{(2n)} = 0$. As a result, only materials with broken inversion symmetry have non-zero $\chi^{(2n)}$ and exhibit even-order non-linear optical properties. Centro-symmetric and isotropic materials such as liquids show only odd-order non-linear effects.

To develop smaller and thinner non-linear media in next-generation optoelectronic devices, much effort has been devoted to solve the phase-mismatch issues in non-linear crystals. While approaches such as quasi-phase matching or creating zero-index metamaterials are viable, using 2D TMDs would eliminate the requirement for phase matching as the crystal thickness ranging from sub-nanometres to several nanometres is far below the coherence length. Naturally, 2D TMDs may serve as ideal candidates for such materials, as the second-harmonic polarisation is confined within the atomic-scale thickness.

5.5.2 Motivation for SHG Spectroscopy

In 2D TMDs, SHG has found several applications from microscopic and spectroscopic techniques to up-conversion lasing. First of all, SHG optical spectroscopy allows for non-destructive crystal structure studies of 2D TMDs, as the amplitude of SHG strongly depends on the crystal structure and layer number. The 1T phase has negligible second-harmonic emission as it has an inversion centre in the crystalline unit cell, while for the 2H and 3R phases, the intensity shows a close correlation with the layer number. For the case of 2H-MoS_2, the SHG intensity in an odd number of layers is much stronger than the response for an even number of layers [33]. This is because the odd-numbered layers with no inversion centre have a net non-linear dipole, while centro-symmetric even layers with an inversion centre do not. This net non-linear dipole in odd-layered 2H-MoS_2 has enabled the probing of SHG due to the finite second-order non-linearity ($\chi^{(2)} \neq 0$). Unlike emission of 2H-MoS_2, the SHG amplitude of 3R-MoS_2 increases with the layer number as the SHG dipoles in each layer are all aligned in one direction, allowing for constructive interference of the SHG light [33]. Secondly, the SHG signal of 2D TMDs has been utilised to study the crystal orientation of monolayer and bilayer flakes [35] or polycrystalline films [36].

SHG generation at a fixed wavelength is commonly used for monolayer and few-layer TMDs to probe crystal inversion symmetry and structure, as reviewed earlier, whereas the intensity of the SHG signal as a function of the two-photon laser excitation energy enables a spectroscopic method to study the excitonic structure. Broken inversion symmetry and rich linear and non-linear optical properties make monolayer TMDs an ideal material for applying this technique. Wang et al. demonstrated the variation of SHG intensity of monolayer WSe_2 with two-photon laser excitation laser scanned from 1.7 to 2.4 eV [37]. SHG intensity varies over 3 orders of magnitude and shows the strongest values at energies of the 1s A exciton (1.75 eV) and 1s B exciton (2.17 eV) [37]. This resonance feature in the SHG spectrum suggests SHG spectroscopy can serve as an optical tool for the determination of exciton states. It is also reported by Seyler et al. that under cryogenic conditions, the SHG intensity of a field-effect transistor device based on monolayer WSe_2

can be tuned over an order of magnitude by an electrostatic field and is strong at the A exciton energy [6].

The SHG signal not only plays an important role in the characterisation of 2D TMDs but also opens up a window for developing atomically thin non-linear devices used for frequency up-conversion. As Li et al. demonstrated that in 2H-MoS_2, the monolayer has much larger SHG than the few-layer and bulk counterparts [38], it is reasonable to explore the SHG conversion efficiency in monolayer TMDs. Recent experimental results have demonstrated that the conversion efficiency represented by the second-order susceptibilities ($\chi^{(2)}$) ranges from 0.1 and 100 nm/V in monolayer TMDs, which is comparable with that in non-linear III–V materials. Recent advances in SHG of 2D TMDs have shown relatively strong second-harmonic responses using a pump laser at appropriate wavelengths, demonstrating their promise in the application as non-linear optical devices. However, limited by the innate thickness of 2D TMDs, the second-harmonic conversion efficiency still needs to be improved for practical applications.

5.5.3 Methods to Improve SHG

One method to improve the SHG intensity is mentioned in Section 5.5.2, that is, using a two-photon excitation pump with energy aligned to the exciton transitions. This is because the quantum mechanical expression for the second-order susceptibility $\chi^{(2)}$ of TMDs includes matrix elements with one occupied and two un-occupied intermediate virtual states. When the virtual transition level is coincident with exciton states in TMDs, the coupling between the second-harmonic radiation and the $\chi^{(2)}$ becomes large. The exciton transition not only facilitates the resonance-enhanced SHG spectrum but also makes the electrical tunability of the frequency-doubled signals feasible. Seyler et al. made a monolayer WSe_2 field-effect transistor and achieved impressive tunability of $\chi^{(2)}$ and the resulting SHG intensity was enabled through electrostatic doping induced by a gate bias [6]. To be specific, the SHG intensity at the A exciton energy can be tuned by over an order of magnitude in a cryogenic environment, albeit the performance is compromised at ambient temperature [39]. This approach utilises electrostatic engineering of oscillator strength at the resonance frequencies of various exciton states. As a result, the effect is inherently narrowband. When the energy of the pump laser is shifted by only tens of milli-electron-volts, the gate-controlled modulation of the SHG dramatically diminishes. The researchers also demonstrated an interesting SHG selection rule, namely the SHG signal is counter-circularly polarised to the fundamental wave [35], which is different from the PL optical selection rule [10].

These methods to enhance the SHG utilise the inherent exciton resonance of TMDs. To further optimise the SHG signal, external strategies should also be incorporated. Optical resonators that can enhance the PL of TMDs can also boost the SHG. It has been reported that the SHG of TMDs can be significantly enhanced by integrating them with optical cavities.

Fryett et al. transferred monolayer WSe_2 onto a silicon photonic crystal cavity [40], as shown in Fig. 5.18a. The choice of photonic crystal cavity is motivated by its modes with a small mode volume and a high quality factor (Fig. 5.18b), since for a non-linear optical switch, the switching power scales as V_m/Q^3 for second-order non-linearity, where V_m is the mode volume and Q is the quality factor. When the device is pumped by ultra-fast laser pulses near this cavity mode at the telecommunication wavelength, an SHG signal in the visible range with near-unity linear polarisation was observed (Fig. 5.18c), which originates from the coupling of the pump photon to the cavity mode. The photonic crystal cavity induces a twofold increase in the average second-harmonic intensity from WSe_2 as compared to monolayer WSe_2 on silicon. After normalisation by comparing the non-linear conversion efficiencies of WSe_2 on the cavity to that on flat silicon, the cavity enhancement is estimated to be 200-fold.

Larger SHG enhancement was reported by Cubukcu et al. via integrating a MoS_2 monolayer into a doubly resonant optical cavity [41]. As demonstrated in Fig. 5.18d, the device is based on a voltage-controlled micro-Fabry–Perot (FP) cavity that consists of a distributed Bragg reflector (DBR) mirror with high reflectivity and a voltage-deformable silver mirror fabricated on a silicon nitride membrane. The non-linear light–matter interaction in MoS_2 is engineered by optomechanical tuning of the FP cavity length via an applied voltage, which, in turn, controls the intra-cavity optical field strengths at the fundamental frequency (ω) and second-harmonic frequency (2ω). The SHG enhancement level is predicted by the field enhancement theoretical model. Since the intra-cavity electric field is related to the quality factor, this model can also be expressed as $(Q_\omega)^2 \times (Q_{2\omega})$, where Q_ω and $Q_{2\omega}$ are the quality factors for the fundamental and second-order waves, respectively. The experimental SHG enhancement factor which is defined as the ratio between the SHG signals from a MoS_2 monolayer in cavity and on a glass substrate at different voltages and wavelengths is shown as a map in Fig. 5.18e. It is shown that the experimental results agree well with the theoretical model. The maximum SHG enhancement (3300-fold) is observed (Fig. 5.18f) in the case where both the fundamental and second-harmonic waves are spectrally tuned with the two cavity modes, that is, in a double-resonance condition.

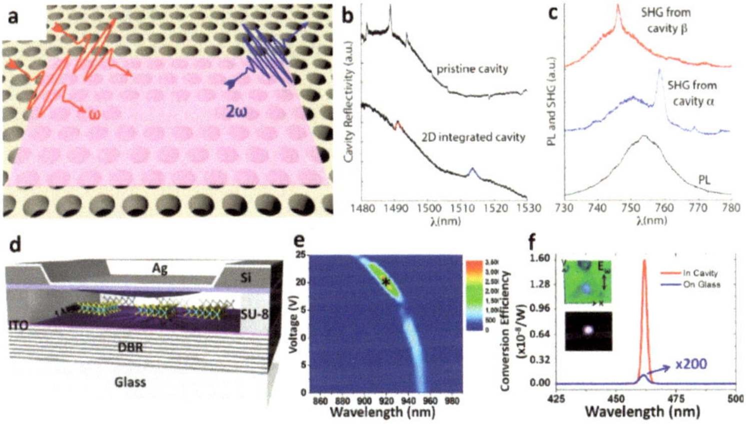

Figure 5.18 SHG enhancement in 2D TMDs by nanostructures. (a) Schematic of WSe_2 transferred on a silicon photonic crystal [40]. (b) Reflectivity of the higher-order cavity modes before and after monolayer WSe_2 transfer [40]. (c) PL from WSe_2 (black) and SHG spectra from mode α (blue) and mode β (red) [40]. (d) Schematic of the device structure formed by the dielectric mirror on a glass substrate and a silver mirror on a suspended membrane [41]. (e) Measured voltage dependence of the total output SHG power enhancement factor normalised to a monolayer on a glass substrate [41]. (f) SHG signal generated from the MoS_2 flake inside the cavity (red solid line) and on a glass substrate (blue solid line) [41]. (d, e, f) Reprinted with permission from Ref. [41]. Copyright (2016) American Chemical Society.

The frequency-doubling process is technologically important for the generation of tunable lasers sources in the visible spectrum, and methods to further improve SHG and reduce loss need to be explored.

5.6 Future Outlook

Recent advances in the development of other 2D semiconductors and van der Waals heterostructures bring about many exciting opportunities. A few examples are worth further investigation. First, the unique valley-dependent properties are intriguing for applications in information processing. An example is coupling 2D TMDs to micro-cavities or plasmonic nanostructures and enhancing the valley polarisation degree of the PL and SHG by making use of the valley optical selection rule in 2D TMDs [10, 42]. Second is achieving strong coupling between light and excitons in 2D TMDs by integrating plasmonic nanostructures with TMDs. The strong coupling can lead to light–matter hybrid states and new quasi-particles, such as exciton polaritons, allowing for the observation of coherent phenomena such as Bose–Einstein condensation as well as enabling novel photonics devices such as polariton lasers [43]. Third, the use of complex excitonic species in optoelectronics devices is not well studied, for example, utilising trions (charged excitons) to transport photo-excited carriers [44]. Here the usage of pump-probe techniques such as transient absorption spectroscopy will help to reveal the exciton dynamics in terms of excitation and relaxation in the light absorption, emission and carrier transport processes [7]. Last but not least is the recent discovery of new 2D semiconductors with anisotropic optical and optoelectronic properties, including phosphorene. Together with 2D TMDs, they provide a material toolbox for 2D semiconductors with gap energies spanning the visible to near-infrared spectral range. Fabricating novel heterostructures by stacking these 2D materials or integrating them with plasmonic nanostructures opens up new opportunities for exploring novel optical properties and developing new optoelectronic devices.

References

1. Wang, Q. H., Kalantar-Zadeh, K., Kis, A., Coleman, J. N. and Strano, M. S. (2012). Electronics and optoelectronics of two-dimensional transition metal dichalcogenides, *Nat. Nanotechnol.*, **7**(11), pp. 699–712.

2. Kumar, A. and Ahluwalia, P. (2012). Electronic structure of transition metal dichalcogenides monolayers 1H-MX_2 (M= Mo, W; X= S, Se, Te) from ab-initio theory: new direct band gap semiconductors, *Eur. Phys. J. B*, **85**(6), p. 186.
3. Carvalho, A., Ribeiro, R. and Neto, A. C. (2013). Band nesting and the optical response of two-dimensional semiconducting transition metal dichalcogenides, *Phys. Rev. B*, **88**(11), p. 115205.
4. Cui, Q., Ceballos, F., Kumar, N. and Zhao, H. (2014). Transient absorption microscopy of monolayer and bulk WSe_2, *ACS Nano*, **8**(3), pp. 2970–2976.
5. Eda, G. and Maier, S. A. (2013). Two-dimensional crystals: managing light for optoelectronics, *ACS Nano*, **7**(7), pp. 5660–5665.
6. Seyler, K. L., Schaibley, J. R., Gong, P., Rivera, P., Jones, A. M., Wu, S., Yan, J., Mandrus, D. G., Yao, W. and Xu, X. (2015). Electrical control of second-harmonic generation in a WSe_2 monolayer transistor, *Nat. Nanotechnol.*, **10**(5), pp. 407–411.
7. Wang, L., Wang, Z., Wang, H.-Y., Grinblat, G., Huang, Y.-L., Wang, D., Ye, X.-H., Li, X.-B., Bao, Q. and Wee, A.-S. (2017). Slow cooling and efficient extraction of C-exciton hot carriers in MoS_2 monolayer, *Nat. Commun.*, **8**, p. 13906.
8. Zeng, H., Liu, G.-B., Dai, J., Yan, Y., Zhu, B., He, R., Xie, L., Xu, S., Chen, X. and Yao, W. (2013). Optical signature of symmetry variations and spin-valley coupling in atomically thin tungsten dichalcogenides, *Sci. Rep.*, **3**, p. 1608.
9. Mak, K. F., He, K., Shan, J. and Heinz, T. F. (2012). Control of valley polarization in monolayer MoS_2 by optical helicity, *Nat. Nanotechnol.*, **7**(8), pp. 494–498.
10. Zeng, H., Dai, J., Yao, W., Xiao, D. and Cui, X. (2012). Valley polarization in MoS_2 monolayers by optical pumping, *Nat. Nanotechnol.*, **7**(8), pp. 490–493.
11. Tonndorf, P., Schmidt, R., Böttger, P., Zhang, X., Börner, J., Liebig, A., Albrecht, M., Kloc, C., Gordan, O. and Zahn, D. R. (2013). Photoluminescence emission and Raman response of monolayer MoS_2, $MoSe_2$, and WSe_2, *Opt. Express*, **21**(4), pp. 4908–4916.
12. Gan, X., Gao, Y., Fai Mak, K., Yao, X., Shiue, R.-J., van der Zande, A., Trusheim, M. E., Hatami, F., Heinz, T. F. and Hone, J. (2013). Controlling the spontaneous emission rate of monolayer MoS_2 in a photonic crystal nanocavity, *Appl. Phys. Lett.*, **103**(18), p. 181119.
13. Wu, S., Buckley, S., Schaibley, J. R., Feng, L., Yan, J., Mandrus, D. G., Hatami, F., Yao, W., Vučković, J. and Majumdar, A. (2015). Monolayer

semiconductor nanocavity lasers with ultralow thresholds, *Nature*, **520**(7545), pp. 69–72.

14. Butun, S., Tongay, S. and Aydin, K. (2015). Enhanced light emission from large-area monolayer MoS_2 using plasmonic nanodisc arrays, *Nano Lett.*, **15**(4), pp. 2700–2704.
15. Lee, B., Park, J., Han, G. H., Ee, H.-S., Naylor, C. H., Liu, W., Johnson, A. C. and Agarwal, R. (2015). Fano resonance and spectrally modified photoluminescence enhancement in monolayer MoS_2 integrated with plasmonic nanoantenna array, *Nano Lett.*, **15**(5), pp. 3646–3653.
16. Wang, Z., Dong, Z., Gu, Y., Chang, Y.-H., Zhang, L., Li, L.-J., Zhao, W., Eda, G., Zhang, W. and Grinblat, G. (2016). Giant photoluminescence enhancement in tungsten-diselenide–gold plasmonic hybrid structures, *Nat. Commun.*, **7**, p. 11283.
17. Chen, X., Park, H.-R., Pelton, M., Piao, X., Lindquist, N. C., Im, H., Kim, Y. J., Ahn, J. S., Ahn, K. J. and Park, N. (2013). Atomic layer lithography of wafer-scale nanogap arrays for extreme confinement of electromagnetic waves, *Nat. Commun.*, **4**, p. 2361.
18. Siegfried, T., Ekinci, Y., Martin, O. J. and Sigg, H. (2013). Gap plasmons and near-field enhancement in closely packed sub-10 nm gap resonators, *Nano Lett.*, **13**(11), pp. 5449–5453.
19. Tan, S. F., Wu, L., Yang, J. K., Bai, P., Bosman, M. and Nijhuis, C. A. (2014). Quantum plasmon resonances controlled by molecular tunnel junctions, *Science*, **343**(6178), pp. 1496–1499.
20. Desai, S. B., Seol, G., Kang, J. S., Fang, H., Battaglia, C., Kapadia, R., Ager, J. W., Guo, J. and Javey, A. (2014). Strain-induced indirect to direct bandgap transition in multilayer WSe_2, *Nano Lett.*, **14**(8), pp. 4592–4597.
21. Akselrod, G. M., Ming, T., Argyropoulos, C., Hoang, T. B., Lin, Y., Ling, X., Smith, D. R., Kong, J. and Mikkelsen, M. H. (2015). Leveraging nanocavity harmonics for control of optical processes in 2D semiconductors, *Nano Lett.*, **15**(5), pp. 3578–3584.
22. Chou, C.-H. and Chen, F.-C. (2014). Plasmonic nanostructures for light trapping in organic photovoltaic devices, *Nanoscale*, **6**(15), pp. 8444–8458.
23. Chen, H., Yang, J., Rusak, E., Straubel, J., Guo, R., Myint, Y. W., Pei, J., Decker, M., Staude, I. and Rockstuhl, C. (2016). Manipulation of photoluminescence of two-dimensional $MoSe_2$ by gold nanoantennas, *Sci. Rep.*, **6**, p. 22296.

24. Bhanu, U., Islam, M. R., Tetard, L. and Khondaker, S. I. (2014). Photoluminescence quenching in gold-MoS_2 hybrid nanoflakes, *Sci. Rep.*, **4**, p. 5575.

25. Conley, H. J., Wang, B., Ziegler, J. I., Haglund Jr, R. F., Pantelides, S. T. and Bolotin, K. I. (2013). Bandgap engineering of strained monolayer and bilayer MoS_2, *Nano Lett.*, **13**(8), pp. 3626–3630.

26. Mak, K. F., Lee, C., Hone, J., Shan, J. and Heinz, T. F. (2010). Atomically thin MoS_2: a new direct-gap semiconductor, *Phys. Rev. Lett.*, **105**(13), p. 136805.

27. Buscema, M., Steele, G. A., van der Zant, H. S. and Castellanos-Gomez, A. (2014). The effect of the substrate on the Raman and photoluminescence emission of single-layer MoS_2, *Nano Res.*, **7**(4), pp. 561–571.

28. Terrones, H., Del Corro, E., Feng, S., Poumirol, J., Rhodes, D., Smirnov, D., Pradhan, N., Lin, Z., Nguyen, M. and Elias, A. (2014). New first order Raman-active modes in few layered transition metal dichalcogenides, *Sci. Rep.*, **4**, p. 4215.

29. Lee, C., Yan, H., Brus, L. E., Heinz, T. F., Hone, J. and Ryu, S. (2010). Anomalous lattice vibrations of single-and few-layer MoS_2, *ACS Nano*, **4**(5), pp. 2695–2700.

30. Kim, J. Y., Kim, J. and Joo, J. (2016). Surface-enhanced Raman scattering for 2-D WSe_2 hybridized with functionalized gold nanoparticles, *Opt. Express*, **24**(24), pp. 27546–27553.

31. Kang, Y., Najmaei, S., Liu, Z., Bao, Y., Wang, Y., Zhu, X., Halas, N. J., Nordlander, P., Ajayan, P. M. and Lou, J. (2014). Plasmonic hot electron induced structural phase transition in a MoS_2 monolayer, *Adv. Mater.*, **26**(37), pp. 6467–6471.

32. Sun, Y., Liu, K., Hong, X., Chen, M., Kim, J., Shi, S., Wu, J., Zettl, A. and Wang, F. (2014). Probing local strain at MX_2–metal boundaries with surface plasmon-enhanced Raman scattering, *Nano Lett.*, **14**(9), pp. 5329–5334.

33. Zhao, M., Ye, Z., Suzuki, R., Ye, Y., Zhu, H., Xiao, J., Wang, Y., Iwasa, Y. and Zhang, X. (2016). Atomically phase-matched second-harmonic generation in a 2D crystal, *Light Sci. Appl.*, **5**(8), p. e16131.

34. New, G. (2011). *Introduction to Nonlinear Optics* (Cambridge University Press).

35. Jiang, T., Liu, H., Huang, D., Zhang, S., Li, Y., Gong, X., Shen, Y.-R., Liu, W.-T. and Wu, S. (2014). Valley and band structure

engineering of folded MoS_2 bilayers, *Nat. Nanotechnol.*, **9**(10), pp. 825–829.

36. Yin, X., Ye, Z., Chenet, D. A., Ye, Y., O'Brien, K., Hone, J. C. and Zhang, X. (2014). Edge nonlinear optics on a MoS_2 atomic monolayer, *Science*, **344**(6183), pp. 488–490.
37. Wang, G., Marie, X., Gerber, I., Amand, T., Lagarde, D., Bouet, L., Vidal, M., Balocchi, A. and Urbaszek, B. (2015). Giant enhancement of the optical second-harmonic emission of WSe_2 monolayers by laser excitation at exciton resonances, *Phys. Rev. Lett.*, **114**(9), p. 097403.
38. Li, Y., Rao, Y., Mak, K. F., You, Y., Wang, S., Dean, C. R. and Heinz, T. F. (2013). Probing symmetry properties of few-layer MoS_2 and h-BN by optical second-harmonic generation, *Nano Lett.*, **13**(7), pp. 3329–3333.
39. Rodrigues, S. P. and Cai, W. (2015). Nonlinear optics: tuning harmonics with excitons, *Nat. Nanotechnol.*, **10**(5), pp. 387–388.
40. Fryett, T. K., Seyler, K. L., Zheng, J., Liu, C.-H., Xu, X. and Majumdar, A. (2016). Silicon photonic crystal cavity enhanced second-harmonic generation from monolayer WSe_2, *2D Mater.*, **4**(1), p. 015031.
41. Yi, F., Ren, M., Reed, J. C., Zhu, H., Hou, J., Naylor, C. H., Johnson, A. C., Agarwal, R. and Cubukcu, E. (2016). Optomechanical enhancement of doubly resonant 2D optical nonlinearity, *Nano Lett.*, **16**(3), pp. 1631–1636.
42. Li, Z., Li, Y., Han, T., Wang, X., Yu, Y., Tay, B. K., Liu, Z. and Fang, Z. (2016). Tailoring MoS_2 exciton-plasmon interaction by optical spin-orbit coupling, *ACS Nano*, **11**, pp. 1165–1171.
43. Liu, W., Lee, B., Naylor, C. H., Ee, H.-S., Park, J., Johnson, A. C. and Agarwal, R. (2016). Strong exciton–plasmon coupling in MoS_2 coupled with plasmonic lattice, *Nano Lett.*, **16**(2), pp. 1262–1269.
44. Lui, C., Frenzel, A., Pilon, D., Lee, Y.-H., Ling, X., Akselrod, G., Kong, J. and Gedik, N. (2014). Trion-induced negative photoconductivity in monolayer MoS_2, *Phys. Rev. Lett.*, **113**(16), p. 166801.

Chapter 6

Integration of 2D Materials with Photonic Nanostructures

Haitao Chen
College of Advanced Interdisciplinary Studies, National University of Defense Technology, Changsha 410073, China
Nonlinear Physics Centre, Research School of Physics and Engineering, Australian National University, Canberra, ACT 2601, Australia
haitaochen0517@outlook.com

This chapter describes the prospects of integration of 2D materials with photonic nanostructures to enable future active nanophotonic devices. We focus on development of controlling light emission from transition metal dichalcogenides (TMDs) by nanophotonic integration. The general principles of enhancing light emission by photonic nanostructures is discussed at the beginning. Then we present two examples of our study of integration of 2D TMDs onto plasmonic metal nanoparticles and dielectric waveguide-grating structures. Future directions of integration of 2D materials with photonic nanostructures are also discussed.

Two-Dimensional Materials in Nanophotonics: Developments, Devices, and Applications
Edited by Yuerui Lu

ISBN 978-981-4800-22-8 (Hardcover), 978-0-429-42877-7 (eBook)
www.jennystanford.com

6.1 Introduction

The development of 2D materials brings up new opportunities for 2D physics and applications. Among them, semiconductor transition metal dichalcogenides (TMDs) such as MoS_2 show quite advanced properties. Especially, in the monolayer limit, they become direct-bandgap semiconductors. The monolayer form demonstrates a range of novel optical effects, including strong excitonic effects, photoluminescence (PL), valley polarisation, single-photon emission and second-harmonic generation (SHG), which show great potential for fabricating novel "flat" active nanophotonic devices in the future.

Besides, the atomically thick 2D TMDs have a naturally passivated surface without any dangling bonds, which makes them suitable to integrate with photonic structures. On the other hand, enabling light sources in silicon photonics is an important requirement for on-chip optoelectronic applications [36, 39]. While silicon alone cannot generate the required light, integration with other direct-bandgap materials is being sought. Integration of germanium or III-V materials on silicon faces technical challenges due to the lattice constant mismatch and different thermal properties [23]. With the desire to overcome such limitations, the integration of 2D TMDs onto silicon photonic structures emerges as a new promising solution [41, 46]. Two-dimensional materials are held together by out-of-plane van der Waals forces and can be transferred onto a silicon substrate without lattice mismatch issues.

However, the light-emitting efficiency of such monolayer TMDs is generally low due to the sub-nanometre light–matter interaction length and defects. To enable their practical applications, new schemes are desirable to enhance the emission efficiency and further control other emission properties such as directionality and polarisation.

Integration with photonic nanostructures is a solution to such challenges, since photonic nanostructures such as plasmonic nanoparticles show great capability to manipulate light–matter interaction at the nanoscale. Indeed, some examples have successfully demonstrated integration of 2D TMDs with different photonic nanostructures to control the PL from these 2D materials, including planar photonic crystal cavities [13, 42], optical micro-cavities

[10, 25], optical ring resonator [45] and so on. Lasing with low threshold has been demonstrated by integrating 2D TMDs with photonic cavities [10, 43].

Among all these work, most of them are based on the Purcell effect. In the following section, we will first briefly introduce how the Purcell effect induces changes in the emission property of a nanoemitter, and then we will present two examples of integration of monolayer TMDs with metal plasmonic nanoparticles and dielectric waveguide-grating structures, respectively.

6.2 Purcell Effect and Emission Enhancement

The spontaneous emission rate of an emitter depends not only on the inherent properties of the materials but also on the electromagnetic environment it interacts with, which is well known as the Purcell effect. Thus, we could modify the emission rate through changing the environment. Purcell made the first proposal to put an atom inside cavities to enhance its magnetic transition in 1946 [32].

To describe the modification of the emission rate, we use the quantified term Purcell factor. It is defined as the ratio between the modified and the free-space emission rate. For an emitter whose transition frequency aligns with the cavity resonance and has a narrower line width compared to the resonance width, when the emitter locates at the maximum field in the cavity mode and the dipole moment aligns with the polarisation of cavity mode, the Purcell factor F_p is [26, 31, 34]

$$F_P = \frac{\Gamma_g}{\Gamma_0} = \frac{3Q\lambda^3}{4\pi^2 V_0} \tag{6.1}$$

where Γ_g and Γ_0 are the emission rate in the cavity and free space, respectively. λ is the wavelength of transition, Q is the quality factor of the cavity and V_0 is the volume of the cavity mode. This equation tells that we need resonators that could confine light into a small volume and keep it for a long time to modify the emission rate significantly. However, the two requirements for resonators seem contradictory since better confinement usually accompanies higher loss. The research of modification of spontaneous emission is mostly about developing a better trade-off between these two aspects [31].

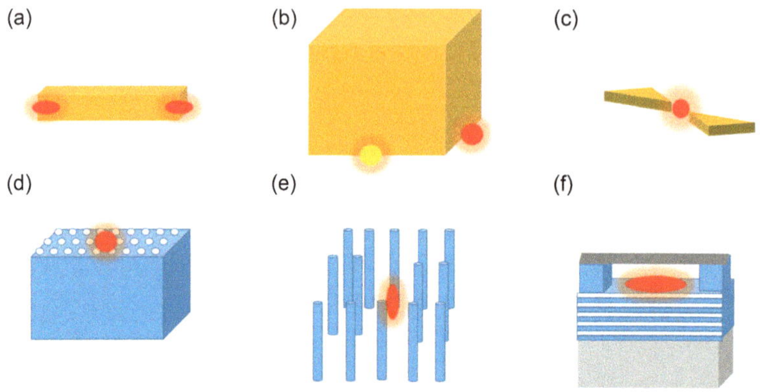

Figure 6.1 Examples of photonic resonators. Schematic designs of plas-monic nanoparticles of different geometries: (a) bar, (b) cube and (c) bowtie. Photonic crystal cavities working on modes of different types: (d) planar, (e) vertical and (f) optomechanical. The coloured dots refer to modes confined by these resonators.

Metal nanoparticles and photonic crystal cavities are the two most commonly used resonators to enhance the emission from emitters. In metal nanoparticles, the light field could be confined beyond the diffraction limit through coupling to electron oscillation, or plasmons, in metals. With the development of nanofabrication techniques, nanoparticles of different geometries and smaller size have been designed to achieve better confinement of light. Figure 6.1 shows a few examples of these designs, such as bar (Fig. 6.1a), cube (Fig. 6.1b) and bowtie (Fig. 6.1c). With plasmonic metal nanoparticles, the optical field could be confined in the scale of down to 10 nm, and even a smaller mode volume could be obtained by confining light in the gaps between nanoparticles [15]. However, the cost we need to pay for such strong confinement is the high damping loss from metals. The quality factor of the plasmonic metal nanoparticles is in the order of 10 [19]. Besides, the PL emission of the emitter could also be quenched when integrating with metal nanoparticles due to coupling to higher-order plasmon modes or charge transfer from emitter to metals. In contrast, a photonic crystal cavity could achieve quality factors as high as millions [28]. But a photonic crystal cavity works on the principle that light is confined due to reflections enabled by the periodic dielectric structures. Thus the mode volume of these cavities is limited by the diffraction limit, which is around $0.1(\lambda_0/n)^3$. Here λ_0 is the free-space wavelength and n is the refractive index of the medium. Various photonics crystal cavities have been developed to perform improved enhancement of emission. Figure 6.1 also shows a few examples of these designed on planar (Fig. 6.1d), vertical (Fig. 6.1e) and optomechanical (Fig. 6.1f) modes. In addition, we could also use photonic structures to concentrate or increase the absorption of the excitation light to enhance the emission from emitters, so photonic structures with multiple resonances could be applied to boost both excitation and emission processes to achieve best overall light harvesting.

What is more, by coupling to different types of modes excited in the plasmonic nanoparticles or dielectric cavities, other radiation properties of emitters, such as directionality and polarisation, could be well controlled, too. Thus, integration of an emitter with photonic nanostructures is a promising way to control the emission properties of emitters and further enable practical optoelectronic applications.

6.3 2D Material Integration with Plasmonic Nanoparticles

As discussed before, emission from quantum emitters could be well enhanced by localised plasmon resonance (LPR) sustained by metallic particles due to the Purcell effect, which offers great opportunities to control the light emission from 2D TMDs by plasmonic metal particles. The emission of TMDs could be generally altered by metal particles in two ways, namely enhancing the local field in the excitation process or the local density of states in the emission process, at the location of the nanoemitter. Whereas the first effect directly translates to an increase of the excitation rate, the latter modifies the spontaneous emission rate. Generally, modification of the spontaneous emission rate is a more robust way for controlling the emission of such TMDs as it does not depend on the pumping scheme, which is especially important for emitting devices that are electronically pumped.

Indeed, the hybrid systems of plasmonic nanoparticles with 2D TMD materials have been a subject of intense interest. Effects such as photo-current enhancement [24], PL enhancement by plasmonic gold particles [1, 6, 18, 27] and nanoshell structure [38] have been demonstrated. A plasmonic bowtie antenna with Fano resonance was also demonstrated to control the emission and reflection of monolayer MoS_2 [21].

Interestingly, the inverse effect, namely PL quenching, has also been reported when investigating a Au-MoS_2 hybrid system [4]. They attributed the quenching to the charge transfer from the monolayer MoS_2 to the gold nanoantennas through the resulting Schottky barrier of 0.4 eV. However, the quenching of PL emission from the monolayer should also happen at very small distances between the antennas and the TMD due to the coupling of the emitted photons into non-radiative plasmonic resonances of the nanoantennas, that is, dark plasmonic modes [3].

Thus it is desirable to conduct a comprehensive study of the coupling between 2D TMDs and plasmonic metal nanoparticles, so we could realise PL control freely. Here we present our demonstration of manipulating the PL of monolayer $MoSe_2$, mainly in the emission process, from quenching to enhancement by changing the distance between the monolayer $MoSe_2$ and the plasmonic gold particles. Our simulation results reveal the coupling mechanism between the materials and plasmonic particles.

6.3.1 Plasmonic-$MoSe_2$ Sample Preparation

To keep the physical picture simple and easy to study, we design a plasmonic gold bar array with a single resonance that overlaps with the PL emission peak of monolayer $MoSe_2$. The reason we choose monolayer $MoSe_2$ is that its emission peak is around 780 nm, which is away from the gold inter-band transition range and excludes the absorption effects from gold.

Rectangular gold nanoparticles with fixed width and height, both of 40 nm, and varying lengths in the range of 70-130 nm are prepared and arranged in a square lattice with a centre-to-centre distance between adjacent elements of 505 nm. The nanoantennas are fabricated by standard electron beam lithography (EBL). After characterising the transmittance spectra of these samples, some are coated with thin-layer silica around 8 nm by physical sputtering. We choose two samples specifically such that both have plasmonic resonances at the PL peaks of monolayer $MoSe_2$, but one without a spacer and one with a spacer. The transmittance profiles of the two sample obtained by polarised white-light spectroscopy are shown in Figs. 6.2a and 6.2b, respectively. Insets of the figures indicate the polarisation directions of the incident beams. The distinct polarisation-dependent transmittance profiles identify the nature of the excitation of the LPR. Note that for the sample with the spacer, as shown in Fig. 6.2b, the plasmonic resonances red-shifts a bit duc to the change of the refractive index environment after coating the spacer.

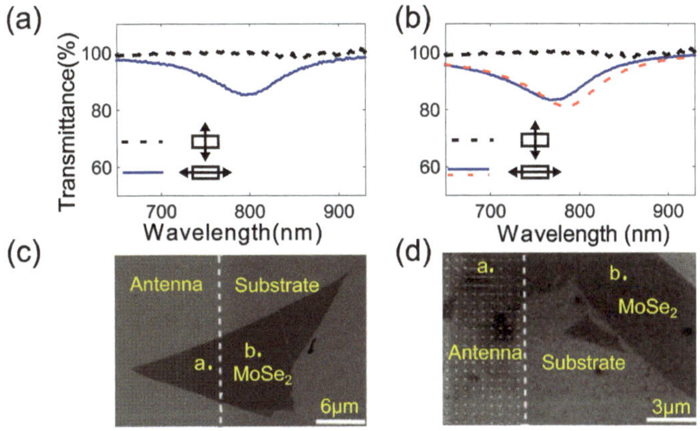

Figure 6.2 (a, b) Transmittance profiles of the sample (a) without and (b) with a spacer with illumination light of different polarisation; the direction of the illuminated light polarisation is illustrated in the insets. In (b), the solid blue and dashed red lines represent the transmittance before and after spacer coating, respectively. (c, d) SEM images of the two samples with and without a spacer, respectively [18].

Next, single layers of $MoSe_2$ samples are mechanically exfoliated from the bulk crystal and transferred onto both samples. For reference purposes, we let the monolayer flakes on both samples sit partly on the antenna array and partly on the silica substrate. The scanning electron microsce (SEM) images of the sample without and with a spacer are shown in Figs. 6.2c and 6.2d, respectively.

6.3.2 PL Characterisation

To study the effects of plasmonic nanoparticles on the PL emission of monolayer $MoSe_2$, we conducted 2D PL scanning. Figures 6.3a and 6.3b show the mapping images for a sample without and with a spacer, respectively. The regions left of the dashed lines are equipped with plasmonic nanoparticles, while the right parts are not. Interestingly, we found that the PL from a the sample without spacer is quenched but it is enhanced for the sample with a spacer, although the plasmonic resonances of both coincide with the peak of the PL emission. To quantify the effects of nanoparticles, we measure the spectral profiles from both samples, as shown in Fig 6.3c,d. The spectral data shown here are taken from points a and b shown in Fig. 6.2c,d.

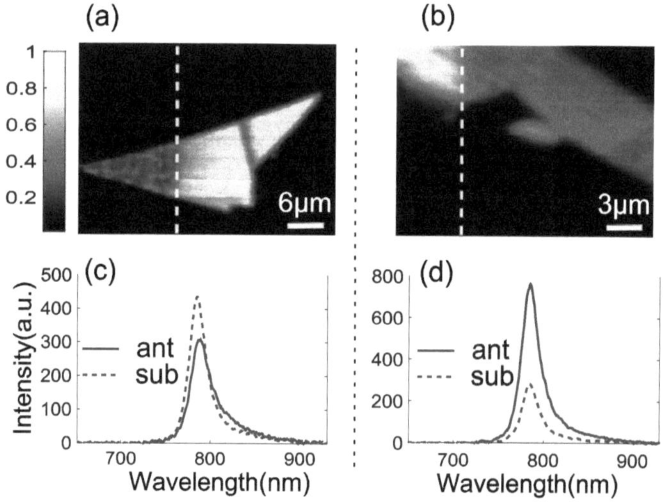

Figure 6.3 (a, b) 2D PL mapping of samples (a) without and (b) with a spacer, respectively; the regions left to the dashed line are pattered with plasmonic antenna. (c, d) PL spectra taken from on- and off-antenna regions for samples (c) without and (d) with a spacer. Solid and dashed lines represent data from on- and off-antenna regions, respectively [18].

To further study the effects of plasmonic nanoparticles, we measured the polarisation-resolved plasmonic effect by inserting a polariser in the detection path. Here we define the plasmonic effect as PL^{ant}/PL^{sub}. This quantity is calculated by taking the average PL value of a small area from the $MoSe_2$-on-antenna region and then subtracting the corresponding background and normalising to the PL from the same area at $MoSe_2$-on-substrate region. It can be seen from the definition that the plasmonic effect would be greater than 1 for PL enhancement but less than 1 for quenching. Figures 6.4a and 6.4.b show the polarisation-resolved plasmonic effects for sample without and with a spacer, respectively. The insets show the relation between the detection polarisation angle and the orientation of the gold bar nanoparticles. The distinct polarisation-dependent behaviours of both samples reflects the intrinsic property of LPR resonance along the long axis, which further confirms that the PL changes are mainly due to the excitation of LPR.

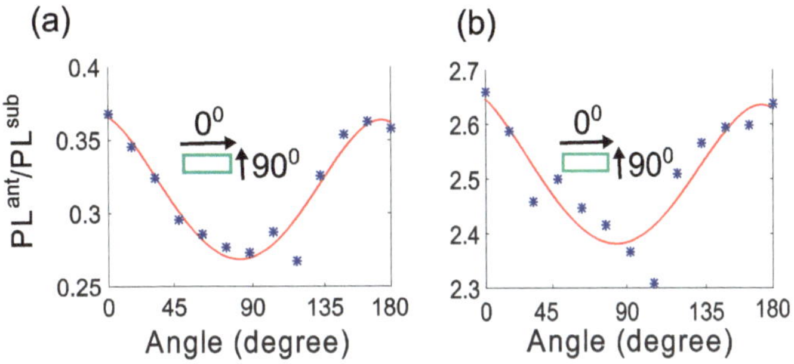

Figure 6.4 (a, b) Antenna effects dependence on the PL emission of different polarisation angles of two samples without and with a spacer, respectively. Asterisks are measured data, and red lines are fitting curves. The insets indicate polarisation direction with respect to the gold rectangular antenna [18].

6.3.3 Numerical Simulation

To further understand the nature of the involved processes, numerical calculations were conducted. We built up the same plasmonic structures as we used in the experiments, and we used electric dipole emitters to mimic the emission from monolayer $MoSe_2$. We studied how a few critical quantities, including the excitation rate, radiative decay rate, quantum yield and the overall PL intensity, evolve with the thickness of the spacer between the dipole emitter and the plasmonic nanoparticles. Figure 6.5 shows the results for dipole emitters of different orientations at different positions. All the results are normalised to the case with no plasmonic nanoparticles. As the overall PL is a product of the quantum yield and the excitation rate, we could conclude, on the basis of results shown in Fig 6.5, that quantum yield is the dominant factors that affects PL emission in our system. From Fig. 6.5b that shows that the quantum yield has a transit when the spacer thickness is around 8 nm, this justified our choice of the spacer thickness in experiments.

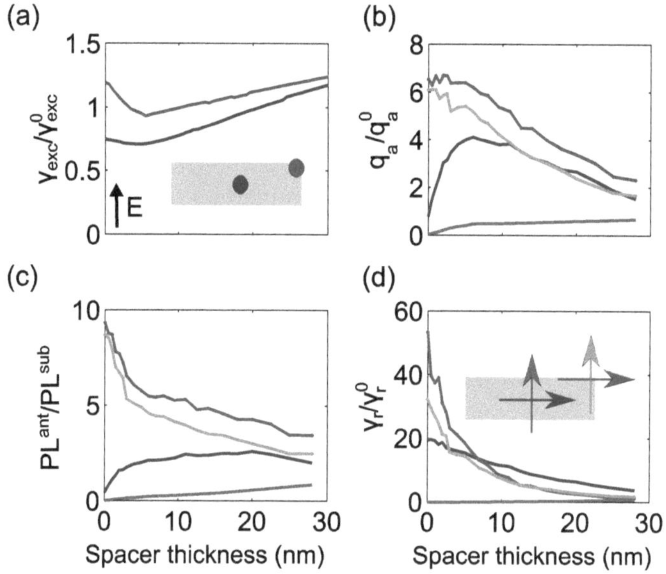

Figure 6.5 Simulated dependences of the (a) excitation rate, (b) quantum yield, (c) PL intensity and (d) radiative decay rate on the spacer thickness for our experimental setup. The inset in (a) shows a dipole emitter located at different positions of the plasmonic antenna. The arrows in the inset of (d) indicate dipole emitters with different orientations [18].

In summary, in these examples, we demonstrated integration of monolayer $MoSe_2$ onto resonant plasmonic nanoparticles experimentally. By changing the spacer thickness between the nanoparticles and the $MoSe_2$ layer, we showed PL control from quenching to enhancement. The numerical simulation results revealed the coupling mechanism between the plasmonic nanoparticles and the 2D materials. We could also conclude from the experimental results that it is important to control the distance between the plasmonic nanoparticles and 2D materials to harvest the PL enhancement brought by the plasmonic resonance.

6.4 2D Material Integration with Dielectric Waveguide Grating

As discussed in Section 6.3, the localised hot spots of plasmonic nanostructures make their coupling to 2D materials highly sensitive to the distance between emitter and structure, often requiring challenging nanometre-precision positioning. Dielectric nanostructures, such as photonic crystal cavities, have also been proposed for enhancing the emission from TMD monolayers [12, 29, 42]. Among them, integration of 2D materials with silicon photonic structures is particularly interesting because it will enable versatile light sources onto silicon photonics platforms [46]. Here we present another example of our study of integrating a WSe_2 monolayer onto a dielectric waveguide grating structure. This multi-resonant structure is silicon-based, scalable and compatible with integration with a silicon photonics platform.

6.4.1 Structure Design and Sample Preparation

By engineering the available photonic modes in the environment, the light–matter interaction strength could be effectively boosted [3]. Generally, the largest enhancement of the overall PL emission can be achieved when utilising photonic structures with multiple resonances that couple to both excitation and emission radiation. We realise such a scheme through integrating a WSe_2 monolayer onto a shallow multi-resonant grating structure inscribed into a planar silicon waveguide, which supports multiple propagating mode resonances.

The side view of our experimental arrangement is shown in Fig 6.6a. A WSe_2 monolayer is positioned on top of a grating etched into a planar waveguide made of amorphous silicon (a-Si). The grating periodicity is selected such that both the excitation laser and the emission couple to available waveguide modes. The grating structure facilitates coupling of the pump light (from free space, as indicated by the arrow) to waveguide mode, which increases the local field intensity at the monolayer position. Thus, the absorption of the pump light by the WSe_2 monolayer is increased and translates to higher excitation efficiency. The emission of WSe_2 also couples into waveguide modes supported by the high-index a-Si layer. The coupling of the WSe_2 layer to the silicon waveguide reduces the radiative lifetime of emission, which, in turn, is extracted efficiently to free space due to the grating structure. Therefore, such an experimental scheme offers boosting the PL emission simultaneously through the excitation and emission processes. Furthermore, as the emission couples into different leaky modes supported by the waveguide grating structure, the emission would be highly directional depending on the dispersion of the modes supported by the structure. Part (i) of Fig. 6.6b shows the designed parameters of the waveguide grating structure (side view), part (ii) shows a scanning electron microscope (SEM) image of the top view of the fabricated grating, and the calculated TE_0, TM_0 and TE_1 mode profiles confined by this structure are shown in part (iii).

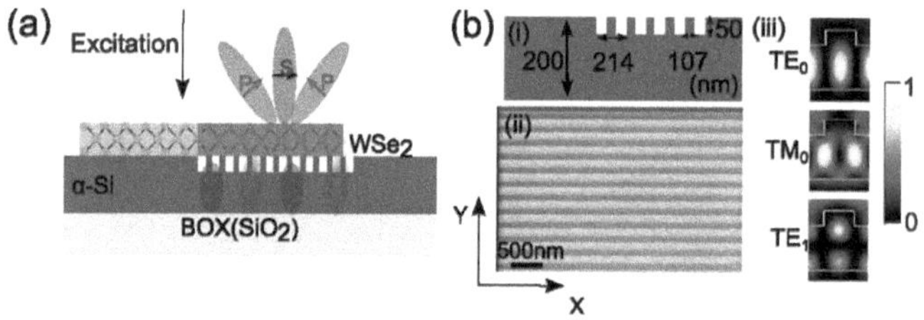

Figure 6.6 (a) Schematic side view of the structure under investigation: a WSe_2 monolayer is located on top of a grating inscribed into the planar waveguide. Geometrical dimensions, of course, are not to scale. (b) Designed parameters of the waveguide grating structure (i) and SEM image of a top view of the grating structure used in experiments (ii); part (iii) shows calculated total field profiles at guided-mode resonant wavelengths of the waveguide grating structure for normal incident plane waves. Reprinted with permission from Ref. [29]. Copyright (2017) American Chemical Society.

Measured transmittance of the fabricated waveguide grating structure is shown in Fig 6.7a for different linear light polarisations relative to the un-patterned area for almost normal incidence. Multiple resonances arising from the excitation of waveguide modes are visible. The black dashed line shows the measured PL emission spectrum of the WSe_2 monolayer overlapping with the resonances. Finite-element light-scattering simulations for normal incident plane waves are done to confirm the nature of the observed resonances; the results are shown in Fig. 6.7b, where we observe a qualitatively good match between experimental and simulation results.

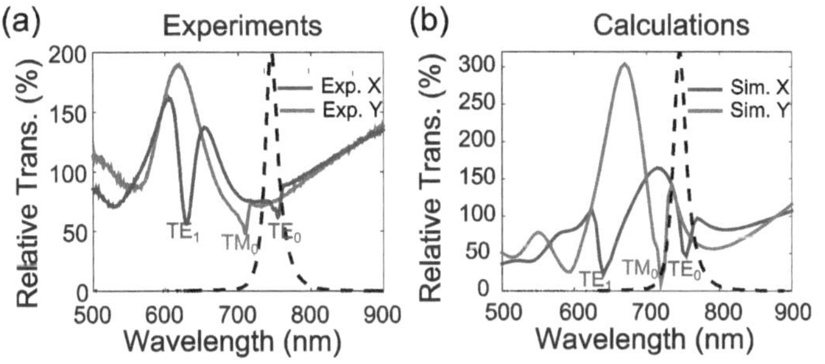

Figure 6.7 (a, b) Measured and calculated transmittance of the waveguide grating structure for different linearly polarised light relative to the un-patterned area, respectively. The dashed line shows the measured emission spectrum of the WSe_2 monolayer. Reprinted with permission from Ref. [29]. Copyright (2017) American Chemical Society.

6.4.2 PL Intensity and Direction Characterisation

After characterising the transmittance of the samples, a monolayer of WSe_2 exfoliated from a bulk crystal is dryly transferred onto the waveguide grating structure. Figure 6.8a shows the optical microscope (left) and 2D PL mapping images (right) of the sample, respectively, where we could observe obvious PL enhancement induced by the grating. By exciting the sample with a 633 nm (around the TE_1 resonance) He-Ne continuous-wave (CW) laser, we observe up to 8 times enhancement of the PL from the on-grating area compared to the off-grating one. The spectra measured from two different positions (on-grating and off-grating regions) are shown in Fig. 6.8b.

To confirm and distinguish the enhancement effects coming from excitation and emission, we also measured the dependence of the enhancement factor on the excitation wavelength, as shown in Fig. 6.8c. The strongest enhancement happened when the sample was excited by a laser wavelength of around 630 nm, which corresponded to the TE_1 resonance of our sample and fulfilled our expectation. The overall enhancement factor was slightly weaker than that in our CW experiments, which was due to a broader spectral and pulsed (a few picoseconds) excitation. This experiment also showed the feasibility of tuning the enhancement factor by varying the excitation wavelength. In addition, time-resolved measurements were conducted to check the emission enhancement, as shown in Fig. 6.8d. For better comparison, here we normalised the emission intensity to its respective maximum and plotted the decay behaviour in a logarithmic scale. We also fitted the decay curve into a biexponential function [44] shown as solid lines. We could observed that the radiative behaviour of WSe_2 for the on-grating region was around twice faster than off-grating one, which proved that we harvested emission enhancement from our samples, too. Thus, we concluded that our multi-resonant Si waveguide grating structure could effectively enhance the PL of a WSe_2 monolayer averagely up to 8 times by combining both the excitation and the emission enhancements.

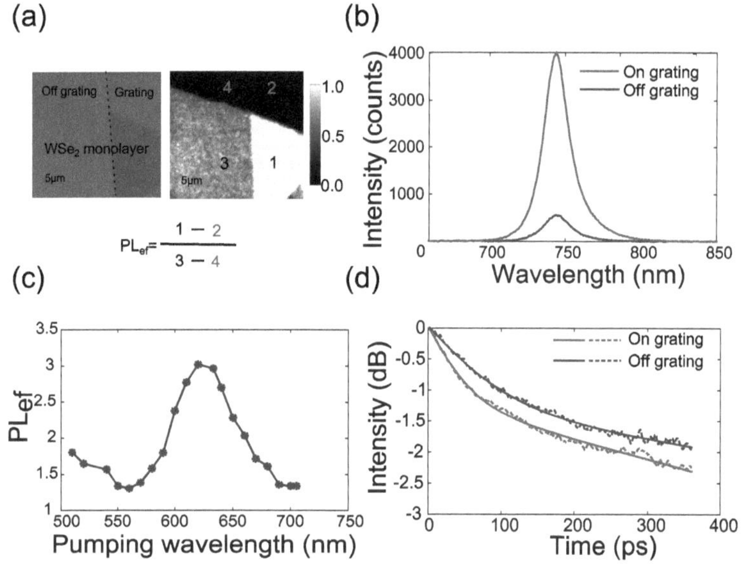

Figure 6.8 (a, left) Optical image of the sample. (a, right) 2D PL mapping of the sample. (b) PL spectra of WSe_2 measured from on-grating and off-grating regions. (c) Dependence of the PL enhancement factor on the excitation wavelength. (d) Time-resolved measurements from on-grating and off-grating regions. Reprinted with permission from Ref. [29]. Copyright (2017) American Chemical Society.

To further explore the coupling between the monolayer WSe_2 and waveguide grating structure, we investigated the far-field angular emission from our system by measuring the back-focal plane image of the emission. We observed quite different emission patterns from the on-grating and off-grating regions. Figure 6.9a shows the back-focal plane image of the total emission from the on-grating regions, where we observed that the emission goes preferentially into four distinct angular regions. In contrast, the back-focal plane image of emission from unpatterned regions, shown in Fig. 6.9d, displayed a typical pattern, with intensity decaying from the centre. Furthermore, we mapped the back-focal plane imaging of the emission of different polarisations. Figures 6.9b and 6.9c show the emission patterns when a polariser oriented across and along the grating ridge was applied in the detection path, respectively. These two images show clearly that the emission direction was polarisation-dependent. Therefore, the emission directionality and intensity in the grating region could be tailored with a polariser to a great degree.

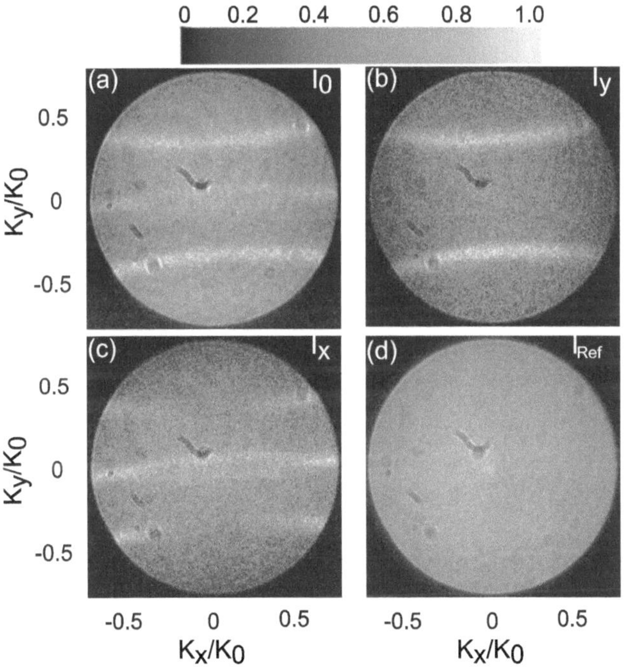

Figure 6.9 Experimental back-focal plane images of emission from a WSe_2 monolayer. (a) The overall emission from on-grating WSe_2 and (b, c) the back-focal plane images of the emission component polarised perpendicular (I_y) and parallel (I_x) to the grating ridge from the on-grating WSe_2 structure, respectively. (d) Back-focal plane image of the total emission from the off-grating WSe_2 structure. Reprinted with permission from Ref. [29]. Copyright (2017) American Chemical Society.

6.4.3 Theoretical Calculation

To investigate further the physical mechanism behind the measured PL characteristics, we performed numerical simulations of the emission of a WSe_2 monolayer coupled to the waveguide grating structure. The WSe_2 monolayer emission was modelled as a superposition of different electric dipole emitters, placed 1 nm above the grating. As the WSe_2 monolayer did not support out-of-plane oriented dipole moments [35], we only simulated in-plane electrical dipoles, oriented across (y direction) and along (x direction) the grating ridge.

The simulated far-field distribution in momentum space, averaged over the positions and orientations, is shown in Fig. 6.10a. Good matching of the singular dipole far-field calculations and the measured momentum space of the emission was obtained, when compared to Fig. 6.9a. By considering only the far-field component perpendicular to the grating ridges (Fig. 6.10b), we reproduced the main features in Fig. 6.9b, in which there is high peak intensity around $k_y \approx \pm 0.3k_0$. Conversely, considering only the intensity component parallel to the grating ridges (Fig. 6.10c) allowed us to reproduce the features of Fig. 6.9c.

The observed polarisation-selective directional emission property is linked to the waveguide modes the emitters couple to. To demonstrate this, we calculated the dispersion relation of TE_0 and TM_0 waveguide modes for a flat-slab waveguide system (air/ 175nm a-Si /SiO_2) folded into the first Brillouin zone with a grating vector $G = 2\pi/214$ nm (Fig. 6.10d). The folded k-space cross section of the waveguide-mode dispersion led to four curves related to ±1 diffraction coupling of radiation to the TE_0 and TM_0 waveguide modes. These four curves matched well with what was observed in the experiments and far-field simulations, which indicates that these are indeed the modes that the monolayer is coupling to.

In summary, in this example, we demonstrated enhanced and polarisation-selective directional emission from monolayer WSe_2 integrated onto a Si waveguide grating structure. The PL enhancement and directionality have been realised by simultaneously coupling the emission and the excitation fields into the resonant modes supported by the structure. By tailoring the resonant frequencies and dispersion of the waveguide modes, we showed great flexibility in controlling the WSe_2 monolayer emission in both intensity and directionality. Simulation results explained our experimental observation well.

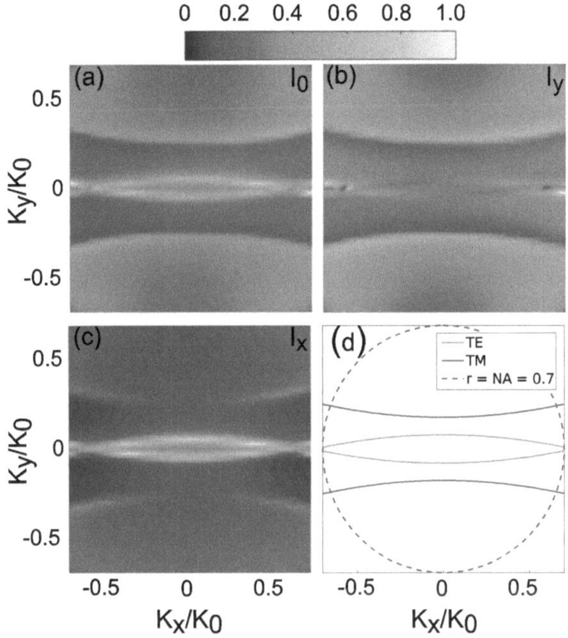

Figure 6.10 Folded dispersion relation of the TEo and TMo waveguide modes at a wavelength of 750 nm for a flat-slab waveguide system, assuming a period of 214 nm in the *y* direction. Reprinted with permission from Ref. [29]. Copyright (2017) American Chemical Society.

6.5 Conclusion and Discussion

In conclusion, we discussed the general prospects and principles of integration of 2D materials with resonant photonic nanostructures. We then presented two typical examples of integration of 2D TMDs onto resonant plasmonic metal nanoparticles and dielectric waveguide grating structures, respectively. In the examples, we showed that photonic nanostructures can control the emission from 2D TMDs in various aspects, including intensity, directionality, polarisation and lifetime. This justifies that integration of 2D materials with photonic nanostructures will play an important role in enabling future 2D material nanophotonic devices.

The field of 2D materials is developing rapidly, as are the new designs of photonic nanostructures. There is a lot more to explore in this direction.

A greater understanding of the coupling between photonic nanostructures and 2D materials is required, thus potentially better controlling the properties of these materials. For example, non-linear coupling between 2D materials, especially second-order non-linearity, has attracted intense interest [7, 9, 11, 47]. Besides, more and more novel materials are emerging, for example, layered WTe_2 (one member of TMDs) has been reported to possess large and non-saturating magnetoresistance [2]. On the other hand, novel photonic nanostructures, especially dielectric ones [20, 39], with multiple types of resonances have brought unprecedented opportunities to study the coupling between photonic structures and 2D materials. What's more, heterostructures formed by stacking different kinds of TMDs [8, 22, 33] or mixing with other 2D materials [5, 16, 30] have shown advanced electronic and optical properties; integration of these heterostructures with photonic nanostructures will further control and enhance their performance. For example, it will be intriguing to see how a resonant photonic structure affects the charge transfer between different materials across the layers. What's more, the properties of 2D materials could be tuned by electric or magnetic fields. It will be interesting to investigate the tunability for hybrid structures consisting of 2D materials and photonic nanostructures. In particular, polariton modes formed through coupling between the TMDs and cavities (strong coupling) have shown lots of interesting physical phenomena recently [14, 17, 37, 40], and there will be plenty of room to explore here.

Acknowledgements

I would like to acknowledge a number of collaborators for their substantial and valuable contribution to the results presented in the chapter. Most of all I would like to thank S. Nanz, A. Abass, J. Yan, T. gao, D. Y. Choi, J. Yang, E. Rusak, J. Straubel, M. Decker, I. Staude, C. Rockstuhl, Y. Lu, Y. S. Kivshar and D. N. Neshev. This work was supported by the Australian Research Council under the Discover Projects and Center of Excellence research programs. I acknowledge the support of the ACT Node of the Australian Nanofabrication Facility.

References

1. Akselrod, G. M., Ming, T., Argyropoulos, C., Hoang, T. B., Lin, Y., Ling, X., Smith, D. R., Kong, J. and Mikkelsen, M. H. (2015). Leveraging nanocavity harmonics for control of optical processes in 2D semiconductors, *Nano Lett.*, **15**(5), pp. 3578–3584.
2. Ali, M. N., Xiong, J., Flynn, S., Tao, J., Gibson, Q. D., Schoop, L. M., Liang, T., Haldolaarachchige, N., Hirschberger, M., Ong, N., et al. (2014). Large, non-saturating magnetoresistance in WTe_2, *Nature*, **514**(7521), pp. 205–208.
3. Anger, P., Bharadwaj, P. and Novotny, L. (2006). Enhancement and quenching of single-molecule fluorescence, *Phys. Rev. Lett.*, **96**, pp. 3–6.
4. Bhanu, U., Islam, M. R., Tetard, L. and Khondaker, S. I. (2014). Photoluminescence quenching in gold-MoS_2 hybrid nanoflakes, *Sci. Rep.*, **4**, p. 5575.
5. Britnell, L., Ribeiro, R., Eckmann, A., Jalil, R., Belle, B., Mishchenko, A., Kim, Y.-J., Gorbachev, R., Georgiou, T. and Morozov, S. (2013). Strong light-matter interactions in heterostructures of atomically thin films, *Science*, **340**(6138), pp. 1311–1314.
6. Butun, S., Tongay, S. and Aydin, K. (2015). Enhanced light emission from large-area monolayer MoS_2 using plasmonic nanodisc arrays, *Nano Lett.*, **15**(4), pp. 2700–2704.
7. Chen, H., Corboliou, V., Solntsev, A. S., Choi, D.-Y., Vincenti, M. A., de Ceglia, D., De Angelis, C., Lu, Y. and Neshev, D. N. (2017). Enhanced second harmonic generation from two-dimensional $MoSe_2$ on a silicon waveguide, *Light. Sci. Appl.*, **6**, p. e17060.
8. Chen, H., Wen, X., Zhang, J., Wu, T., Gong, Y., Zhang, X., Yuan, J., Yi, C., Lou, J., Ajayan, P. M., et al. (2016). Ultrafast formation of interlayer hot excitons in atomically thin MoS_2/WS_2 heterostructures, *Nat. Commun.*, **7**, p. 12512.
9. Day, J. K., Chung, M.-H., Lee, Y.-H. and Menon, V. M. (2016). Microcavity enhanced second harmonic generation in 2D MoS_2, *Opt. Mater. Express*, **6**(7), pp. 2360–2365.
10. Dufferwiel, S., Schwarz, S., Withers, F., Trichet, A., Li, F., Sich, M., Clark, C., Nalitov, A., Solnyshkov, D., Malpuech, G., Novoselov, K., Smith, J., Skolnick, M., Krizhanovskii, D. and Tartakovskii, A. (2014). Exciton-

polaritons in van der Waals heterostructures embedded in tunable microcavities, *Nat. Commun.*, **6**, pp. 8579–8579.

11. Fryett, T. K., Seyler, K. L., Zheng, J., Liu, C.-H., Xu, X. and Majumdar, A. (2016). Silicon photonic crystal cavity enhanced second-harmonic generation from monolayer WSe_2, *2D Mater.*, **4**(1), p. 015031.
12. Galfsky, T., Sun, Z., Considine, C. R., Chou, C.-T., Ko, W.-C., Lee, Y.-H., Narimanov, E. E. and Menon, V. M. (2016). Broadband enhancement of spontaneous emission in two-dimensional semiconductors using photonic hypercrystals, *Nano Lett.*, **16**(8), pp. 4940–4945.
13. Gan, X., Gao, Y., Fai Mak, K., Yao, X., Shiue, R.-J., van der Zande, A., Trusheim, M. E., Hatami, F., Heinz, T. F., Hone, J., et al. (2013). Controlling the spontaneous emission rate of monolayer MoS_2 in a photonic crystal nanocavity, *Appl. Phys. Lett.*, **103**(18), p. 181119.
14. Gartstein, Y. N., Li, X. and Zhang, C. (2015). Exciton polaritons in transition-metal dichalcogenides and their direct excitation via energy transfer, *Phys. Rev. B*, **92**(7), p. 075445.
15. Halas, N. J., Lal, S., Chang, W.-S., Link, S. and Nordlander, P. (2011). Plasmons in strongly coupled metallic nanostructures, *Chem. Rev.*, **111**(6), pp. 3913–3961.
16. Hamm, J. M. and Hess, O. (2013). Two two-dimensional materials are better than one, *Science*, **340**(6138), pp. 1298–1299.
17. Hu, F., Luan, Y., Scott, M., Yan, J., Mandrus, D., Xu, X. and Fei, Z. (2017). Imaging exciton–polariton transport in $MoSe_2$ waveguides, *Nat. Photonics*, **11**, pp. 356–360.
18. Chen, H., Yang, J., Rusak, E., et al. (2016). Manipulation of photoluminescence of two-dimensional $MoSe_2$ by gold nanoantennas, *Sci. Rep.*, **6**, Article no. 22296.
19. Khurgin, J. B. (2015). How to deal with the loss in plasmonics and metamaterials, *Nat. Nanotechnol.*, **10**(1), pp. 2–6.
20. Kuznetsov, A. I., Miroshnichenko, A. E., Brongersma, M. L., Kivshar, Y. S. and Luk'yanchuk, B. (2016). Optically resonant dielectric nanostructures, *Science*, **354**(6314), p. aag2472.
21. Lee, B., Park, J., Han, G. H., Ee, H.-S., Naylor, C. H., Liu, W., Johnson, A. C. and Agarwal, R. (2015). Fano resonance and spectrally modified photoluminescence enhancement in monolayer MoS_2 integrated with plasmonic nanoantenna array, *Nano Lett.*, **15**(5), pp. 3646–3653.
22. Lee, C.-H., Lee, G.-H., Van Der Zande, A. M., Chen, W., Li, Y., Han, M., Cui, X., Arefe, G., Nuckolls, C., Heinz, T. F., et al. (2014). Atomically thin p–n

junctions with van der Waals heterointerfaces, *Nat. Nanotechnol.*, **9**(9), pp. 676–681.

23. Liang, D., Roelkens, G., Baets, R. and Bowers, J. E. (2010). Hybrid integrated platforms for silicon photonics, *Materials*, **3**(3), pp. 1782–1802.
24. Lin, J., Li, H., Zhang, H. and Chen, W. (2013). Plasmonic enhancement of photocurrent in MoS_2 field-effect-transistor, *Appl. Phys. Lett.*, **102**(20), p. 203109.
25. Liu, X., Galfsky, T., Sun, Z., Xia, F., Lin, E.-C., Lee, Y.-H., Kéna-Cohen, S. and Menon, V. M. (2015). Strong light–matter coupling in two-dimensional atomic crystals, *Nat. Photonics*, **9**(1), pp. 30–34.
26. Merzbacher, E. (1997). *Quantum Mechanics*, 3rd edn. (Wiley).
27. Najmaei, S., Mlayah, A., Arbouet, A., Girard, C., Léotin, J. and Lou, J. (2014). Plasmonic pumping of excitonic photoluminescence in hybrid MoS_2 – Au nanostructures, *ACS Nano*, **8**(12), pp. 12682–12689.
28. Noda, S., Fujita, M. and Asano, T. (2007). Spontaneous-emission control by photonic crystals and nanocavities, *Nat. Photonics*, **1**(8), pp. 449–458.
29. Chen, H., Nanz, S., Abass, A., et al. (2017). Enhanced directional emission rom monolayer WSe2 integrated onto a multiresonant silicon-based photonic structure, *ACS Photonics*, **4**(12), pp. 3031–3038.
30. Novoselov, K., Mishchenko, A., Carvalho, A. and Neto, A. C. (2016). 2D materials and van der Waals heterostructures, *Science*, **353**(6298), p. aac9439.
31. Pelton, M. (2015). Modified spontaneous emission in nanophotonic structures, *Nat. Photonics*, **9**(7), pp. 427–435.
32. Purcell, E. M. (1946). Spontaneous emission probabilities at radio frequencies, *Phys. Rev.*, **69**, pp. 37–38.
33. Rigosi, A. F., Hill, H. M., Li, Y., Chernikov, A. and Heinz, T. F. (2015). Probing interlayer interactions in transition metal dichalcogenide heterostructures by optical spectroscopy: MoS_2/WS_2 and $MoSe_2/WSe_2$, *Nano Lett.*, **15**(8), pp. 5033–5038.
34. Sakurai, J. J. M. (1994). *Modern Quantum Mechanics*, revised edn. (Addison-Wesley).
35. Schuller, J. A., Karaveli, S., Schiros, T., He, K., Yang, S., Kymissis, I., Shan, J. and Zia, R. (2013). Orientation of luminescent excitons in layered nanomaterials, *Nat. Nanotechnol.*, **8**(4), pp. 271–276.

36. Sell, D., Yang, J., Doshay, S., Zhang, K. and Fan, J. A. (2016). Visible light metasurfaces based on single-crystal silicon, *ACS Photonics*, **3**(10), pp. 1919–1925.

37. Sidler, M., Back, P., Cotlet, O., Srivastava, A., Fink, T., Kroner, M., Demler, E. and Imamoglu, A. (2017). Fermi polaron-polaritons in charge-tunable atomically thin semiconductors, *Nat. Phys.*, **13**(3), pp. 255–261.

38. Sobhani, A., Lauchner, A., Najmaei, S., Ayala-Orozco, C., Wen, F., Lou, J. and Halas, N. J. (2014). Enhancing the photocurrent and photoluminescence of single crystal monolayer MoS_2 with resonant plasmonic nanoshells, *Appl. Phys. Lett.*, **104**(3), p. 031112.

39. Staude, I. and Schilling, J. (2017). Metamaterial-inspired silicon nanophotonics, *Nat. Photonics*, **11**(5), pp. 274–284.

40. Sun, L., Wang, Q., Zhang, B., Shen, X. and Lu, W. (2016a). Room temperature exciton polaritons in thin WS_2 flakes, *Frontiers in Optics (Optical Society of America)*, p. FF5H.2.

41. Sun, Z., Martinez, A. and Wang, F. (2016b). Optical modulators with 2D layered materials, *Nat. Photonics*, **10**(4), pp. 227–238.

42. Wu, S., Buckley, S., Jones, A. M., Ross, J. S., Ghimire, N. J., Yan, J., Mandrus, D. G., Yao, W., Hatami, F., Vučković, J., Arka, M. and Xiaodong, X. (2014). Control of two-dimensional excitonic light emission via photonic crystal, *2D Mater.*, **1**(1), p. 011001.

43. Wu, S., Buckley, S., Schaibley, J. R., Feng, L., Yan, J., Mandrus, D. G., Hatami, F., Yao, W., Vučković, J. and Majumdar, A. (2015). Monolayer semiconductor nanocavity lasers with ultralow thresholds, *Nature*, **520**(7545), pp. 69–72.

44. Yan, T., Qiao, X., Liu, X., Tan, P. and Zhang, X. (2014). Photoluminescence properties and exciton dynamics in monolayer WSe_2, *Appl. Phys. Lett.*, **105**(10), p. 101901.

45. Ye, Y., Wong, Z. J., Lu, X., Ni, X., Zhu, H., Chen, X., Wang, Y. and Zhang, X. (2015). Monolayer excitonic laser, *Nat. Photonics*, **9**, pp. 733–737.

46. Youngblood, N. and Mo, L. (2016). Integration of 2D materials on a silicon photonics platform for optoelectronics applications, *Nanophotonics*, **6**(6), doi:https://doi.org/10.1515/nanoph-2016-0155.

47. Zeng, J., Yuan, M., Yuan, W., Dai, Q., Fan, H., Lan, S. and Tie, S. (2015). Enhanced second harmonic generation of MoS_2 layers on a thin gold film, *Nanoscale*, **7**(32), pp. 13547–13553.

Chapter 7

Strong Light–Matter Interactions in 2D Materials Assisted by On-chip Optoelectronic Devices

Xuetao Gan, Liang Fang and Jianlin Zhao
MOE Key Laboratory of Material Physics and Chemistry under Extraordinary Conditions, and Shaanxi Key Laboratory of Optical Information Technology, School of Science, Northwestern Polytechnical University, Xi'an 710072, China
xuetaogan@nwpu.edu.cn

7.1 Introduction

Recently, 2D materials with a layered structure have emerged as one of the most interesting media for constructing optoelectronic devices because of their unique and plentiful electrical and optical properties [1–3]: (1) The family of 2D materials exhibit a rich variety of electronic properties, including semi-metallic graphene, insulating boron nitride, semiconducting transition metal dichalcogenides (TMDs) and black phosphorus [4]. The optical bandgaps of semiconducting 2D materials could be engineered by

Two-Dimensional Materials in Nanophotonics: Developments, Devices, and Applications
Edited by Yuerui Lu

ISBN 978-981-4800-22-8 (Hardcover), 978-0-429-42877-7 (eBook)
www.jennystanford.com

changing their layer numbers, modulating alloy composition or stacking into a heterostructure, which covers spectral ranges from ultra-violet to infrared throughout the visible range. (2) While 2D materials only have mono- or few-atom thickness, light–matter interactions inside them could be extremely strong. For instance, monolayer graphene absorbs the single-passed light with an efficiency of 2.3% over a wavelength range from visible to mid-infrared [5]. Due to the reduced dielectric screening and effective many-body interaction, monolayer molybdenum disulfide (MoS_2) could yield an absorption coefficient of ~10% around the excitonic resonances [6]. Also, relying on the strong confinement of carriers in the 2D regime, the carrier density could be tuned readily via vertical electrical field or chemical doping, producing considerable modification of optical responses. (3) 2D materials' surface are naturally passivated without any dangling bonds. It is possible to construct vertical heterostructures using different 2D materials via the weak van der Waals forces, which therefore gets rid of issues of lattice mismatch and enables band structure engineering [7]. To achieve various active optoelectronic devices, p-i-n, p-n, and Schottky junctions are easy to build intentionally.

Determined by the unique electronic band structure and stacking-order-dependent interlayer coupling, 2D materials also have outstanding non-linear optical properties. For instance, in a single layer of graphene, because of the relatively low density of states, its optical absorption in infrared range becomes saturable under high-intensity radiation, enabling graphene-based saturable absorbers for tunable mode-locking lasers. Furthermore, relying on the zero-bandgap and linear dispersion of graphene, strong four-wave mixings (FWMs) and third harmonic generations (THGs) are observed in mono- and few-layer graphene flakes with the broadband on-resonance [8]. The comparison between FWM from graphene and a gold film implies a high third-order non-linearity of 10^{-7} esu in graphene, which is about 7 orders of magnitude larger than that in silica. High third-order non-linearities have been demonstrated in few-layer black phosphorus as well [9]. For atomic TMDs, due to the inversion symmetry broken in varied thicknesses, extraordinary second-harmonic generations (SHGs) were experimentally demonstrated. Bulk MoS_2 has been well known for its absence of the SHG due to the inversion symmetry. However, in atomically thin MoS_2 an odd number of layers presents strong SHGs, while the even number of layers still cannot yield second-order non-linear processes [10, 11]. This could be attributed to the space group transformations from D_{3d} to D_{3h} for MoS_2 slices with even and odd numbers of layers. In addition, in monolayer WS_2 and WSe_2, the measured second-order non-linear susceptibility is in the order of 10^3 pm/V, which is about 3 orders of magnitude higher than those in common bulk materials [12]. This unique second-order non-linearity in monolayer TMDs could be explained as follows: The photon energy of the SHG signal is on-resonance with the exciton energy formed by an electron-hole pair. This exciton resonance-enhanced SHG could also be easily tuned electrically due to the low dielectric screening [13]. These interesting non-linear optical responses in 2D materials have great potential to extend their optoelectronic applications into the non-linear regime, such as coherent light source generations, all-optical signal processing and up-conversion photo-detectors.

As mentioned before, the light–matter interactions in 2D materials are much stronger than bulk materials. A number of experimental studies have verified the 2.3% interband transition absorption in monolayer graphene over a wide wavelength range [5, 14]. Comparatively, in a thick film of indium gallium arsenide, the thickness required to get an absorbance of 2.3% over the light at 1.55 μm is about 20 nm. However, to make an optoelectronic device utilising monolayer graphene, the 2.3% absorption is too weak to realise high performance. It is well known that electrically tunable absorption in graphene could be exploited to implement optical modulators. Even if graphene's 'Pauli blocking' were achieved perfectly, the modulation depth over the single-passed light is only 0.1 dB, which is far away from the normal value of 10 dB to satisfy the real optical signal processing. In addition, the atomic thickness of 2D materials limits their SHG implementations strongly, which can only be observed with a pulsed pump laser with extremely high peak intensity. However, continuous-wave (CW)-pumped SHGs with simple, low-power and low-cost light sources, such as semiconductor laser diodes, are more desirable for device operations. Therefore, it is essential to strengthen light–matter interactions in 2D materials, which is important to achieve high-contrast modulations, high-responsive photo-detections, enough internal gains for light emissions and low-power non-linear processes.

There are a variety of methods which have been demonstrated to improve light–matter interactions in 2D materials. For example, monolayer graphene was inserted into optical cavities assembled by two distributed Bragg reflectors or by two silver film reflectors [15]. An integrated graphene photo-detector was reported with an enhancement factor of 26. In these Fabry–Perot (FP) cavities, an optical signal could be reflected for a long time to couple with the inserted 2D materials effectively. In an optical cavity, the sustained time period τ of the optical field is determined by the quality (Q) factor with a function of $Q = \omega\tau$, where ω is the optical frequency of resonant mode. Hence, the enhanced light–matter interaction in the FP cavity is proportional to the Q-factor. Unfortunately, the fabrication of a hybrid 2D material-FP cavity is complex and hinders the achievement of a high Q-factor. To allow the insertion of 2D materials, the two reflectors have to be assembled separately. Also, the depositions of reflectors suffer from the incompatibility with the 2D materials, leading to non-uniform and low-quality films.

The surface plasmonic mode around metal nanostructures is another attractive architecture to enhance the coupling between 2D materials and the optical field [16]. In this structure, the optical field is strongly confined in a scale of tens of nanometres near the metal-air interface, yielding a high density of photon state. The 2D materials have a thickness of only a few atoms, consequently enabling strong couplings between 2D materials and surface plasmonic modes of the metal structures. There are two major problems to exploit metal nanostructures in enhancing light–matter interactions: the excitation of the surface plasmon polariton is very difficult with the requirement of a large wave vector of the incident light; metal material's intrinsic optical absorption yields a strong Ohmic loss to the optical field, especially for the propagation modes, which lowers the enhancement efficiency significantly. While plasmon-enhanced graphene photo-detection and light emission from monolayer MoS_2 have been reported, their enhancement factors are low and the device fabrications with nanometre precision still have challenges. Recently, another important and effective technique to realise strong light–matter interaction in 2D materials attracts much interest, which involves a plethora of dipole-type polaritonic excitations [17, 18]. Specifically, electrically tunable and highly confined plasmon polaritons were predicted and observed in graphene for optoelectronics, bio-sensing and other mid-infrared applications. In hexagonal boron nitride, low-loss infrared-active phonon-type polaritons exhibit hyperbolic behaviour for some optical frequencies. In atomically thin TMDs, reduced dielectric screening gives rise to optically prominent excitons with large binding energy and polaritonic modes. These polaritons are very promising for the applications in special frequencies, which support the excitation of polaritons. It cannot be considered as a universal way to comply with strong light–matter interactions in 2D materials to facilitate their practical applications. In the past few years, these enhancement techniques have been studied widely in the aspects of fundamental physics and device applications. Readers are suggested to read the aforementioned references if they are interested.

In this chapter, the authors would like to emphasise chip-integrated nanophotonic structures, including waveguides, photonic crystals and resonators, which could provide a much better platform to enhance light–matter interactions in 2D materials and construct high-performance optoelectronic devices. Compared to the aforementioned techniques, several advantages of on-chip nanophotonic devices could be concluded: (1) The planar platform of the on-chip devices is compatible with the physical structure of 2D materials, enabling their facile integration by wafer-scale transferring or epitaxy growing 2D materials onto the photonic wafer directly [19]. The integration processes is much simpler. (2) The integration of 2D materials could be performed after the fabrication of photonic devices, which would not degrade their performance at all. (3) It is easy to construct 2D material-based electrical devices for photo-detectors, modulators or light-emitting devices on the planar platform of the photonic chip. (4) Optical signals in chip-integrated nanophotonic structures are guided in a sub-micro-metre slab. Strong evanescent fields leak out of the photonic slab by vertical total internal reflection. When the atomically thin 2D materials are coated onto these photonic structures, their locations have strong evanescent fields to ensure effective coupling, which is about 10%-40% of the maximum optical field in the structure. For instance, 100% light absorption in monolayer graphene is possible. (5) A variety of photonic structures could be chosen to enhance light–matter interactions via different concepts, providing such options as compact footprints or broadband operations. For a waveguide or a resonator, when light propagates inside them, the interaction length or time with the 2D materials is extended significantly compared to the vertically incident light in a single-pass way.

Before the demonstration of enhanced light–matter interactions in the integrated architecture of 2D materials and on-chip nanophotonic devices, it is also valuable to state another important motivation: Integrated with chip-integrated nanophotonic devices, 2D materials have great potential to function as active media for on-chip optical interconnects. Recently, it has been widely recognised that on-chip optical interconnects are a viable alternative to copper wire-based electrical interconnects to extend the trend predicted by Moore's law. On-chip optical interconnects could readily offer greater bandwidth, lower power and absence of electromagnetic interference and signal cross-talk. Specifically, silicon photonics, which has been well developed in the past two decades, provides the most promising platform for on-chip optical interconnects [20, 21]. The key driving force behind silicon photonics is the ability to use complementary metal-oxide semiconductor (CMOS)-compatible technology to fabricate photonic elements, leading to high-volume production at low cost [22]. Additionally, silicon's high refractive index allows the photonic devices to have compact dimensions; the integrated photonic circuits could be constructed in a large scale with high density. Silicon's indirect bandgap of ~1.1 eV makes its transparency for the telecom-band optical signal. As a result, silicon by itself offers an almost complete suite of high-quality passive photonic devices, including low-loss waveguides, polarisation controllers, power dividers and optical filters [23–25].

Unfortunately, the achievements of silicon-based high-performance active devices for multi-functionalised optical interconnects still have challenges [26]. Silicon's indirect bandgap with a value of ~1.1 eV hinders the realisations of on-chip light source and photo-detectors at the telecom band. Currently, the most promising way to introduce efficient electrically pumped light sources is to place III-V gain materials onto silicon directly [27, 28], where III-V layers either wafer-bond or epitaxially grow on the silicon wafer scale. Germanium as another group IV material has a considerable absorption coefficient of the near-infrared wavelength regime for the realisation of high-performance photo-detectors on silicon photonic chips [29, 30]. Unfortunately, these hybrid integration processes with III-V or germanium layers are difficult to be compatible with CMOS technology due to the requirement of high temperature. Also the lattice and thermal mismatch would introduce threading dislocations, degrading the material quality and device performances [31]. What's more, integrations of light sources or photo-detectors on silicon photonic devices require multiple and repeated processes of material deposition, lithography, dry etching and ion implantations, leading to the complex fabrication processes with inevitably low production yields and high cost. Optical modulators are another fundamental active component in optical interconnects. While silicon has no linear Pockels effect to achieve a highly efficient electro-optical modulator, the plasma dispersion effect due to the carrier density variation in silicon could give rise to enough refractive index perturbation to achieve a significant phase shift or intensity modulation in photonic devices [32]. Though the achievements of silicon modulators are not required to consider embarrassing growing process of hybrid active materials, their fabrications also suffer from the complex processes of selective lithography, etching and ion implantations, which are still challenges for practical and low-cost manufacture of high-performance modulators. Lastly, it is worth noting that the intrinsic semiconducting properties of silicon, IIIV materials and germanium would set up the upper limit of these active devices' performances.

Two-dimensional materials have a wide range of material properties. It is also possible to assemble 2D materials with different layer numbers and compositions. These attributes allow for their realisations of chip-integrated active optoelectronic devices, which present potential advantages over bulk III-V or germanium materials, such as less fabrication processes, broadband spectral range and high-speed operation. The naturally passivated surface of 2D materials without any dangling bonds facilitates the device integrations as well. For the electrical contacts of 2D materials' optoelectronic devices, the atomically thin materials provide a variety of opportunities to realise the Ohmic contact, getting rid of the complex ion implantation and selective metal deposition [33]. For example, graphene as a semi-metal material could have low contact resistance with the electrodes, which could also be exploited as a spacing layer to flatten the contact between other 2D materials and metal electrodes. Monolayer boron nitride has been transferred onto a MoS_2 transistor as a tunnelling barrier layer to improve the electrical contacts [34]. Also, high-carrier mobility and bipolar behaviours in graphene and black phosphorus provide unique possibilities to enable optoelectronic devices to operate at high speed [33]. As a result, integrating 2D materials onto on-chip photonic structures to construct optoelectronic devices would be benefited from ease of fabrication, compact footprint and efficient light–matter interactions, which promises a novel architecture for optical interconnects. The integration also indicates great potential to develop practical applications of 2D materials, relying on the enhanced light–matter interaction by photonic structures, as well as the compatibility with the fabrication of electrical integrated circuits.

In this chapter, to illustrate the effectively enhanced light–matter interactions in 2D materials by chip-integrated nanophotonic devices, the authors review the integrations of 2D materials with photonic wire waveguides, ring resonators and photonic crystal nanocavities, which are three of the most important elements for on-chip optical interconnects. Figure 7.1a–d displays the schematics of different configurations to couple light with monolayer graphene, which are vertical illuminations by single-passed light, integrations with waveguide, integrations with micro-ring and integrations with photonic crystal nanocavity, respectively. In Section 7.2, enhanced light–matter interactions in a broadband spectral range would be introduced by integrating monolayer graphene onto a waveguide. The enhancement was proved by calculating the graphene-induced waveguide propagation loss, which is much larger than the light attenuation when light passes through the monolayer graphene singly. Even more effective enhancements are then demonstrated in Sections 7.3 and 7.4, which utilise the micro-ring resonators and photonic crystal cavities. Compared to the waveguide integration, these optical resonators are much compact, and the enhanced light–matter interactions are endowed by the long sustained time of photons in the resonant modes. The strong interactions between 2D materials and the optical field are characterised by the degradation of Q-factors in the optical resonators. Other optical responses in 2D materials, including light emission, Raman scattering and non-linear processes, could also be strongly enhanced. It is worth noting that relying on these strongly enhanced light–matter interactions, a number of chip-integrated 2D material optoelectronic devices have been realised with much better performance than their corresponding devices in the vertical illumination regime, including high-contrast modulators, high-response photo-detectors and low-threshold micro-lasers. In this chapter, the authors will only focus on enhancements of optical responses in 2D materials; readers could find these devices' operations and performances elsewhere.

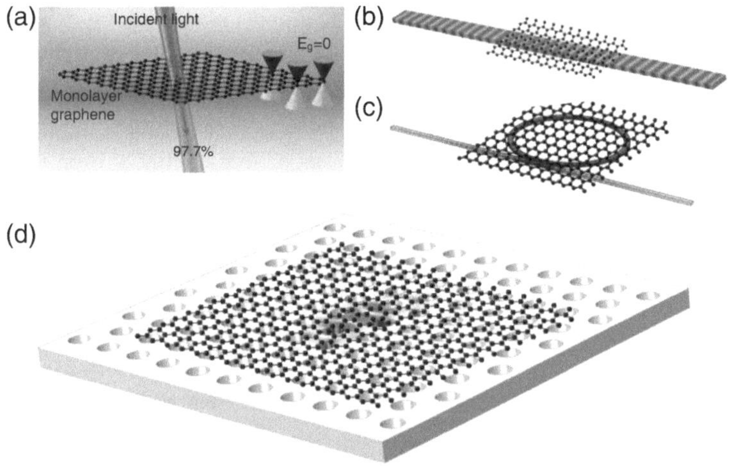

Figure 7.1 Different configurations to realise light–matter interactions in monolayer graphene. (a) Graphene is illuminated by normal incident light. (b–d) Graphene integrated with waveguide, micro-ring and photonic crystal cavity.

7.2 Enhanced Optical Absorption in Hybrid Graphene-Waveguide Structures

Chip-integrated photonic waveguides, as the basic transmission line of optical signals, are fundamental bodies to construct various passive and active devices, including couplers, splitters, modulators and photo-detectors. A variety of materials have been utilised to fabricate photonic waveguides, such as silicon, silicon oxide, silicon nitride, III-V compounds, etc., though silicon waveguides are much more favourable for integrated circuits. Single-mode waveguides with low transmission losses and high confinements are desirable for signal processing. Generally, to reduce the scattering losses induced by the sidewall roughness, waveguides are buried in a cladding layer with a low refractive index. According to the total internal reflection, the waveguide cross section for single-mode operation is determined by the refractive index contrast between waveguide core and cladding layer. For instance, because of silicon's high refractive index at the telecom band (~ 3.4), its high-quality waveguides could be designed with a cross section on scales of sub-micro-metres.

In hybrid structures of waveguide and 2D materials, light–matter interactions are enhanced by extending the coupling length of the evanescent field and 2D materials, as shown in Fig. 7.1b. To study that, monolayer graphene is chosen as a represented material to integrate with a silicon waveguide, as schematically shown in Fig. 7.2a. The top cladding oxide layer on the silicon waveguide should be thinned down to couple the guiding mode to graphene via the leaky evanescent field. By controlling this oxide layer thickness, different coupling strengths are available for specific optoelectronic devices. For instance, in a graphene-integrated photo-detector, the graphene layer is preferred to contact the waveguide directly for effective light absorption. If the graphene layer is exploited to make a thermal modulator, a thin spacing layer may be required to reduce the insertion loss. To study that quantitatively, a straightforward way is calculating the waveguiding mode distribution of the assembled structure. Techniques based on finite-element analysis or the finite-difference time-domain could be coded to simulate the electromagnetic fields in waveguides. Figure 7.2b displays a simulated electrical field distribution of the graphene-waveguide mode at 1550 nm using a finite-element method (COMSOL Multiphysics). Here, the waveguide dimension is chosen as 220 nm × 500 nm for the transverse electrical (TE) mode operation, and silicon's refractive index is 3.4. The cladding oxide layer has a thickness of 10 nm over the silicon waveguide and a refractive index of 1.48. The refractive index of graphene was 2.38 + 1.68*i* for light around 1550 nm. The field distribution along the middle of the structure is displayed as well. The calculated optical intensity at the location of the graphene layer is about 10.6% of the maximum intensity.

In monolayer graphene, it is well known that the attenuation of vertically incident light is about 2.3%. The extended interaction length between waveguide mode and graphene could result in strong attenuation of light propagation. Because of graphene's absorption, the effective refractive index of the calculated guiding mode in Fig. 7.2b is a complex value, $n_{eff} = 2.341917 - 0.002116i$. The propagation loss of the hybrid waveguide structure could be estimated as ~0.074 dB μm^{-1} from the imaginary part of the refractive index by $4 \times 10^{-5}\pi \text{Im}(n_{eff})/\lambda \ln(10)$, where λ is the operation wavelength. Hence, an integrated graphene-waveguide structure with a length of 100 μm, the attenuation of light propagation is 7.4 dB, which is much higher than the 0.1 dB attenuation in the normal incidence configuration. With similar calculations, the propagation losses in graphene-waveguide structures with different structures could be analysed. In Fig. 7.2c, by changing the thickness of the top oxide layer, propagation losses of the hybrid waveguides are plotted. For the absence of the oxide layer, the graphene coupled with the waveguide mode directly, the attenuation rate could be high as ~0.091 dB μm^{-1}. However, if the spacing oxide layer is as thick as 100 nm, the propagation loss is reduced significantly to a value ~0.015 dB μm^{-1}. Notably, the thickness of the oxide layer could be controlled experimentally using a technique of chemical mechanical polishing, which consequently gives rise to desired light absorption in graphene.

The enhanced light–matter interaction between graphene and the evanescent field of the waveguide could also be modulated by shaping the waveguide's cross section. For instance, if the waveguide has a thinner geometry, according to guiding mode theory, a much stronger evanescent field would leak at the top surface of the waveguide to interact with graphene more effectively. In Fig. 7.2d, a plot of waveguide propagation losses versus different waveguide heights is shown. The waveguide width is maintained as 500 nm, and the spacing oxide layer is set as 10 nm thick. Much stronger light attenuation could be obtained for a waveguide with a thinner shape, which exceeds ~0.17 dB μm^{-1} for a height of 100 nm. Of course, if the waveguide is so thin that the optical field cannot be confined well, the intrinsic propagation loss in the silicon waveguide has to be considered, which may even be larger than that caused by graphene absorption. For thicker waveguides, the dependence of propagation loss on the waveguide height is not so strong though. On the other hand, by varying the width of the silicon waveguide with a constant waveguide height, the leaky evanescent field could also be modified to tune the coupling strength between graphene and the waveguide mode.

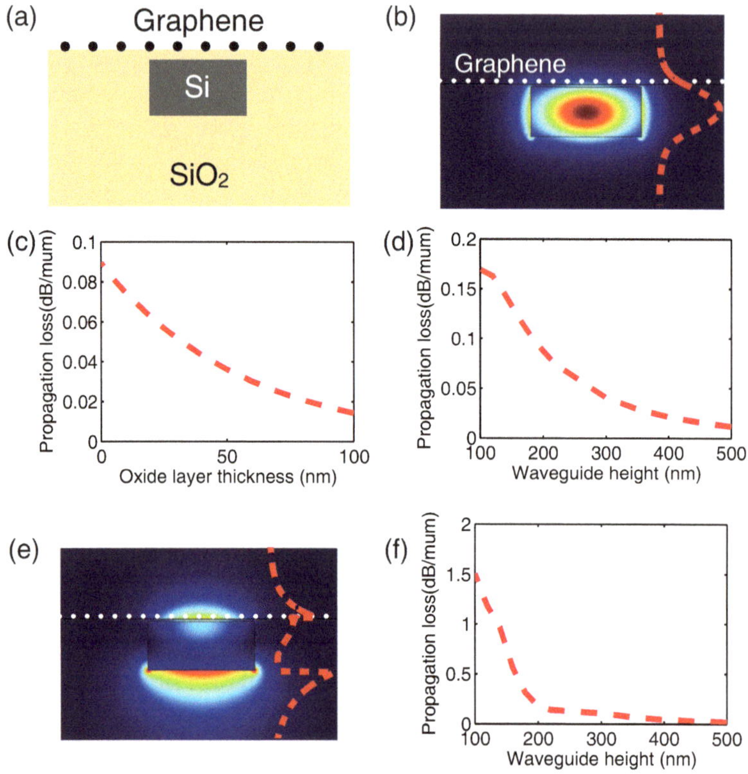

Figure 7.2 (a) Cross section of a hybrid graphene-waveguide structure. (b) Simulated electrical field distributions of the TE guiding mode in the hybrid device at a wavelength of 1550 nm, where the red dashed line represents the intensity distribution through the middle of the hybrid structure. (c) Calculated propagation losses as a function of the spacing oxide layer thicknesses for the TE guiding mode. (d) Propagation losses of the guiding modes in hybrid graphene-waveguide structures by changing the silicon waveguide's height but with a constant width of 500 nm. (e) Simulated electrical field distribution of the TM guiding mode at 1550 nm in the hybrid waveguide structure. (f) Propagation losses versus the waveguide heights for the TM guiding modes.

In photonic waveguides, a guiding mode with transverse magnetic (TM) polarisation is also allowed, which could be achieved by controlling the polarisation of the input light experimentally. By employing guiding modes with different polarisations, the light–matter interaction in graphene could be specified. Figure 7.2e displays a TM mode distribution at 1550 nm of the graphene-waveguide employed in Fig. 7.2b. One can find the electrical field in the TM mode has a more evanescent field leaking at the waveguide's top surface, which therefore would result in larger propagation loss. Figure 7.2f plots the transmission attenuation versus the waveguide heights, which is an implication of the coupling strength between the optical field and graphene. It reveals for the same waveguide cross section that the TM guiding mode could improve the graphene-induced propagation loss by almost 1 order of magnitude, especially for a waveguide thinner than 200 nm. Note that because of the moderately strong leaky evanescent field in the TM guiding mode, the intrinsic loss is high due to the surface scattering and material's non-linear absorptions. The waveguides for TM guiding mode operations are designed to have a height larger than the width to lower the intrinsic loss. In this case, the light interaction with the top graphene is inevitably weakened. An optimisation of waveguide structure is necessary to realise high-performance optoelectronic devices.

Experimentally, a straightforward way to determine the propagation loss or the absorption coefficient of a graphee-waveguide hybrid is the comparison of waveguide transmissions before and after the graphene integration [35, 36]. However, there are many influence factors to cause measurement uncertainties. For instance, waveguide-coupling efficiencies might be varied for each measurement due to the modifications of position misalignment or polarisation control. To avoid that, an accurate method is to utilise a Mach-Zehnder interferometer (MZI) circuit assembled by two waveguides [37], as schematically shown in Fig. 7.3a. The two arms of the interferometer are covered by a graphene layer with different lengths. Different from the conventional methods with direct measurements of input and output optical powers, the interferometric architecture determines the optical absorptions by measuring the extinction ratio (ER) of the interference fringes in the transmission spectrum. The transmission power P_0 of a light at wavelength A could be expressed by

$$P_0 = (P_{\text{in}} / 4)[e^{-\alpha l_1} + e^{-\alpha l_2} + 2e^{-\alpha(l_1+l_2)/2} \cos(2\pi n_{eff} \Delta L / \lambda)] \quad (7.1)$$

Here, α is the linear absorption coefficient in the graphene-waveguide hybrid. The graphene lengths over the two arms are l_1 and l_2, and ΔL is the path length difference between the two arms. The effective index of n_{eff} is the one of the waveguide without graphene's modification. Also, the intrinsic propagation losses in the bare waveguide are not considered, which are much smaller than graphene-induced attenuation. The input power P_{in} is assumed to be divided into the two arms with the same ratio. The ER of a transmission spectrum is given by

$$ER = \frac{P_{\max}}{P_{\min}} = \left(\frac{1+e^{\alpha\Delta l/2}}{1-e^{\alpha\Delta l/2}}\right) \approx \left(\frac{4}{\alpha\Delta l}\right)^2 \tag{7.2}$$

where $\Delta l = l_1 - l_2$. This approximation is valid when αl is small. Hence, by measuring the ERs from the transmission spectra, the absorption coefficient α in the graphene-waveguide structures could be estimated, and the absolute input and output optical powers are not necessary to be measured.

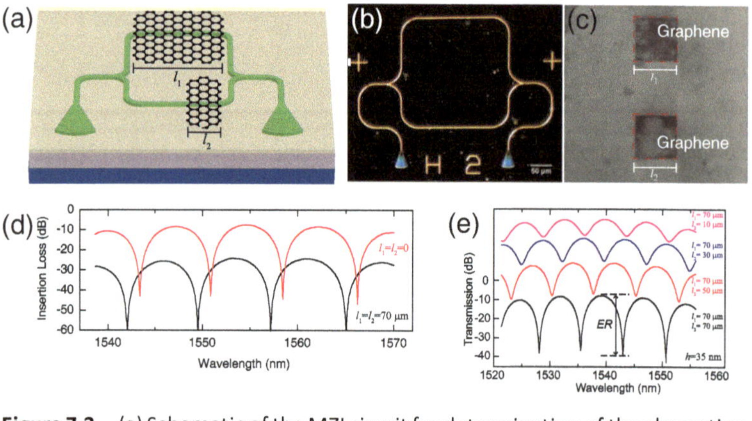

Figure 7.3 (a) Schematic of the MZI circuit for determination of the absorption coefficient in the graphene-waveguide structure. (b) Optical microscope image of the fabricated device with graphene layers covered on both arms of the MZI. (c) SEM image of the fabricated device, where graphene layers are indicated by the red dashed box. (d) Measured transmission spectra of the graphene-integrated and bare MZIs. (e) Transmission spectra of the graphene-integrated MZI, where the graphene layer coated on one arm is etched into short lengths gradually. Adapted from Ref. [37], with the permission of AIP Publishing.

An MZI structure was fabricated on a silicon-on-insulator wafer with the top silicon layer of 115 nm thickness. To obtain a planarised platform in the MZI, a negative electron beam resist hydrogen silsesquioxane (HSQ) was spin-coated onto the device with post-baking, which is a spacing layer between graphene and the waveguide as well. Its thickness could be modified by wet-etching with a buffered oxide etchant. Monolayer graphene grown by chemical vapour deposition (CVD) was exploited to integrate with the MZI. A polymethyl methacrylate (PMMA) polymer layer was used to assist the transfer, followed by selective lithography and oxygen plasma etching to pattern graphene's length. An optical microscope image of the fabricated device is displayed in Fig. 7.3b. The MZI has an arm difference of ΔL = 100 μm. Because the total thicknesses of the HSQ and silicon oxide layer is not specially designed to achieve high optical contrast for monolayer graphene, it is difficult to recognise them optically. Alternatively, using scanning electron microscopy, clear graphene areas could be observed due to the different electron conductivity between graphene and HSQ, as indicated in Fig. 7.3c.

To couple light into the waveguide, two grating couplers were designed at both ends of the MZI. The transmission spectra were obtained by tuning a narrow-band laser over a spectral range. Simultaneously, the transmission powers were recorded using a photo-detector. In the first measured device, the two arms were coated by graphene layers with the same lengths, that is, $l_1 = l_2$. Because of the negligible intrinsic loss of the bare silicon waveguide, the main attenuations are contributed by the graphene-waveguide hybrid sections. Hence, the same lengths of l_1 and l_2 will yield a high ER similar to that obtained from a device without graphene, as shown in Fig. 7.3d. Graphene on one arm was then etched gradually to have varied lengths, which leads to imbalanced graphene absorptions in the two arms. Figure 7.3e plots the interference fringes when one arm's graphene was etched to 50, 30 and 10 μm. The thickness of the HSQ spacing layer is about 35 nm. The ER decreases gradually with the increased Δl, which is consistent with the predicted results from Eq. 7.2. The absorption coefficient of the graphene-waveguide hybrid was estimated as 0.106 dB μm^{-1}. Devices with different HSQ thicknesses were characterised as well, showing similar conclusions about the ER [37]. A lower absorption coefficient was obtained for a thicker HSQ spacing layer. These experiments agree well with the analysis from the simulation results shown in Fig. 7.2.

7.3 Effective Light–Matter Interactions in Micro-Ring-Integrated 2D Materials

Micro-ring resonators on photonic chips could be considered as a photonic waveguide which is looped back on itself. When the optical path length of the micro-ring is exactly a whole number of wavelengths, a constructive interference would be reached and a resonance occurs. Micro-ring resonators have a traveling-wave resonant mode, which is different from the standing-wave mode in FP cavities or photonic crystal cavities. Normally, micro-ring resonators are side-coupled with a bus waveguide to access their resonant modes, which yields periodical resonant dips in the waveguide transmission spectrum. Micro-ring resonators have been widely employed for chip-integrated optical interconnects and sensing due to their compact size, high quality factors and compatibility with the integration of other passive and active devices. Similar as other planar photonic devices, optical fields confined in micro-rings have a strong evanescent field on the top surface, arising from the total internal reflection, providing a platform to enhance light–matter interactions in the integrated 2D materials. In the micro-ring 2D material-integrated system, the micro-ring's resonance condition would be modified by the interactions between the evanescent field and 2D materials; simultaneously, optical responses in the 2D materials could be controlled by the recirculating fields in the micro-ring, such as absorption, light emission and non-linear processes. In this section, only the micro-ring-enhanced linear optical absorption will be introduced. For the resonator-enhanced light emission and non-linear processes, the authors would like to discuss them in detail in the next section involving photonic crystal cavities. Compared to micro-ring resonators, photonic crystal cavities have a much stronger localised optical field to enable strongly enhanced light–matter interaction.

Micro-ring resonators have a long circumference in a scale of tens of micro-metres. It is possible to tailor the 2D materials' dimension to cover various fractions of the micro-ring to control their interaction with the optical field [38]. As shown in Fig. 7.4a, a waveguide-coupled micro-ring is partially covered by a graphene layer. Figure 7.4b displays the cross section of one of the integrated structures. Two graphene layers separated by a thin alumina layer are deposited on a 220-nm-thick silicon micro-ring. This structure promises the electrical gating of graphene's absorption for modulating the transmission spectra of the micro-ring. Surely, it could also be used to study the micro-ring-enhanced optical absorption in graphene.

In a bare micro-ring resonator, the optical field confinements are purely contributed by the total internal reflection, and the *Q*-factors are determined by the material absorption, sidewall roughness, bending radiation loss and coupling loss to the bus waveguides. After the integration of a graphene layer, its light absorption will lower the *Q*-factor by another loss channel. Pioneering work was performed by simulating the resonant modes of the structure shown in Fig. 7.4a,b. A preliminary analysis was conducted to characterise the propagation constant and mode profile of a waveguide with the same cross section as the graphene micro-ring. By setting the graphene layers to have different chemical doping levels or the absence of graphene, different effective refractive indices were obtained. Graphene's absorption coefficient as well as the refractive index are determined by the Fermi level according to a simple theoretical model based on both inter- and intra-band transitions [39]. By switching the chemical potential between 0.3 eV and 0.5 eV, the calculated sheet conductivities are $1.75 \times 10^5 Sm^{-1}$ and $3.58 \times 10^3 Sm^{-1}$, respectively. The corresponding dielectric constants are estimated as 10.44 and 11.91, respectively. The simulations indicate the effective refractive indices for the low- and high-doping levels are $2.2543 - 8.42 \times 10^{-5}i$ and $2.2546 - 3.79 \times 10^{-3}i$, which are different from the value of 2.2576 in the bare silicon micro-ring [38]. Because of graphene's extremely tiny thickness and relatively small conductivity, it only introduces an imaginary part in the propagation constant without altering the mode profile in a significant way. This conclusion is consistent for other photonic structures coupled with graphene via the evanescent wave. The varied effective refractive indices would give rise to resonant wavelength shifts and *Q*-factor variations.

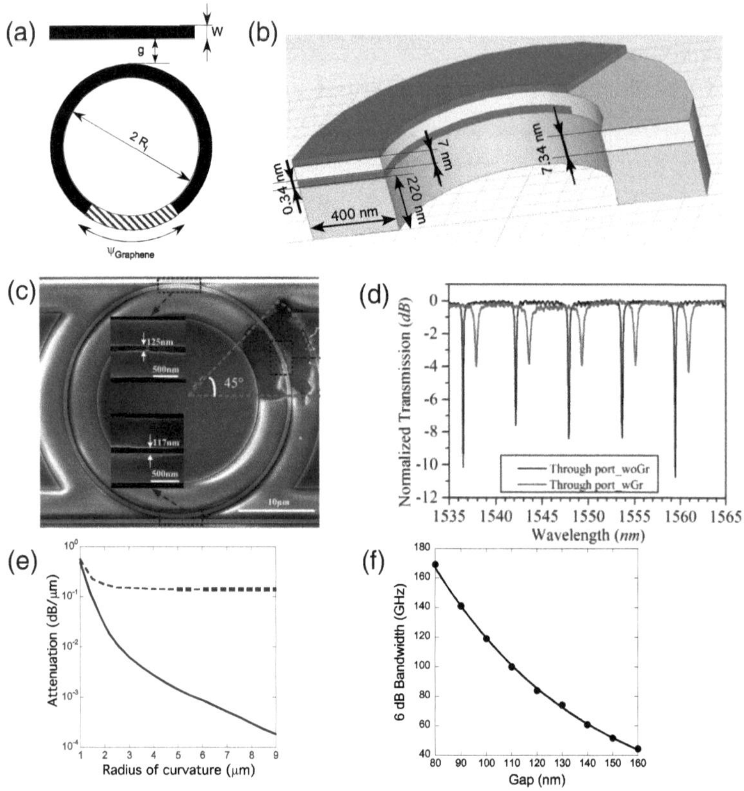

Figure 7.4 (a) Schematic of a graphene-integrated micro-ring, where a side-coupled bus waveguide is included and the coverage angles of the graphene layer could be controlled. (b) Cross section of the micro-ring integrated with two graphene layers. (c) SEM image of a graphene-coated silicon micro-ring, showing a coverage angle of 45°. (d) Transmission spectra of the micro-ring before and after graphene deposition. (e) Bending loss of a micro-ring in the absence and presence of two graphene layers. (f) A 6 dB bandwidth of the resonant dip versus the gap between ring and waveguide in a graphene-micro-ring device. Adapted with permission from Refs. [38] and [40]. Copyright (2012, 2016) Optical Society of America.

A similar device was fabricated by transferring one layer of CVD graphene onto a silicon micro-ring [40], as shown in Fig. 7.4c. The micro-ring was accessed via a through port and a drop port to study the transmission and reflection spectra of the integrated device. With a broadband amplified spontaneous emission (SE) light source and an optical spectrum analyser, the transmission spectra of the micro-ring before and after graphene integration were measured, as shown in Fig. 7.4d. Significant reduction of the ER and broadening of the resonance were observed because of the increased propagation loss induced by monolayer graphene. While the presence of graphene induces no detectable variations of the free spectral range of the multiple resonant modes, the resonant dips have obvious red shifts. These experimental results are consistent with these theoretical analyses. Extracted from the measured transmission spectra, the enhanced linear absorption coefficient of in-plane monolayer graphene was estimated as 0.23 dB/μm.

In this micro-ring-graphene hybrid system, the enhanced optical absorption in graphene could be understood as the variations of transmission in the side-coupled bus waveguide. As mentioned before, the losses or transmission variations of the micro-ring depend on its geometry as well, including bend radius and coupling gap between ring and waveguide. It is worth investigating the effect of graphene's absorption on the microring's transmission when different ring parameters are designed. Figure 7.4e illustrates the bending loss in the absence and presence of two graphene layers. In the absence of graphene, the bending loss tends to decrease to zero when the radius increases since the negligible scattering loss. However, after the integration of graphene, the conclusion is different. For a large radius, the loss approaches a limited value of 0.13 dB/μm due to the mode loss induced by graphene absorption. For a radius smaller than 2 μm, the main loss results from the bending loss. Figure 7.4f displays the line widths of the resonant peaks with respect to the gaps between ring and waveguide in a range of 80-160 nm. It reveals the Q-factors of the graphene-integrated micro-ring increase as the gap becomes larger, which is expected. The transmission variation, however, has no so obvious dependence on the coupling gaps, which is also determined by the graphene length. The extension of the graphene layer in the ring, as indicated in Fig. 7.4a, could be controlled to achieve critical coupling in micro-ring with a maximum ER.

The micro-ring-enhanced light–matter interaction in graphene was exploited to facilitate the non-linear parametric frequency conversions, which have been detailed demonstrated in Ref. [41]. In this work, a graphene monolayer was integrated onto a silicon micro-ring with a radius of 10 μm. A highly doped graphene layer was chosen to relieve its strong optical absorption. The Q-factor of the micro-ring decreases slightly after the integration of graphene, from 9100 to 9000. Two lasers at wavelengths of 1557 nm and 1567 nm are coupled into the hybrid structure as the pump and signal, which are on-resonance with the micro-ring's resonant modes. An idler of the FWM process around 1547 nm was obtained. The conversion efficiency was estimated as −37 dB. A control measurement of the FWM is repeated in a bare silicon micro-ring by setting the powers of the pump and signal to be equivalent to those in the graphene-coated micro-ring. The control experiment indicated an enhancement factor of 6.8 dB, thanks to graphene's strong non-linearity.

Evanescent field coupling between a MoS_2 monolayer and a silicon nitride micro-ring was studied as well [42]. Monolayer MoS_2 has an optical bandgap in the visible range around 650 nm. There is no optical absorption or emission at the telecom band, which hinders the disclosure of the enhanced optical response in silicon-based micro-ring resonators. Silicon nitride is a wide-bandgap semiconductor transparent to both visible and infrared wavelengths. Photonic devices in silicon nitride could be fabricated with the standard CMOS processing as well. To expose the strongly enhanced optical responses in MoS_2 by micro-rings, resonant modes in the visible range should be chosen. A device was fabricated by transferring a mechanically exfoliated monolayer MoS_2 onto a silicon nitride micro-ring, as shown in Fig. 7.5a. A thin layer of silicon dioxide was deposited on the silicon nitride micro-ring to assist the deposition of MoS_2. A MoS_2 monolayer was coated over the micro-ring with a distributed azimuthal angle of 45°. Of course, monolayer MoS_2 with different coverage angles could be integrated on a micro-ring to perform a series of control experiments.

The MoS_2 micro-ring device was characterised first using its photo-luminescence (PL). Figure 7.5b displays a spatially resolved PL image with a 532 nm excitation laser. Strong PL intensity was observed from the suspended MoS_2 area as a result of the high quantum yield. On the micro-ring part where there is no MoS_2, PL signals were observed as well. This non-local PL signal indicated the excitation laser and PL emission couple into the ring's resonant modes.

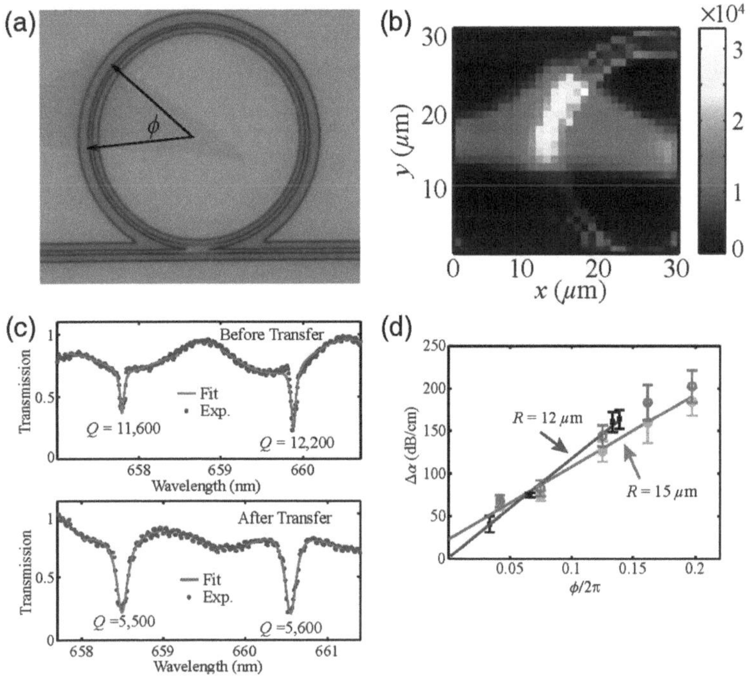

Figure 7.5 (a) Optical microscope image of a silicon nitride micro-ring integrated with *a* MoS_2 monolayer. (b) Spatially resolved PL image of the integrated device, where the spectra were integrated over a range between 600 nm and 720 nm. (c) Transmission spectra of the ring resonator before and after the deposition of MoS_2. The Q-factors were extracted from Lorentzian fittings. (d) MoS_2-induced losses for different coverage angles of different radii of the ring. Reprinted from Ref. [42], with the permission of AIP Publishing.

Then transmission spectrum of the MoS_2 micro-ring was investigated by tuning a laser in the range of 650-660 nm, which overlaps with the exciton absorption peak of monolayer MoS_2. It is revealed that monolayer MoS_2 has an absorbance around 5%-10% at the exciton resonance, which is much higher than monolayer graphene. The monolayer MoS_2 would modify the ring's Q-factors significantly. One of the representative results was obtained from a device with a 15° distributed angle of coverage. Figure 7.5c shows the transmission spectra before and after MoS_2 deposition. Two resonant modes with different Q-factors are included in this spectral range. The monolayer MoS_2 alters the mode effective index of the micro-ring via its real and imaginary dielectric functions. In the transmission spectra, resonant wavelength shifts and Q-factor reductions were observed in both resonant modes. The reduction of the intrinsic Q-factor could be used to extract the change in absorption per length α_{MoS2} of the ring caused by the monolayer MoS_2. The relation between the Q-factor and the total loss in the micro-ring is determined by $Q = \frac{10}{\ln(10)} \frac{2\pi n_g}{\lambda_0 \alpha}$. Here, n_g is the effective refractive index of the guided mode, and λ_0 is the resonant wavelength. Therefore, by calculating the variation of Q-factors, the loss due to graphene's absorption could be estimated. Figure 7.5d presents the coefficient α_{MoS2} with different coverage angles for different radii of rings. A weighted linear fit was performed to each radius, resulting in $\alpha_{MoS2} \sim 0.085$ dB μm^{-1} ($\sim$0.1157 dB μm^{-1}) for resonators with a radius of 15 μm (12 μm). These values are comparable with the absorption coefficient in graphene-waveguide structures.

7.4 Photonic Crystal Cavity-Enhanced Optical Responses in 2D Materials

Planar photonic crystals (PPCs) are formed by etching periodic air-holes in a dielectric slab with a lattice constant of sub-microns. When light propagates through the PPC slab, it would suffer from a Bragg reflection induced by the periodic refractive index perturbation, which results in a photonic forbidden band for a certain wavelength range [43]. When defects are introduced by missing, shifting, or shaping air-holes, the periodicity of the PPC is broken and a PPC cavity is formed. If light with a wavelength located in the photonic forbidden band is coupled into the defect, it will be confined in-plane by the periodic air-holes around with nearly 100% reflectance. At the same time, the light would be confined vertically by the total internal reflection. This confinement of light in three dimensions gives rise to resonant modes with high quality (Q) factors, which are typically in the order of 10^4. Normally, the defects in PPC cavities only scale over several air-holes, that is, the resonant modes are confined in cubic wavelength volumes ($V_{mode} \sim \lambda^3$). Hence, the electrical fields of resonant modes in PPC cavities are extremely high, which are proportional to the ratio of Q/V_{mode}. PPC cavities have been widely used to implement cavity quantum electrodynamics, ultralow threshold laser and high-efficiency harmonic generations.

In optical interconnects, PPC cavities are important elements to achieve optical filters, optical buffers and memories.

According to guiding wave theory, the evanescent field of the PPC mode decays in an exponential function out of the slab. In the integrated 2D material-PPC cavity, 2D materials have only a few nanometres and therefore could couple with the cavity mode strongly. A number of spectroscopy studies have been implemented on 2D material-PPC cavities and demonstrated strongly enhanced light–matter interactions in 2D materials over the optical absorption, Raman scattering, light emission and non-linear processes [44-46].

7.4.1 Enhancing Graphene's Absorption Using a Photonic Crystal Cavity

Figure 7.6a displays an optical microscope image of a fabricated graphene-PPC cavity hybrid. Here, the graphene flake is mechanically exfoliated and wetting transferred onto the PPC cavity with the assistance of a PMMA polymer layer. The PPC cavity is fabricated in a gallium phosphide (GaP) membrane, which could operate in the visible spectral range for studying the enhanced light emission and Raman scattering in 2D materials. The PPC cavity is designed with a defect of linearly missed three holes (L3). The compact sizes of PPC cavities allow a graphene flake to cover several of them simultaneously. Because of graphene's strong stiffness, the transfer process would not destroy its flat structure, as illuminated by the atomic force microscope (AFM) image shown in Fig. 7.6b. The graphene layer suspends over the air-holes without any breakage, which could maintain graphene's electrical properties as well in the optoelectronic devices.

To study the coupling between 2D materials and PPC cavity mode, a vertically cross-polarised microscope setup is employed to monitor the reflective behaviours of the cavity modes before and after the integration of graphene [44], as schematically shown in Fig. 7.6c. A polarised beam splitter is placed before the objective lens to control the excitation source and the reflection signal to have perpendicular polarisations. A half-wave-plate is inserted between the polarised beam splitter and objective lens to match the polarisation direction of the incident laser and the sample orientation. Normally, the objective lens is 50× or 100× to focus the laser into a spot size smaller than 2 μm. The incident laser is a super-continuum laser with a visible to near-infrared spectral range. The reflection signal from the PPC cavity is collected and analysed by a spectrometer with a cooled charge-coupled device (CCD) camera. Note that in an optical interconnects chip, PPC cavities are fabricated to couple with a side PPC waveguide, enabling a high coupling efficiency around 90%. Here, the vertically coupling regime of the cross-polarisation setup can only provide a coupling efficiency lower than 10%. However, it supports the measurements over a device only with a size of several micro-metres, facilitating the device fabrications and straightforward measurements. Also, as it will be demonstrated later that the vertical scattering of PPC cavity could be considered as a coupling waveguide, which simplifies the analysis of enhanced light–matter interaction in the reflection regime.

The measured reflection spectra of the device shown in Fig. 7.6a are displayed in Fig. 7.6d. Before the integration of graphene, a resonant mode at a wavelength of 1477.3 nm is obtained with a narrow and high resonant peak, as indicated by the red dotted line. Integrated with a graphene monolayer, the resonant peak red-shifts to 1479.1 nm with a broadened line width, and the reflection intensity is quite weak. By fitting the resonant peaks with Lorentzian functions, the Q-factors of the PPC cavity are 2640 and 360 before and after graphene integration, respectively. An attenuation of the cavity reflection is calculated as well by $-10\log_{10}(R_g/R_0)$, as shown in the inset of Fig. 7.6d. Here R_g and R_0 are reflection spectra of the PPC cavity with and without graphene. A remarkable attenuation of 20 dB is obtained at a resonant wavelength of 1477.3 nm. Hence, the single layer of carbon atom could make the cavity nearly opaque, which is a strong contrast to the 0.1 dB attenuation achieved over a single-passed light illumination. A 1.8 nm red-shifted resonant wavelength also indicates strong coupling between graphene and the resonant mode.

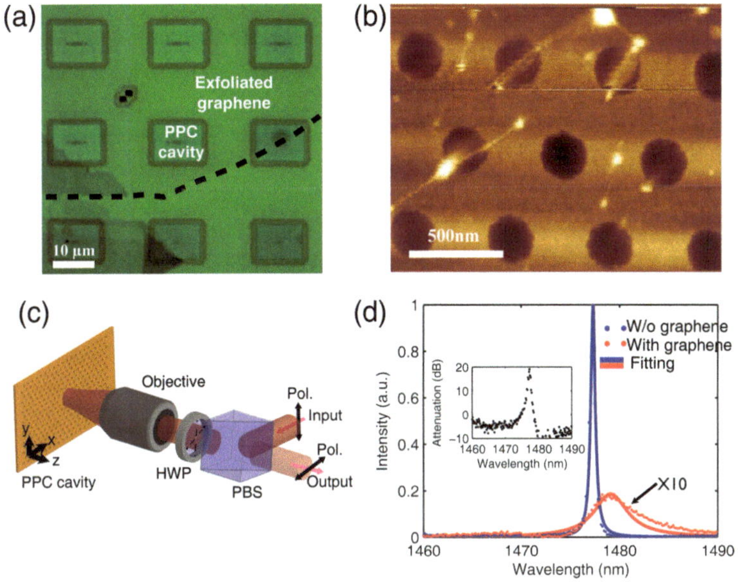

Figure 7.6 (a) Optical microscope image of PPC cavities coated with monolayer graphene, where the graphene boundary is indicated by the dashed line. (b) AFM image of the graphene-PPC cavity hybrid. (c) Schematic of the cross-polarisation microscope setup. (d) Measured reflection spectra of the PPC cavity before and after graphene integration. Inset is the calculated attenuation spectrum of the cavity reflections. Adapted with permission from Ref. [44]. Copyright (2012) American Chemical Society.

To explain these experimental observations, a theory model could be proposed, as schematically shown in Fig. 7.7a. An optical cavity integrated with a graphene layer couples to a tunnelling waveguide to permit light to couple into and out of the cavity. The waveguide has forward- and backward-propagating modes, mode 1 and mode 2, and the coupling rates with the cavity resonance are κ_{ca} and κ_{cb}, respectively. The incoming (outgoing) light of mode 1 and mode 2 is denoted by $a_{\text{in}}(a_{\text{out}})$ and $b_{\text{in}}(b_{\text{out}})$, respectively. The bare cavity has a resonant frequency ω_0 and an energy decay rate κ_c' due to other lossy modes and material absorption. The total energy decay rate $\kappa_c = \kappa_{ca} + \kappa_{cb} + \kappa_c'$ gives the quality (Q) factor of the intrinsic cavity as $Q = \omega_0/(\kappa_c' + \kappa_{ca} + \kappa_{cb})$. When perturbed by the presence of a graphene sheet, the cavity mode undergoes a shift of $\Delta\omega$ in the resonant frequency and experiences an energy loss induced by graphene with a decay rate of κ_{cg}.

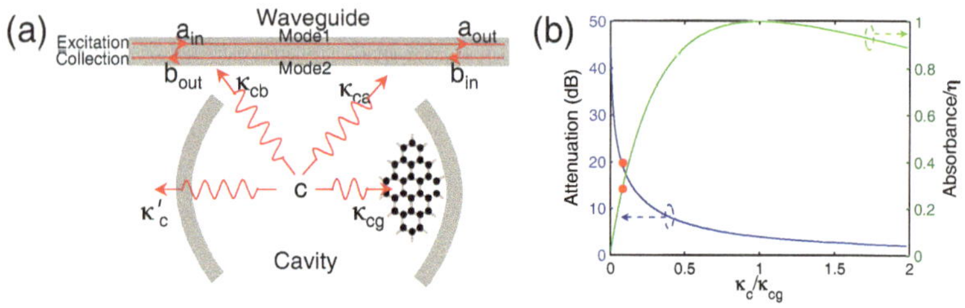

Figure 7.7 (a) Schematic of the coupling model in hybrid graphene cavity, where the graphene-integrated cavity is coupled with a tunnelling waveguide. (b) Calculated cavity's reflection attenuation and graphene's absorbance with respect to the coupling strengthen between graphene and resonant mode. Adapted with permission from Ref. [44]. Copyright (2012) American Chemical Society.

For weak coupling between the cavity and waveguides, the evolution of resonant mode in the graphenecavity system is governed by the temporal coupled-mode equations[?]

$$\frac{\mathrm{d}c}{\mathrm{d}t} = -(i(\omega_0 + \Delta\omega - \omega) + \kappa_{ca}/2 + \kappa_{cb}/2 + \kappa_{c'}/2 + \kappa_{cg}/2) + \sqrt{\kappa_{ca}}a_{in} + \sqrt{\kappa_{cb}}b_{in} \tag{7.3}$$

The outgoing fields of the waveguides a_{out} and b_{out} are related to the incoming field by

$$a_{out} = -a_{in} + \sqrt{\kappa_{ca}}c \tag{7.4}$$

$$b_{out} = -b_{in} + \sqrt{\kappa_{cb}}c \tag{7.5}$$

To be consistent with these experimental implementations, only mode 1 has incoming field, that is, $a_{in} \neq 0, b_{in} = 0$. Also, for the reflection measurements with a crossed polariser, the excitation mode (mode 1) a_{in} and collected mode (mode 2) b_{out} are approximately Gaussian spatial modes, which are defined by the objective lens. Considering the symmetric confinement of the resonant mode, its coupling coefficient with the Gaussian modes of a_{in} and b_{out} should be equal. By assuming a coupling efficiency η between the microscope modes and the cavity radiation field, the cavity mode couples with the microscope modes with rates of $\kappa_{ca} = \kappa_{cb} = \eta\kappa_c$.

Solving for c from Eq. 7.3 in steady state and substituting the result into Eqs. 7.4 and 7.5, one can obtain the system reflection $R(\omega)$ and transmission $T(\omega)$ coefficients at frequency ω

$$R(\omega) = \left|\frac{b_{\text{out}}}{a_{\text{in}}}\right|^2 = \frac{\eta^2\kappa_c^2}{(\omega_0 + \Delta\omega - \omega)^2 + (\kappa_c/2 + \kappa_{cg}/2)^2} \quad (7.6)$$

$$T(\omega) = \left|\frac{a_{\text{out}}}{a_{\text{in}}}\right|^2 = \frac{(\omega_0 + \Delta\omega - \omega)^2 + (\kappa_{cg}/2 + \kappa_c'/2)^2}{(\omega_0 + \Delta\omega - \omega)^2 + (\kappa_c/2 + \kappa_{cg}/2)^2} \quad (7.7)$$

The loss of the cavity system (1 —T— R) arises from the intrinsic cavity loss and graphene absorption, and the graphene absorption $A(\omega)$ could be extracted as

$$A(\omega) = \frac{\eta\kappa_c\kappa_{cg}}{(\omega_0 + \Delta\omega - \omega)^2 + (\kappa_c/2 + \kappa_{cg}/2)^2} \quad (7.8)$$

For the bare PPC cavity without graphene, the same expressions apply with the changes of $\kappa_{cg} = 0$ and $\Delta\omega = 0$.

By fitting the experimentally obtained reflection spectra to Eq. 7.6, one can obtain the characteristic parameters of $\omega_0 = 1.28 \times 10^3$ THz, $\kappa_c = 1.9 \times 10^{-4}\omega_0$, $\Delta\omega = -1.24 \times 10^{-3}\omega_0$, and $\kappa_{cg} = 2.4 \times 10^{-3}\omega_0$. The resonance frequency shift $\Delta\omega$ and additional energy decay rate κ_{cg} are comparable in magnitude and are much higher than κ_c. It shows that monolayer graphene dominates the photon loss in the PPC cavity with a factor of $\kappa_{cg}/(\kappa_{cg} + \kappa_c) \approx 92\%$, which is in contrast to the 2.3% absorption of the single-passed light.

From this theoretical model, one can deduce more indications about the modulation of cavity reflection and cavity-enhanced graphene absorption as a function of the intrinsic cavity loss rate κ_c, the waveguide coupling efficiency η and the graphene loss rate κ_{cg}. From Eq. 7.6, it is easy to deduce the effect of a graphene layer on the on-resonance reflection of the cavity. We find a ratio for the reflections with and without graphene of $\left(1 + \frac{\kappa_{cg}}{\kappa_c}\right)^2$. The ratio shows the reflection of graphene-PPC cavity system can be dramatically modulated by altering the parameter $\frac{\kappa_{cg}}{\kappa_c}$. The blue line in Fig. 7.7b plots the predicted value for the relative cavity attenuation induced by graphene, $-10\text{Log}_{10}(R_g/R_0)$, as a function of κ_c/κ_{cg}, which is the ratio of the loss rate of the unloaded cavity to that with graphene. For a cavity with low κ_c or high Q, the attenuation from a single layer of graphene can be as high as 40 dB (at the resonance frequency of the bare PPC cavity). A modulation of this magnitude should be possible with existing PPC cavities ($Q \sim 10^6$). Even for a low-quality PPC cavity with $\kappa_c \approx \kappa_{cg}$ the reflection still drops by about 5 dB.

Unlike the modulation of PPC cavity reflection, the cavity-enhanced on-resonance absorption of graphene $\frac{\eta\kappa_c\kappa_{cg}}{(\kappa_c/2+\kappa_{cg}/2)^2}$ has a maximum value of η, which is reached by optimising the system decay rates to satisfy the condition $\kappa_{cg} = \kappa_c$, as shown in the green line curve of Fig. 7.7b. This maximum absorption condition could be understood as follows: if κ_c κ_{cg}, photons decay via the cavity losses before being absorbed by graphene; alternatively, if $\kappa_c < \kappa_{cg}$, photons cannot be efficiently coupled into the cavity in the first place. Therefore, matching the graphene-induced loss to the intrinsic cavity loss yields maximum absorption in the graphene. For efficient absorption, the above equations show a linear dependence on η. This coupling efficiency can exceed 45% using the side-coupled cavity from a waveguide [47], or tapered fibre coupling [48]. By employing efficient coupling and choosing $\kappa_c \sim \kappa_{cg}$, the model indicates that it is possible to reach 45% absorption into a monolayer graphene, which can be improved further by placing a reflector under the cavity. The maximum absorption in graphene can reach 50% if the waveguide-cavity coupling is responsible for the entire intrinsic loss of the cavity, that is, $\kappa_{ca} + \kappa_{cb} = \kappa_c$. However, in cavities possessing traveling-wave resonant mode, such as ring resonators, there is no backward coupling, which enables maximum absorption of 100% in graphene. This coupled graphene-PPC cavity model could be applied to other 2D material-resonator systems to study the enhancement of interaction between electromagnetic field and 2D materials.

These experimental and theoretical results indicate monolayer graphene could control a PPC cavity's resonance effectively via its refractive index and absorption. Simultaneously, the PPC cavity's localised resonant mode would enhance graphene's absorption strongly. These two abilities in the integrated graphene-PPC cavity system promise the improvement of graphenebased optoelectronic devices, such as modulators and photo-detectors [49–51].

7.4.2 Enhanced Raman Scattering in Graphene by a Photonic Crystal Cavity

The PPC cavity-enhanced absorption in graphene indicates strong interaction between the electrical field of the resonant mode and the grahene layer. The enhanced optical responses in graphene could be validated from other optical processes, for example, Raman scattering, as displayed in Fig. 7.8. A GaP-PPC cavity with a shorter resonant wavelength is employed to integrate with a CVD-grown graphene layer. A scanning electron microscope (SEM) image is shown in Fig. 7.8a. Before and after graphene's deposition, reflection spectra of the PPC cavity are measured, as shown in Fig. 7.8b. With graphene, the Q-factor of the cavity drops from 860 to 330 and the resonant wavelength is red-shifted from 635.7 nm to 636.3 nm. Calculated from Eq. 7.6, the experimental data indicate the parameters of $\Delta\omega = -0.9 \times 10^{-3}\omega_0$ and $\kappa_{cg} = 1.9 \times 10^{-3}\omega_0$. Note that the perturbation induced on this device by graphene is weaker than that on the device shown in Fig. 7.6. This difference can be attributed to the shorter resonant wavelength of the current device, yielding a weaker evanescent field coupling with graphene.

The Raman spectra from the graphene-PPC cavity hybrid are measured on the cross-polarisation microscope setup as well. Differently, the excitation source is changed into a narrow-band laser with a tunable wavelength range of 635 to 639 nm, covering the resonant wavelength of the graphene-PPC cavity system. The Raman signal scattered from the device is collected by the objective lens and directed to the spectrometer via the polarised beam splitter. Before the reflection signal couples into the spectrometer, a long-pass filter is employed to get rid of the reflected excitation laser. Because of the cross-polarisation configuration, the measured Raman signal is the perpendicular component to the polarisation direction of the excitation laser. Figure 7.8c displays a measured Raman spectrum of the device excited at 637 nm. Two prominent features at 1578 cm^{-1} and 2640 cm^{-1} are observed, corresponding to the G and 2D bands, respectively. Note that the 2D peak is weaker than the G peak due to the high doping level of the CVD graphene during the wetting transfer process.

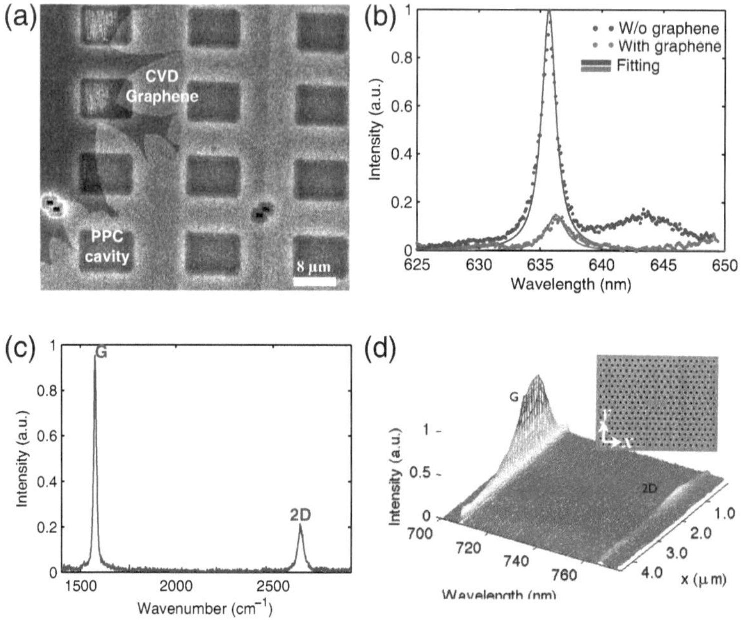

Figure 7.8 (a) SEM image of a device with CVD graphene integrated onto PPC cavities. (b) Measured reflection spectra of the PPC cavity before and after CVD graphene deposition. (c) Raman spectrum from the CVD graphene layer excited by a 637 nm laser, showing two peaks at 1578 cm^{-1} and 2640 cm^{-1} for G and 2D peaks, respectively. The weaker 2D peak compared to the G peak is caused by chemical doping. (d) Raman spectra of graphene when the hybrid device is spatially scanned along the *x* axis. Adapted with permission from Ref. [44]. Copyright (2012) American Chemical Society.

To examine the cavity-enhanced Raman scattering from the integrated graphene-PPC cavity hybrid, Raman spectra are acquired continuously when the sample is moved along the *x* axis by a step of 100 nm, as indicated in Fig. 7.8d. It reveals when the laser spot focuses on the defect region of the PPC cavity, where the resonant mode localises, both G and 2D Raman peaks are much stronger. Since the defect region of the PPC cavity has a flat substrate without air-holes, the reflectance of optical signal should be higher at this region, which may contribute to stronger Raman scattering. To eliminate this uncertainty factor, the excitation laser wavelength was then tuned across the resonant wavelength of the graphene-PPC cavity hybrid and the Raman spectra are acquired continuously, as shown in Fig. 7.9a. Both of the G and 2D peaks have the same pump wavelength dependence, presenting high intensities around the resonant wavelength. By extracting the peak intensities at different pump wavelengths, the Raman peaks show Lorentzian curves centred at 636.3 nm with a *Q*-factor of ~330, as shown in the inset of Fig. 7.9a. These two Lorentzian lines are consistent with the cavity resonance shown in Fig. 7.8b. In the two curves, the flat baseline arises from incident laser exciting on the graphene directly, not through the cavity mode. Normalising the peak Raman signals to the baseline, a 2.8- and a 3.3-fold increase of the G and 2D bands are obtained, respectively.

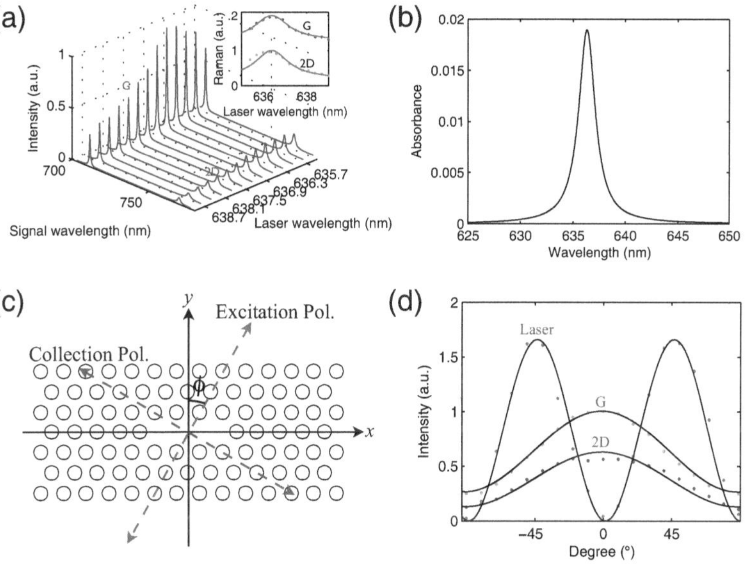

Figure 7.9 (a) Pump wavelength dependence of the G and 2D Raman peaks. Inset plots the Lorentzian fittings for the peak intensities of G and 2D peaks. (b) Enhancement spectrum of graphene's absorption by the cavity mode. (c) Schematic of the excitation and collection for graphene Raman scattering, which have perpendicular polarisation directions. (d) Polarisation dependence of cavity-enhanced Raman signal, showing 90° periodicity of the laser reflection and 180° periodicity of the Raman peaks. Adapted with permission from Ref. [44]. Copyright (2012) American Chemical Society.

As mentioned at the beginning of this subsection, this wavelength-dependent Raman scattering matches the expected cavity field enhancement, which also increases the optical absorption. To verify that, the enhanced absorption in graphene could be derived from the experimentally measured cavity spectra (shown in Fig. 7.8b) as well as Eq. 7.8. By assuming the coupling efficiency $\eta \approx 4\%$ between the cavity's far-field radiation and the Gaussian mode of the objective lens, the light absorbed by the graphene layer could be estimated, which has a maximum absorbance of 1.88% for the on-resonance light. On the other hand, for a graphene monolayer on a flat GaP substrate, the light absorbance could be estimated as follows: In the visible-to-infrared spectral range, graphene suspended in air has a frequency-independent absorbance of 2.3%. In the spectral range of the experiments, the GaP material is transparent and has a refractive index around 3.1. The absorbance A_0 of the graphene sheet on semi-infinite bulk GaP without a photonic crystal can be calculated from fractional changes of reflectance and transmission obtained by solving Maxwell's equations

$$1-\frac{T_g}{T_s}=\frac{2}{n_s+1}\pi\alpha \tag{7.9}$$

$$\frac{R_g}{R_s}-1=\frac{4}{n_s^2-1}\pi\alpha \tag{7.10}$$

$$T_s + R_s = 1 \tag{7.11}$$

$$T_g + R_g + A_0 = 1 \tag{7.12}$$

where T_s (R_s) and T_g (R_g) are the transmission (reflection) coefficients of the GaP substrate and graphene on the substrate, respectively, and n_s is the refractive index of GaP. The reflectance from the GaP substrate satisfies $R_s=\left(\frac{n_s-1}{n_s+1}\right)^2$. From these equations, we obtained $A_0 \approx 0.238\pi\alpha \approx 0.54\%$. Therefore, the light absorption in a graphene monolayer enhanced by the PPC cavity is about 3.41 times of the absorption by a graphene monolayer on a GaP substrate. With this enhancement factor, one can plot the absorption spectrum of a graphene monolayer on the PPC cavity, as shown in Fig. 7.9b. The field enhancement which leads to this increase in absorption for the graphene also gives rise to the enhanced Raman scattering efficiency, which is consistent with the results shown in Fig. 7.9a.

This cavity-enhanced Raman scattering arises from the effective interactions between the cavity's near field and graphene monolayer. In the cross-polarisation setup, there is a half-wave-plate which could be used to control the far-field coupling efficiencies of the excitation laser and collection signal. Therefore, by rotating the half-wave-plate, the polarisation dependence of graphene's Raman scattering could be extracted, and the cavity enhancement could be confirmed further. In the measurement, if the half-wave-plate makes an angle of $\phi/2$ with respect to the y axis defined in Fig. 7.8d, then the excitation and detection axes will be rotated by an angle of ϕ, as shown in Fig. 7.9c. For the fundamental mode of the employed PPC cavity with three missing air-holes, only the y component of incoming or outgoing electrical fields can couple to the resonant mode. Considering the appropriate projection for the incoming and outgoing fields for the reflection of the laser excitation, we expect an angular variation of $\cos^2\phi\sin^2\phi$. As can be seen in Fig. 7.9d, this prediction agrees very well with the experiment. The cavity field used to enhance the graphene Raman scattering is proportional to the input coupling, which has a $\cos^2\phi$ dependence. The cavity field excites graphene's Raman signal through its near field, with normalised mode field components polarised along x and y directions, E_x and E_y. The Raman scattering has parallel and perpendicular polarisation dependence factors with respect to the excited field, which are assumed as $f_{\parallel}$ and $f_{\perp}$ Considering the polarisation dependence of the output coupling, the collected Raman signal excited by E_x is proportional to $\cos^2\phi|E_x|^2(f_{\parallel}\cos^2\phi + f_{\perp}\sin^2\phi)$, and that excited by E_y is proportional to $\cos^2\phi|E_y|^2(f_{\parallel}\sin^2\phi + f_{\perp}\cos^2\phi)$. The experimental results for the polarisation dependences of G and 2D Raman bands are fitted to $0.76\cos^2\phi + 0.25$ and $0.45\cos^2\phi(\cos^2\phi + 1.6\sin^2\phi) + 0.12$, as shown in Fig. 7.9d. The fitting results indicate the ratios of $\frac{|E_y|^2 f_{\parallel} + |E_x|^2 f_{\perp}|}{|E_y|^2 f_{\perp} + |E_x|^2 f_{\parallel}|} = 1$ = 1 and 1.6 for the G and 2D bands, respectively. From the simulation of the cavity mode, $\int \frac{|E_y|^2}{|E_x|^2}\,\mathrm{d}x\mathrm{d}y = 1.92$ in the plane of the graphene location could be obtained. Combining with this near-field distribution, the polarisation dependence factors are estimated to have ratios of $f_{\parallel}/f_{\perp}$ = 1 and 6.47 for the G and 2D bands, respectively. These results indicate that the G band has unpolarised scattering, while the 2D band exhibits strong polarisation dependence. The cavity-enhanced Raman scattering could also serve as a precise method for fundamental studies of 2D materials. For instance, high-precision Raman scattering could assist in optical studies of grain boundaries and edge states.

7.4.3 Controlling Spontaneous Emission of 2D Materials Using a Photonic Crystal Cavity

According to the Wigner-Weisskopf approximation, the radiation decay rate of an emitter is described by Fermi's 'golden rule' which contains the local density of electromagnetic states. Thence, the characteristics of spontaneously emitted light depend strongly on the dielectric environment of the emitter. Both the total emission rate and the direction-dependent spectra would be affected by the dielectric environment. The last two sub-sections have demonstrated that the localised resonant mode of a PPC cavity could couple with the integrated 2D materials strongly. This enhanced local density of the photon states in the cavity mode could also be exploited to effectively control the emission behaviour in 2D materials.

Monolayer TMDs, that is, MX_2 (M = Mo, W; X = S, Se, Te), exhibit PL quantum yields which are enhanced by a factor more than 10^4 compared to the bulk crystal [52, 53], which could be attributed to the transition from the indirect bandgap to the direct bandgap when the layer number decreases to one. However, the PL efficiency of monolayer MoS_2 is still very low at $\sim 10^{-2}$ because the non-radiative recombination rate $1/\tau_{nr}$ far exceeds the spontaneous emission (SE) rate $1/\tau_r$ [52]. It is reported that for a MoS_2 monolayer on SiO_2 substrate, τ_{nr} and τ_r are estimated as ~100 ps and ~10 ns, respectively, at room temperature [52, 54]. Though monolayer WSe_2 is reported to have a much higher PL quantum yield, its excitonic PL is only 40 times higher than that in monolayer MoS_2. As discussed in the last paragraph, by exploiting the strong Purcell effect in optical cavities with cubic wavelength-scale mode volume and high Q-factor, it is possible to achieve significantly improved PL efficiency of monolayer MX_2.

Pioneering work was conducted over a MoS_2 monolayer integrated with a PPC cavity [45], which achieved an SE rate enhancement factor of nearly 70. The MoS_2 monolayer was mechanically exfoliated from the bulk material onto a PMMA polymer film, as shown in Fig. 7.10a. By employing the wet transfer process, the MoS_2 monolayer was placed onto a GaP-PPC cavity precisely. Figure 7.10b displays an optical microscope image of the finished device. While the MoS_2 monolayer has no optical contrast on the GaP substrate, its position could be recognised by the side thicker film. The employed GaP-PPC cavity has a defect of three linearly missed air-holes, as indicated by the SEM image in Fig. 7.10c. Around the air-suspended GaP-PPC cavity, there are trenches with a width about 300 nm. The PPC cavity was fabricated in a 138-nm-thick GaP membrane with lattice spacing a = 165 nm and an air-hole radius r = 0.3a, yielding resonant modes in a wavelength range of 600–700 nm to overlap the PL spectrum of the MoS_2 monolayer.

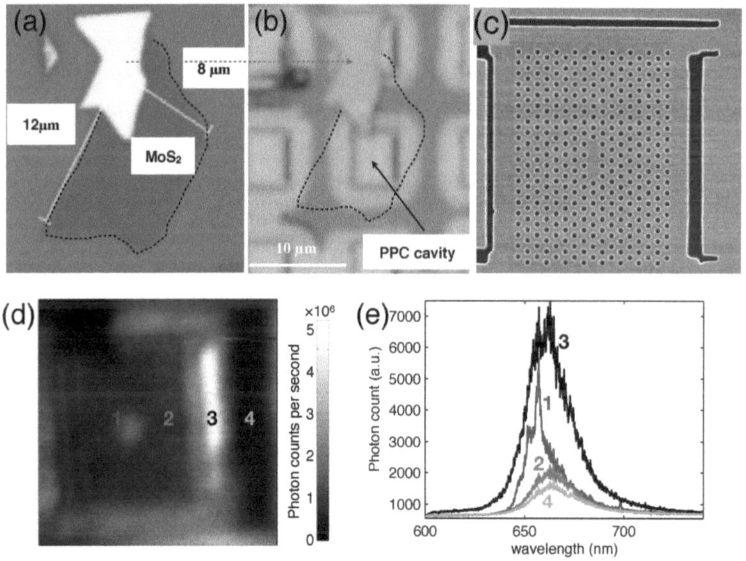

Figure 7.10 (a) Optical microscope image of the mechanically exfoliated MoS_2 monolayer. (b) Optical microscope image of an integrated monolayer MoS_2-PPC cavity, where the MoS_2 flake is correlated with the image shown in (a). (c) SEM image of the employed PPC cavity. (d) PL spatial mapping image of the monolayer MoS_2-PP cavity. (e) PL spectra from different regions, which are marked in (d). Adapted from Ref. [45], with the permission of AIP Publishing.

The monolayer MoS_2-PPC cavity was then characterised on a micro-PL confocal microscope with a 532 nm CW excitation laser, which was focused to a beam diameter of ~400 nm and with a power of ~50 μW. The vertically scattering PL signal from the sample was collected by the objective lens and filtered from the excitation laser using a dichroic mirror. The PL signal was analysed by a spectrometer or recorded by an avalanched photo-diode. To examine the modification on MoS_2's PL by the PPC cavity, the device was spatially scanned in a step of 50 nm on a piezo-actuated stage. The measured PL photon flux by the photo-diode was plotted into a spatially resolved image, as shown in Fig. 7.10d. This spatial mapping PL image is acquired in the same region as that indicated by the SEM image. By correlating it with the SEM image, four individual emission profiles of monolayer MoS_2 could be observed relying on different substrates, as marked in Fig. 7.10c. To study the emission behaviours, the spectrally resolved PL over the four regions are acquired as well, as shown in Fig. 7.10d. The result reveals the PL collected from region 3, where the MoS_2 sheet is suspended over a 300-nm-wide trench, is significantly brighter than that obtained from region 4 on the bulk GaP membrane. This is expected due to the suppression of PL quantum yield by the substrate [52]. Also, the high-index GaP slab would sharply reduce the PL collection efficiency because of the total internal reflection. On both regions 3 and 4, the MoS_2 monolayer emits the same fluorescence spectrum centred around 660 nm due to the direct electronic bandgap [52].

In region 2, the MoS_2 monolayer couples with the periodic air-holes of the PPC lattice. According to the theoretically calculated photonic band structure of the employed PPC, the emission band of the MoS_2 monolayer overlaps with its photonic forbidden band, which therefore inhibits MoS_2's in-plane emission channel. Thereby, the SE should be re-directed into near-vertical k-vectors within the light cone of the GaP slab [55]. The combination of emission redistribution and the higher collection efficiency from the PPC lattice would enhance the PL collection into the vertical direction via the suppression of emission into in-plane PPC modes. Hence, the collected photon flux from region 2 is brighter than that from the bulk GaP membrane, as indicated in the PL spectra. However, the PL collected from the cavity defect (region 1) shows even brighter emission than that from region 2. Comparing the spectra acquired in region 1 to those obtained from regions 2-4, one can conclude this enhancement mainly arises from a greatly amplified photon flux of the two peaks at the wavelengths of 655.4 nm and 656.9 nm, which are resonant wavelengths of the PPC cavity according to the measured reflection spectrum.

To further analyse the cavity enhancement effect, PL spectra with different polarisations were analysed from region 1 by rotating a polariser in the PL collection path of the microscope setup. Figure 7.11a displays the measurement results, where ϕ denotes the angle between the cavity y axis and the polarisation direction of the polariser. It is clear that the two peaks at 655.4 nm and 656.9 nm have strong polarisation dependence, which is consistent with the polarisation-dependent far-field coupling of the resonant modes in the L3 cavity. These two peaks are confirmed as the resonant modes of the PPC cavity using a 3D finite-difference time-domain simulation as well. The emission peak at the longer wavelength around 697 nm was confirmed as the resonant mode as well. In the monolayer MoS_2-PPC cavity coupling system, on the cavity defect, two effects contribute to the dramatic reshaping of the MoS_2 SE. First, the SE coupled into the in-plane PPC modes is suppressed due to the photonic bandgap. Second, cavity modes would enhance the SE rate by the Purcell effect due to their high local density of the photon state.

This cavity-enhanced SE rate in monolayer MoS_2 could be analysed by modelling the MoS_2 sheet as a collection of excitonic dipole emitters. The exciton recombination rate could be described as the sum of the radiative and non-radiative recombination rates, $\Gamma = \Gamma_r + \Gamma_{nr}$. In the experiments, by changing the excitation laser power, the PL intensities at the resonant wavelengths were monitored as well, as shown in Fig. 7.11b. A linear dependence on the excitation power was obtained, verifying that the SE processes are below the saturation rate of the MoS_2 emitter. Hence, the emission power P_{PL} would be proportional to $P_{in}A\Gamma_r/(\Gamma_r + \Gamma_{nr})$, where P_{in} is the excitation power and A is the absorbance of monolayer MoS_2 at the excitation wavelength. Because $\Gamma_{nr} >> \Gamma_r$ in MoS_2 and the finite collection angle of the objective lens, one can approximate the experiments by $P_{PL} \propto \eta\Gamma_r / \Gamma_{nr}$, where η is the collection efficiency of the PL emission by the objective lens. Then, the excitons are considered as an ensemble of emitters μ in the MoS_2 on a bulk substrate, which have a natural SE rate of $\Gamma_0(\lambda)d\lambda$ with a transition rate corresponding to the spectral range from λ to $\lambda + d\lambda$. When the MoS_2 sheet couples to the PPC cavity, the modified SE distribution is governed by [55]

$$\Gamma(\lambda)d\lambda = \Gamma_0(\lambda)d\lambda[F_{c,0}L(\lambda)|\psi|^2 + F_{PC}] \tag{7.13}$$

Here, $L(\lambda) = 1/[1+4Q^2\left(\frac{\lambda}{\lambda_{c,0}} - 1\right)^2]$ denotes the cavity's Lorentzian spectrum with $\lambda_{c,0}$ as the resonant wavelength, and $\psi = E \cdot \mu/|E_{\max}||\mu|$ denotes the spatial and angular overlaps between the emitter dipole μ. and the cavity field E. The factor $F_{c,0} = \frac{3}{4\pi^2}\frac{Q}{V_{\text{mode}}}\left(\frac{\lambda_{c,0}}{n}\right)^3$ is the maximum SE enhancement factor of the cavity mode when the emitter dipole is on resonance with the cavity and spatially aligns with the cavity field. The term F_{PC} accounts for the suppression of the SE rate by the PPC lattice and modes other than the cavity mode.

To fit the measured PL spectrum from the cavity-coupled MoS_2, the collection efficiencies of the cavity modes and averaged PPC leaky modes should be considered, apart from the above-analysed Purcell enhancement over the SE rate. Therefore, the cavity-coupled MoS_2 emission spectrum $I_\phi(\lambda)$ with different polarisations ϕ can be calculated by integrating the SE rate given in Eq. 7.13 over the spatial and in-plane orientation densities of the emitter dipoles $\rho(r, \lambda,$ and $\mu)$

$$I_\phi(\lambda) = \Gamma_0(\lambda)\int d\mu d^2r[\eta_{c,0}F_{c,0}L(\lambda)|\psi|^2 \sin(\phi) + \eta_{PC}F_{PC}]\rho(r,\lambda,\mu) \tag{7.14}$$

Here, $\eta_{c,0}$ and η_{PC} are coupling efficiencies of the PL emissions coupled with the cavity modes and averaged PPC leaky modes. Due to the primarily linear polarisation dependence of the cavity modes at 655.4 nm and 656.9 nm, the PL spectra shown in Fig. 7.11a with polarisations of $\phi = 0°$ and 90° present the off- and on-resonance emissions. The spectrally resolved cavity enhancement of the collected emission from the two spectra are calculated, as shown in Fig. 7.11d, which is described by

$$\frac{I_{90^\circ}(\lambda)}{I_{0^\circ}(\lambda)} = \frac{\eta_{c,0}}{\eta_{PC}}\frac{\int d\mu d^2r F_{c,0}L(\lambda)|\psi|^2 \rho(r,\lambda,\mu)}{\int d\mu d^2r F_{PC}\rho(r,\lambda,\mu)} + 1 \tag{7.15}$$

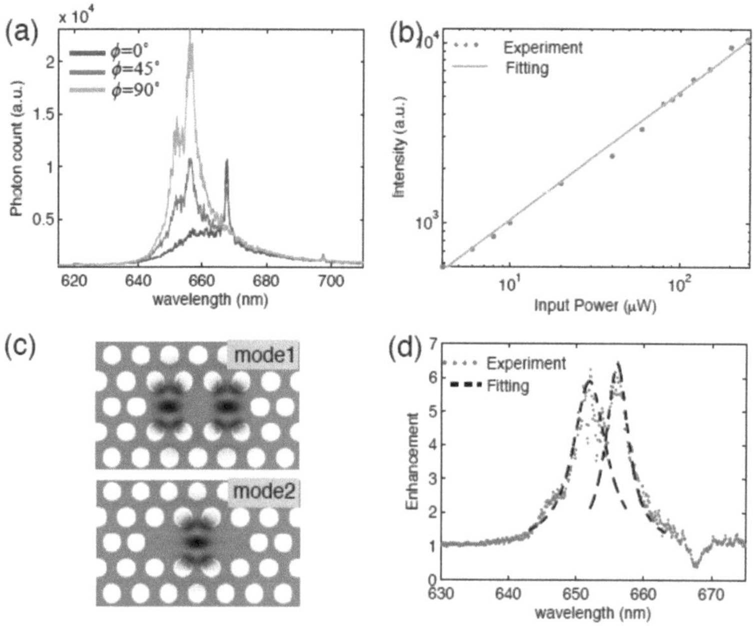

Figure 7.11 (a) Polarisation-dependent PL spectra from the cavity location of the monolayer MoS_2-PPC cavity, where the angle ϕ corresponds to the angle between cavity's y axis and polarisation direction of the polariser. (b) Pump power dependence of the PL emission intensity extracted at the spectra peak. (c) Simulated electrical field distributions of the resonant modes. (d) Enhancement factor obtained from the PL spectra at different polarisations, where the dashed line is the theoretical fitting by considering the Purcell effect and coupling efficiency. Adapted from Ref. [45], with the permission of AIP Publishing.

For simplicity, the two resonant modes at 655.4 nm and 656.9 nm are denoted as mode 1 and mode 2, respectively. By fitting the two peaks to Lorentzian line shapes, the Q-factors of the two modes are 220 and 320, respectively. The cavity fields of modes 1 and 2 are simulated, as shown in Fig. 7.11c. The two modes have mode volumes of 0.63 and $0.33(\lambda/n)^3$, where n is the refractive index of GaP. By integrating the far-field radiations of a dipole spectrally on-resonance or slightly off-resonance with the cavity mode, one can obtain the coupling efficiencies ratios $\frac{\eta_{c,0}}{\eta_{PC}}$ for modes 1 and 2 are 87% and 73%, respectively. Here, the numerical aperture of the objective lens is 0.95. The integral over the angle of the dipole u with respective to the cavity field **E** equals to 1/2 due to the random orientations of dipoles on the 2D MoS_2 sheet. The spatial density of the dipoles corresponds to the excitation of a uniform MoS_2 area by a Gaussian beam with a full width at half maximum of about 400 nm in the $x-y$ plane. Over this excitation area, the spatial integrals of $(|\mathbf{E}||\mu|/|\mathbf{E}_{max}||\mu|)^2$ are 0.169 and 0.079 for modes 1 and 2, respectively, as calculated from their simulated cavity fields. With the calculated V_{mode} and Q-factors derived from the experimental spectra, the maximum Purcell factors $F_{c,0}$ for modes 1 and 2 are estimated about 26.5 and 73.8, respectively. The suppression factor F_{PC} is estimated by simulating the emission power ratio of a dipole on the L3 cavity defect and on the bulk membrane [56]. The emission frequency of the dipole is chosen in the photonic bandgap of the PPC but off-resoance with the cavity mode. The obtained F_{PC} approximately equals 0.4, which is close to the values found in similar PPC structures [55, 56]. Combining these calculations and the Lorentzian functions $L(\underline{\lambda})$ of modes 1 and 2, the theoretical model described in Eq. 7.15 shows a good fit to the experimentally obtained enhancement spectrum, as shown in Fig. 7.11d.

The modifications of the SE rate in monolayer TMD by coupling the emitters to the cavity mode or the PPC lattice were also demonstrated by integrating a WSe_2 monolayer onto a GaP-PPC cavity [57]. The WSe_2 monolayer was mechanically exfoliated from the bulk material and transferred precisely onto the PPC cavity, as indicated in the SEM image of Fig. 7.12a. The PPC cavity is designed to have resonant wavelengths around 750 nm to match the emission peak of the WSe_2 monolayer, and the cavity has a defect of linearly missed four holes. In the fabricated device, the Q-factor is not high either, around 250, because the excitonic absorption of monolayer WSe_2 locates in this wavelength range. Figure 7.12b displays the obtained PL spectra from different regions of the device. When the pump laser at 660 nm was focused on the WSe_2 material on a PPC lattice or on a planar GaP substrate, broadband PL emissions were obtained. However, on the defect region of the PPC cavity, an extra narrow emission peak was observed over the broadband PL emission. This narrow peak is verified as the cavity-enhanced emission by the polarisation dependence and compared with the reflection spectra, as shown in Figs. 7.12c and 7.12d, respectively.

It has been analysed that the PPC lattice would suppress the coupled emitter's SE rate due to the photonic forbidden band. However, in Fig. 7.12b, the emission photon flux from the WSe_2 on the PPC lattice is much stronger than that from the material on the planar substrate, which was observed as well in Fig. 7.10e. As discussed before, due to the inhibited coupling with the in-plane PPC mode, the emission channel is re-directed into the vertical k-space out of the light cone, which therefore makes the emitted radiation break the total internal reflection of the high-refractive-index GaP slab. In addition, in WSe_2 suspended on a periodic air-holes, the quantum yield over the air-holes is enhanced, and the emission is strongly patterned by the sub-wavelength grating.

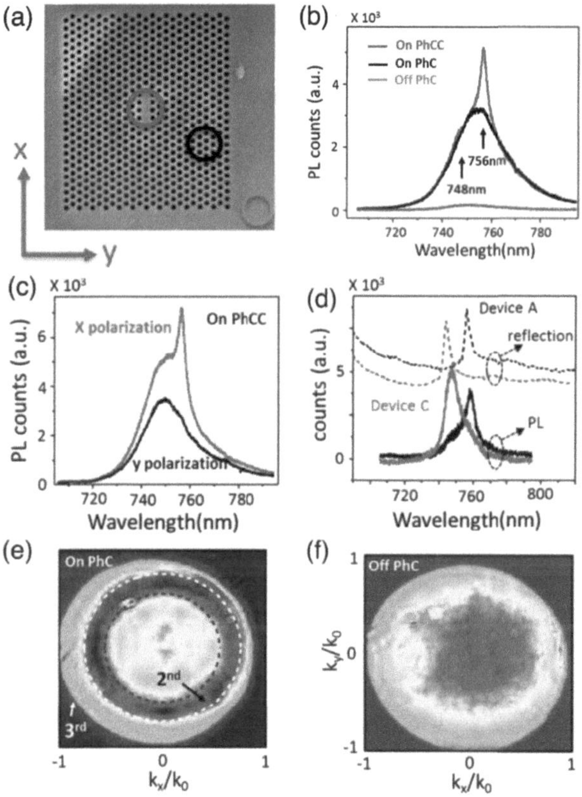

Figure 7.12 Modifications on SE of a WSe_2 monolayer using a PPC cavity. (a) SEM image of a monolayer WSe_2-PPC cavity. (b) PL spectra acquired from different locations of the integrated device. (c) Polarisation-dependent PL spectra obtained from the cavity region. (d) Comparisons between PL peaks and the cavity's reflection peaks. (e, f) Normalised PL intensity distribution over in-plane momentum space measured by momentum-resolved microscopy, when the monolayer is placed on an (e) unpatterned GaP substrate and (f) PPC pattern. Adapted from Ref. [57]. © IOP Publishing. Reproduced with permission. All rights reserved.

This effect could be recognised from the PL's far-field radiation, as shown in Fig. 7.12e. By acquiring the emission intensity image on the back focal plane of the objective lens, the in-plane momentum ($k_{x,}$ k_y) distributions of the far-field light emission were obtained. The in-plane momentum is related to the emission polar angle θ by $\sin(\theta) = \sqrt{k_x^2 + k_y^2}/k_0$, and $k_0 = 2\pi/\lambda$ is the wave number related to the emission wavelength. Modified by the PPC lattice, the exciton emission is highly directed in a ring within the polar angle range of 34° and 48°, corresponding to the in-plane momentum between $0.55k_0$ and $0.75k_0$. Differently, emission from the WSe_2 monolayer located on a planar GaP substrate has a far-field radiation with the in-plane momentum concentrates in the centre, as shown in Fig. 7.12f, presenting a Lambertian-like behaviour.

These PPC cavity-enabled Purcell effects over monolayer TMD indicate this photonic architecture could provide a strong electromagnetic environment to control light–matter interaction in 2D materials. It could expand the scope of solid-state cavity quantum electrodynamics to atomically thin materials, promising a range of optical devices. As an immediate application, low-threshold monolayer semiconductor lasers were achieved in a monolayer TMD-PPC cavity [58, 59], as displayed in Fig. 7.13.

In a previous work [45, 57], PPC cavities were only employed to achieve the SE enhancements in monolayer MoS_2 or WSe_2. However, no stimulated emission was observed even with high pump power. The mainly limited factor is the relatively low Q-factor in the employed PPC cavity, around 300. The intrinsic losses in the integrated devices inhibit the PL emission to get enough optical feedback in the optical cavity. Xu et al. modified the structure of the GaP-PPC cavity by thinning the GaP slab by 50 nm, which optimised the cavity design as well as reduced the sidewall roughness of the air-holes [58]. The Q-factor was improved significantly to a value about 10^4. A WSe_2 monolayer was transferred onto the cavity using a PMMA polymer sacrificial layer, as shown in the device images of Fig. 7.13a. The experimental implementation was similar as that arranged in the previous work. With a 632 nm CW pump laser, the PL emissions from the devices were acquired. To improve the quantum yield of WSe_2, the device is cooled to a low temperature. The PL spectra from the device at the temperature of 80 K show a sharp peak at 793.7 nm with a half-maximum line width of 0.3 nm. To validate the extremely strong peak arising from the cavity mode, polarisation dependence was obtained, which is consistent with the fundamental mode of the PPC cavity. One characteristic of a laser is the non-linear 'kink' occurring around the lasing threshold in the light-in and light-out plot, that is, the L-L curve. In Fig. 7.13b, the L-L curves are presented from the device operated at 130 K and 80 K, respectively. Both of them show the non-linear kink at the laser threshold region. In Fig. 7.13b, a set of power-dependent data for SE away from the cavity resonance is also shown. No kink signature was observed. The fitting curves in the plots were calculated by the cavity laser rate equation, giving rise to a β-factor of 0.19. This large β-factor indicates about 19% of the total SE is couple to the cavity mode, promising a low lasing threshold of 27 nW (~1 W cm^{-2}). Another hallmark feature to distinguish lasing behaviour is to monitor the evolution of the peak line width as a function of the pump power. When the pump power is increased in the range below the threshold, the observed line width narrows from 0.75 nm to 0.5 nm.

Monolayer TMDs have large exciton-binding energies (in the order of 100 meV) due to the reductions of dielectric screening, which have not been achieved in other material groups. They therefore provide a unique opportunity to work as an efficient active medium to achieve room-temperature lasing relying on the stronger excitonic transitions. Excitingly, a low-threshold monolayer laser operated at room temperature was realised by integrating a $MoTe_2$ monolayer with a silicon PPC cavity [59]. Better than the WSe_2-based monolayer laser, this $MoTe_2$ laser was coupled to a nanobeam PPC cavity, which had a higher Q-factor exceeding 5600 as well as a much smaller V_{mode} ~$0.48(\lambda/n)^3$, as shown in Fig. 7.13c. Thence, this device provides much higher Q/V_{mode} to engineer the emission rate in monolayer $MoTe_2$. In the experiments, two laser modes were obtained at 1052 nm and 1132 nm because the PL spectrum of $MoTe_2$ overlaps with the two resonant modes simultaneously. The lasing features were validated by the light-in and light-out plot and the peak line width with respect to the increased pump power, as shown in Fig. 7.13d,e. A β-factor of 0.1 was obtained, and the threshold power density was estimated as 6.6 W cm^{-2}, which is much lower than that of room-temperature excitonic lasers in ultra-violet wavelengths. This low-threshold laser was contributed to the small volume of the active-gain medium, which was estimated as 2440 nm $\times$ 0.7 nm $\times$ 365 nm = (85.4 nm)^3.

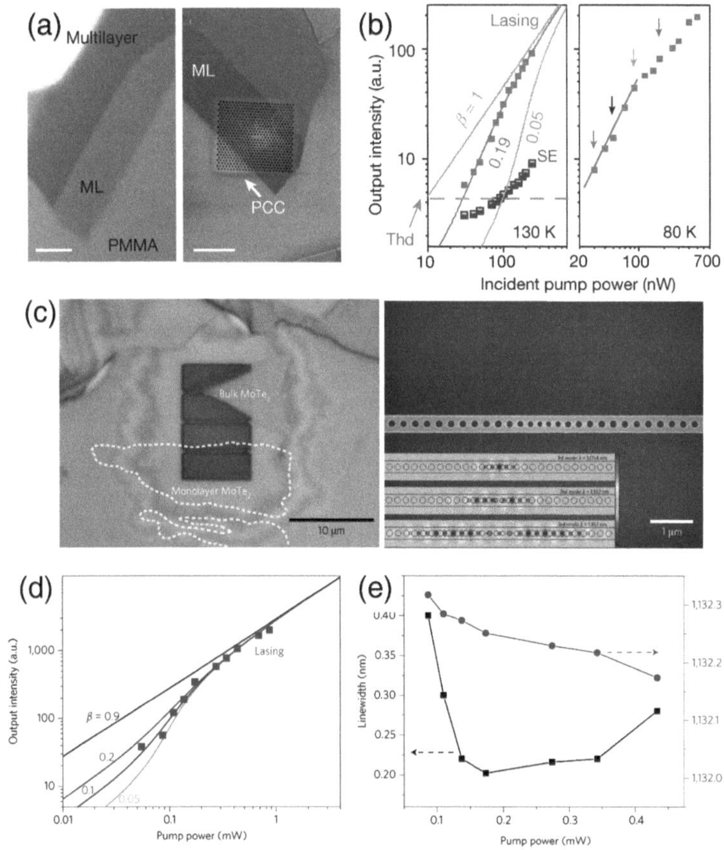

Figure 7.13 Low-threshold monolayer semiconductor lasers integrated on a PPC cavity, (a) Optical microscope and SEM images of monolayer WSe_2 integrated on a PPC cavity for low-threshold lasing. (b) Light-in and light-out power plot of WSe_2's PL emission measured at the temperatures of 130 K and 80 K. (c) Optical and SEM images of a $MoTe_2$ monolayer integrated on a nanobeam cavity lor low-threshold lasing at room temperature. (d) Light-in and light-out power ptot of $MoTe_2$'s PL emission. (e) Line widths and peak wavelengths of the PL emissions from $MoTe_2$ as a function of pump power. (a, b) Reprinted by permission from Springer Nature Customer Service Centre GmbH: Springer Nature, *Nature*, Ref. [58], Copyright (2015). (c, d, e) Reprinted by permission from Springer Nature Customer Service Centre GmbH: Springer Nature, *Nature Nanotechnology*, Ref. [59], Copyright (2017).

The achieved $MoTe_2$-based monolayer laser has advantages of room-temperature operation, ultra-low threshold, smallest volume and smallest size. Besides these, a more attractive virtue of this work was the integration with a silicon-based chip, while it only worked at the near edge of silicon's transparent wavelength. It presents appealing potentials for silicon photonics to solve the challenges of integrating high-efficient light sources on silicon chips, which is crucial for on-chip optical interconnect. Unfortunately, $MoTe_2$'s exciton transition (1100 nm) is far away from the telecom bands, such as 1300 nm or 1550 nm, hindering its realistic application for a on-chip nanolaser source. Black phosphorus as a new emerging 2D material has a direct bandgap in its mono- and few-layer flakes. The bandgap strongly depends on the number of layers, showing a shifted emission wavelength from ~800 nm in the monolayer to 1600 nm in four layers. It is very attractive to integrate few-layer black phosphorus as a gain medium on a silicon PPC cavity to realise a nanolaser at the telecom band. An attempt was carried out by Ota et al. [60]. The experimental results are displayed in Fig. 7.14.

Because of black phosphorus's instability, it is essential to fabricate the integrated device in a glove box without oxygen and humidity. PL spectra centred at around 1100 nm, 1450 nm and 1600 nm were obtained from the exfoliated black phosphorus with various thicknesses. The employed PPC cavities were fabricated in a gallium arsenide membrane with three missing air-holes in a triangular PPC lattice. Resonant modes centred around 1100 nm were obtained with Q-factors over 5000 before the transfer of black phosphorus. Bilayer black phosphorus was integrated onto the cavity. Because of the strong material absorption, the Q-factor of the PPC cavity degraded to a value of ~260. The finished device is displayed in Fig. 7.14a. Black phosphorus's emission behaviours before and after the transfer were measured as well, presenting a decreased PL intensity as well as a red-shifted peak centre. To study the cavity enhancement, PL spectra were taken from the defect cavity region, as shown in Fig. 7.14b. The broad emission peak centred at 1100 nm was from black phosphorus, while the sharp peak at 1070 nm was from the emission coupled to the fundamental cavity mode. The spatial location and polarisation dependences of the cavity-enhanced PL in black phosphorus were examined as well, as shown in Fig. 7.14c,d.

However, laser oscillation was not observed eventually from any of several samples, even with the pumps of strong CW and pulsed laser and cryogenic temperature. One of the possible explanations of the failed lasing behaviour is that the quantum yield in atomically thin black phosphorus is quite low compared to that in monolayer TMDs, which cannot provide a sufficient material gain. The degradation of black phosphorus under high-power laser illumination is another possible factor.

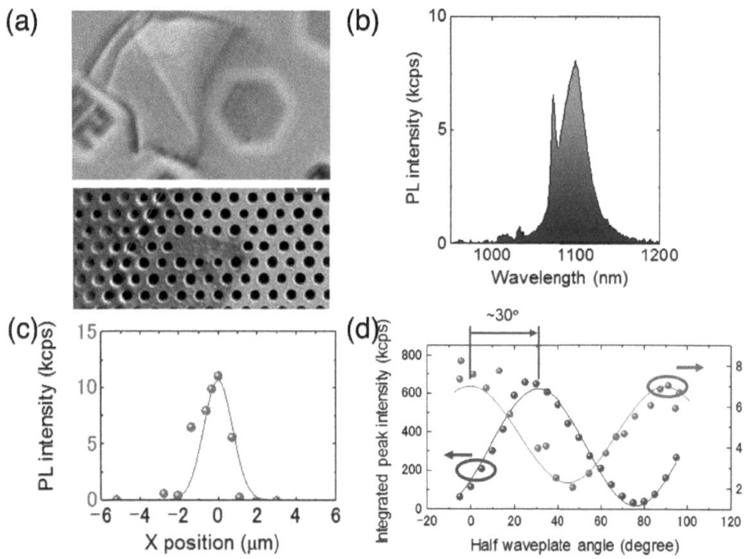

Figure 7.14 Enhanced light emission from a PPC cavity integrated with few-layer black phosphorus. (a) Optical microscope and SEM images of a PPC cavity integrated with few-layer black phosphorus. (b) PL spectrum from cavity-coupled black phosphorus, showing an enhanced peak at the resonant wavelength. (c) Position dependence of the cavity-enhanced PL from black phosphorus, with a localised profile around the cavity defect. (d) Polarisation dependence of the cavity-enhanced PL from black phosphorus. Adapted from Ref. [45], with the permission of AIP Publishing.

The PPC cavity-enabled strong Purcell effects can not only be used to enhance PL emission from 2D materials but also promise the improved efficiency of light-emitting devices (LEDs) assembled from 2D materials. The distinct electronic properties and excitonic effects of monolayer TMDs facilitate the development of atomically thin, electrically driven light emitters [7]. In addition, the possibility to stack different 2D materials vertically provides another more efficient architecture to electrically pump emission from monolayer TMDs [61]. Again, because of the atomically thin gain materials, the total photon fluxes from 2D materials' LEDs are very low. By integrating a PPC cavity with an LED assembled by monolayer graphene, boron nitride and monolayer WSe_2, the electroluminescence (EL) spectrum, polarisations and emission directions of monolayer WSe_2 were modified by the resonant modes [62]. The schematic of the assembled LED integrated with a PPC cavity is shown in Fig. 7.15a. The active light-emitting monolayer WSe_2 was sandwiched by two tunnelling contacts formed by graphene/mono- or bilayer hexagonal boron nitride. A GaP-PPC cavity was precisely placed onto the emitting region of the LED.

A fabricated device is displayed in Fig. 7.15b. When a bias voltage was applied across the two graphene layers, the electrons and holes are tunnelled into the WSe_2 layer, which then form excitons with a finite time. The radiative combination of the electrons and holes gives rise to EL. The EL spectra from the cavity and off-cavity are shown in Fig. 7.15c. The EL spectrum collected from cavity has an extra peak (~759 nm) apart from the EL main peak (~740 nm) from the intrinsic emitting layer. It implies the efficient coupling between the cavity mode and the electrical dipole in WSe_2. To further confirm the cavity-enhanced EL, the polarisation dependence of the two peaks from the cavity was analysed by placing a polariser in the light collection path, as displayed in Fig. 7.15d. The emission from the WSe_2 monolayer arises from the isotropic exciton recombination, and therefore the ~740 nm peak has no polarisation. However, the resonant modes in the PPC cavity have polarisation-dependent far-field radiations. After the EL coupled to the cavity mode, the far-field emissions are collected via the modulation of the cavity mode. The emission at the resonant peak ~759 nm has a strong polarisation with a degree of 84%. In addition to the polarisation analysis, spatially resolved EL was implemented as well to verify the cavity enhancement effect. By virtue of the polarisation dependence of the cavity-enhanced EL, two spatially mappings of the devices were made with polariser perpendicular and parallel to the cavity's far-field radiation. By comparing the difference between the two mapping images, strongly localised EL was obtained at the cavity centre, as displayed in Fig. 7.15e.

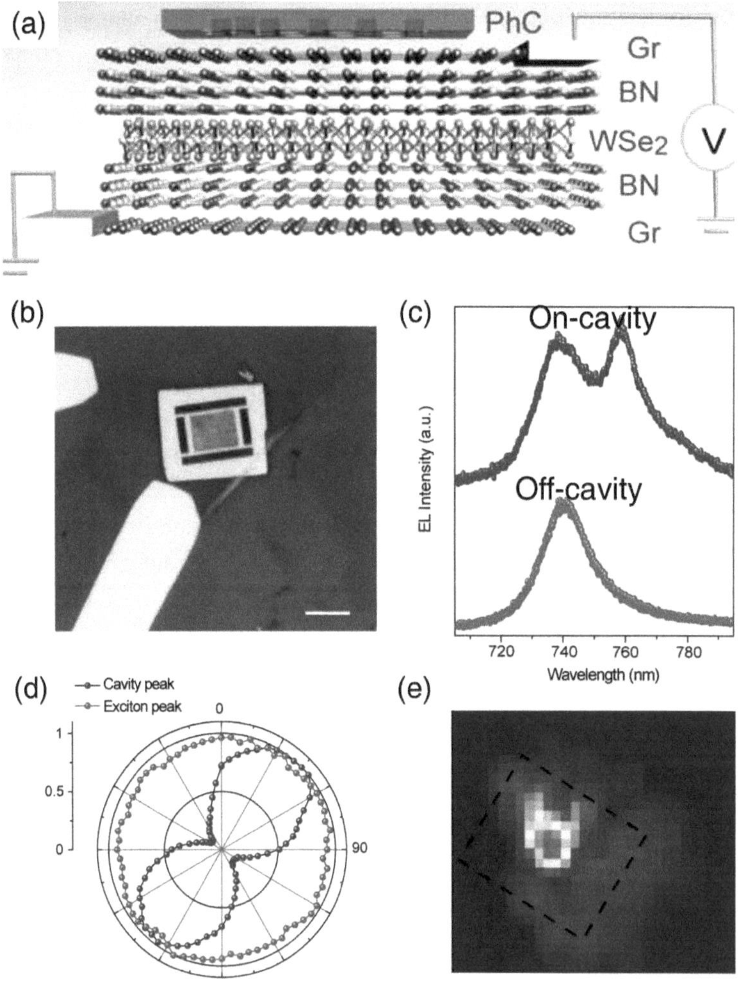

Figure 7.15 PPC cavity-enhanced EL from a monolayer WSe_2 LED. (a) Schematic of the heterostructure LED, where the active material is monolayer WSe_2, and the graphene/boron nitride provides tunnelling carriers, (b) Optical microscope image of a finished device, (c) EL spectra acquired from the cavity and off-cavity, showing a cavity-enhanced peak, (d) Polarised EL from the cavity-integrated LED. (e) Spatial position-dependent EL. Adapted with permission from Ref. [62]. Copyright (2016) American Chemical Society.

7.4.4 Strong Non-linear Processes in 2D Materials Enabled by a Photonic Crystal Cavity

The photons in the resonant modes of PPC cavity could circulate for a long time due to the high Q-factor. Also, the optical field is localised in a mode volume of cubic wavelength. Hence, the optical power density in a PPC cavity is very high, which has been widely employed to enhance non-linear processes. In the integrated 2D material-PPC cavity, the optical intensity at the location of 2D materials could be more than 40% of the maximum intensity in the resonant mode. Since the non-linear processes are proportional to the quadratic and cubic functions of optical intensity for the second and third non-linear processes, the non-linear responses in PPC cavity would be enhanced by factors of $Q/V^2{}_{\text{mode}}$ and $Q/V^3{}_{\text{mode}}$, respectively. Hence, the PPC cavity is a promising platform to enhance 2D materials' non-linear processes.

7.4.4.1 Optical bistability and four-wave mixing in a graphene-PPC cavity system

Figure 7.16 describes an experiment exploiting a silicon PPC cavity to make graphene's third-order non-linearity more pronounced [46]. As demonstrated in Section 7.4.1, strong light absorption in graphene would lower the cavity's Q-factor intensively, degrading the enhancement of non-linear processes. To get over the problem, heavily doped graphene could be chosen to integrate with the PPC cavity, as demonstrated in Ref. [46]. In the transfer process of CVD graphene, chemical etching of the growing substrate (copper foil) is required. And the graphene layer is attached to a polymer layer, which would be removed by acetone. This process normally involves various acids and organic solvents and inevitably yields chemical residuals in the graphene layer. Hence, in the wetting transfer process, graphene is doped in a high chemical level. If graphene's Fermi level is higher than the half-photon energy, its light absorption would be partly inhibited due to Pauli blocking. Figure 7.16a displays the SEM image of a device. Here, the PPC cavity is side-coupled with a PPC waveguide to achieve a high coupling efficiency. The excitation laser and non-linear signals are coupled into and out of the PPC waveguide using lensed fibres at both side of the silicon chip. After the integration of the graphene layer, the cavity's Q-factor was decreased from 23,000 to 7500 due to the limited absorption. The measurement was implemented with a low input power of 2.5 μW to avoid possible non-linear processes.

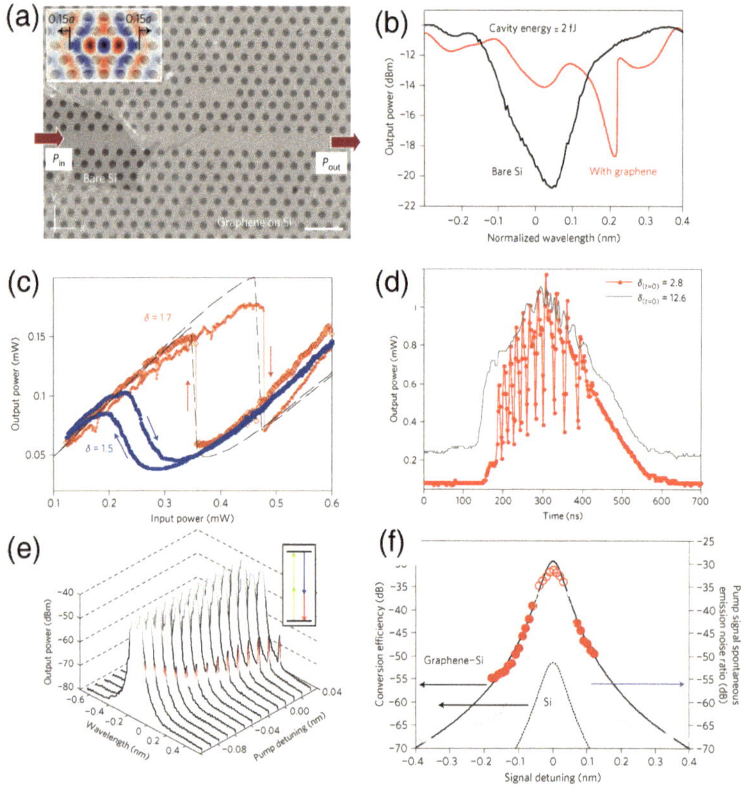

Figure 7.16 PPC cavity-enhanced non-linear processes in graphene. (a) SEM image of a CVD graphene-integrated PPC cavity. (b) Resonant spectra of the PPC cavity acquired before and after graphene's integration. (c) Optical bistability hysteresis in the graphene-PPC cavity hybrid by scanning the input power upwards and downwards, (d) Temporally regenerative power oscillations in the cavity output when the input power changes in the form of the grey line. (e) Measured transmission spectrum with the signal laser fixed –0.16 nm away from the cavity resonance wavelength, while scanning the pump laser detuning from –0.1 to 0.03 nm. (f) Observed and simulated conversion efficiencies for the PPC cavity with and without graphene. Reprinted by permission from Springer Nature Customer Service Centre GmbH: Springer Nature, *Nature Photonics*, Ref. [46], Copyright (2012).

A non-linear photo-thermal effect induced by the graphene layer was first studied on the graphene-PPC cavity hybrid, as shown in Fig. 7.16b. With an input power of 0.6 mW, an obvious red-shifted resonant wavelength was observed in the graphene-PPC cavity hybrid as well as an asymmetric line shape. This result indicates a consistent and large resonance red-shift rate of 1.2 nm mW^{-1}. In contrast, the bare cavity presented a negligible red-shifted wavelength, and the resonant line shape is a symmetric Lorentzian type. With a larger pump power, the red-shifted rate of the bare silicon cavity was estimated as a small value of 0.1 nm mW^{-1}. The strong resonant wavelength shifts from the graphene-PPC cavity hybrid were next investigated for a low-threshold optical bistability. By sweeping the excitation laser power upwards and downwards, the output powers of the device were recorded, presenting a hysteresis loop. For laser wavelengths tuned from the resonant wavelength differently, the bistability curves were obtained, as shown Fig. 7.16c. Here the laser detuning δ is defined as $\delta = (\lambda_{\text{laser}} - \lambda_{\text{cavity}})/\Delta\lambda_{\text{cavity}}$ where $\Delta\lambda_{\text{cavity}}$ is the line width of the intrinsic cavity without graphene. The results imply the power threshold of the bistability is about 100 μW. In Fig. 7.16c, the dashed lines show the numerical predictions of coupled-mode theory for the integrated cavity, including the first-order estimations of the graphene-modified thermal, linear and non-linear losses, and free-carrier parameters. It is worthy to note that because of the strongly enhanced optical intensity in the resonant modes, the light absorption in graphene also has a component of two-photon absorption. The extremely high third-order non-linearity in graphene makes its two-photon absorption are much higher than that in silicon, while the optical field in the silicon slab is much more than that in graphene. Similar as those reported in silicon photonics, two-photon absorption would induce free-carrier density increment and overall enhanced thermally red-shifted wavelength.

Optical bistability is an implication of regenerative oscillation, which has been normally predicted in resonators with strong Kerr effects. In the graphene-PPC cavity system, when the input laser power is well above the bistability threshold, the output optical power has two-state bistable switching and becomes oscillatory, as demonstrated in Fig. 7.16d. With a theoretical analysis with coupled-mode theory, the regenerative oscillations were formed between the competitions of phonon and free-carrier populations, corresponding to the slow thermal red shift and fast free-carrier plasma dispersion blue-shifts. From the measured temporal oscillations for the de-tuned wavelength range of $\delta = 0.68$ to 1.12, a threshold power of 0.4 mW and a period of 9.4 ns were obtained.

In addition to the cavity-enhanced photo-thermal effect and free-carrier generation, the Kerr non-linearity in graphene could also be employed to realise efficient FWMs in the graphene cavity. A Q-factor of 7500 corresponds to a resonant peak line width of 0.2 nm. To satisfy the resonance enhancement, the pump and signal lasers have wavelengths within this line width and incident powers of 0.6 mW. Figure 7.16e shows a series of idler measurements of the degenerated FWM with variously detuned pump wavelengths. The parametric idler was clearly observed as a sideband to the cavity resonance, with the pump wavelength detuned from −0.1 nm to 0.03 nm. Figure 7.16f shows the calculated conversion efficiency of the measured FWM process as well as theoretical predictions. A clear enhancement factor could be observed due to the graphene integration. A Kerr coefficient of $n_2 = 4.8 \times 10^{-17}\ \text{m}^2\text{W}^{-1}$ was derived from the graphene-PPC cavity ensemble, which is an order of magnitude larger than that in bare silicon devices. Using the simulated optical field distribution in the graphene-PPC cavity hybrid as well as the absolute third-order non-linearity in graphene and silicon, an effective third-order non-linearity and Kerr coefficient were calculated. An estimated value of $n_2 = 7.7 \times 10^{-17}\ \text{m}^2\text{W}^{-1}$ was obtained, closely matching with the experimentally obtained value.

7.4.4.2 Continuous-wave-pumped SHG in a few-layer GaSe-PPC cavity

Second-order non-linearities in various materials have more promises in device applications than the third-order non-linearity because of the high efficiency and low pump-power requirement. SHGs as one of the most developed second-order non-linear processes have already been widely utilised to generate new coherent sources, high-resolution microscopy and precise surface spectroscopy. SHGs could only be realised in materials without inversion centro-symmetry. In 2D materials, determined by the unique interlayer coupling and excitonic behaviours, SHGs have interesting features. Combining with the outstanding optoelectronic properties of 2D materials, SHGs could expand their optoelectronic applications into the non-linear regime. Unfortunately, the atomic thickness of 2D materials limits the effective light–matter interaction, and the implementations of SHGs in 2D materials normally require pulsed laser with high peak intensities. For practical applications, CW-pumped SHGs with simple, low-power and low-cost light sources, such as semiconductor laser diodes, would be highly desirable. By integrating 2D materials with high second-order non-linearity onto a PPC cavity, it is possible to achieve the CW-pumped SHGs relying on the strongly confined optical field.

Figure 7.17a shows an optical microscope image of a PPC cavity integrated with few-layer GaSe [63]. GaSe is a well-known non-linear crystal over a wide wavelength range. It has pronounced second-order non-linearity in its atomic layer, which is 1-2 orders of magnitude larger than that of monolayer MoS_2. In addition, SHGs in GaSe have no layer dependence, which facilitates device fabrication. To evaluate the thickness of the transferred GaSe flake, an AFM technique was employed, as displayed in Fig. 7.17b. A thickness of 7.8 nm was estimated, corresponding to the layer number of 9 by assuming the monolayer thickness of 0.85 nm. By using a vertically cross-polarisation confocal setup, the resonant peaks of the PPC cavity before and after GaSe integration were acquired. The resonant wavelength red-shifts from 1536.5 nm to 1551.0 nm. Because of GaSe's high refractive index, it also weakens the confinement of the resonant mode, lowering the Q-factor from 2000 to 1750. To study on the cavity-enhanced SHG, a narrow-band telecom-band laser was tuned to the wavelength of 1551 nm to excite the cavity's resonant mode. The generated second-harmonic signal from the device was collected, which is examined by a spectrometer mounted with a CCD camera. The reflected resonant signal was filtered out from the second-harmonic signal using a short-pass dichroic mirror. Figure 7.17c displays the measured spectrum. With a 0.5 mW pump laser, a strong peak presented at a wavelength of 775.5 nm, corresponding to the SHG of the pump laser exactly.

To verify the SHG pumped by the low-power CW laser, its power dependence was evaluated by varying the laser power, as shown in the log-log plot of Fig. 7.17d. The dashed line represents a linear fitting with a slope of 2.01, confirming the quadratic power dependence of the cavity-enhanced SHG. For the exploited PPC cavity and objective lens, the far-field collection efficiency of the resonant mode is about 40%. By measuring the cavity reflection of an on-resonance illumination, the coupling-in efficiency of the pump laser was estimated as 6%. Therefore, even for a 0.1 mW laser focused on the PPC cavity, the power coupled into the cavity is lower than 10 μW, which can still generate a detectable second-harmonic signal. The low-power CW-pumped SHG in GaSe could be explained by the strongly enhanced optical field in the cavity. To verify this, the dependence of the SHG on the pump laser wavelength was measured, which was tuned through the cavity resonance. As indicated in Fig. 7.17e, as the laser wavelength was tuned away from the resonance, the SHG became weak and achieved a state which the signal is too weak to measure. For an on-resonance pump, the optical field coupled into the cavity is enhanced by a factor proportional to Q, and the enhancement over the second-order non-linear process has a factor proportional to Q^2. Hence, the SHG spectrum shown in Fig. 7.17e would reduce like a line shape described by the squared Lorentzian line shape used to fit the fundamental resonance.

The spatially resolved cavity-enhanced SHG was investigated as well to validate the localised enhancement over the resonant mode. The device was mounted on a 2D piezo-actuated stage with a moving step of 100 nm, and the generated second-harmonic signals at different locations were measured using a photo-multiplier tube (PMT). The cavity-enhanced SHG is pumped by the near-field of the cavity's resonant mode. Efficient SHGs can only be observed when the pump laser is coupled into the cavity. Therefore, the spatial position of the detected SHG is determined by the cavity's coupling-in region. For the employed cavity, light could vertically couple into the cavity effectively around the modification region, which has a scale about 2.5×2 μm^2. In the SHG spatial mapping, a similar area with a strong signal was observed, as shown in Fig. 7.17f. Outside the cavity's mode-coupling region, the GaSe layer is only pumped by the vertically illuminated laser, which is too weak to yield the observed SHG for a CW pump.

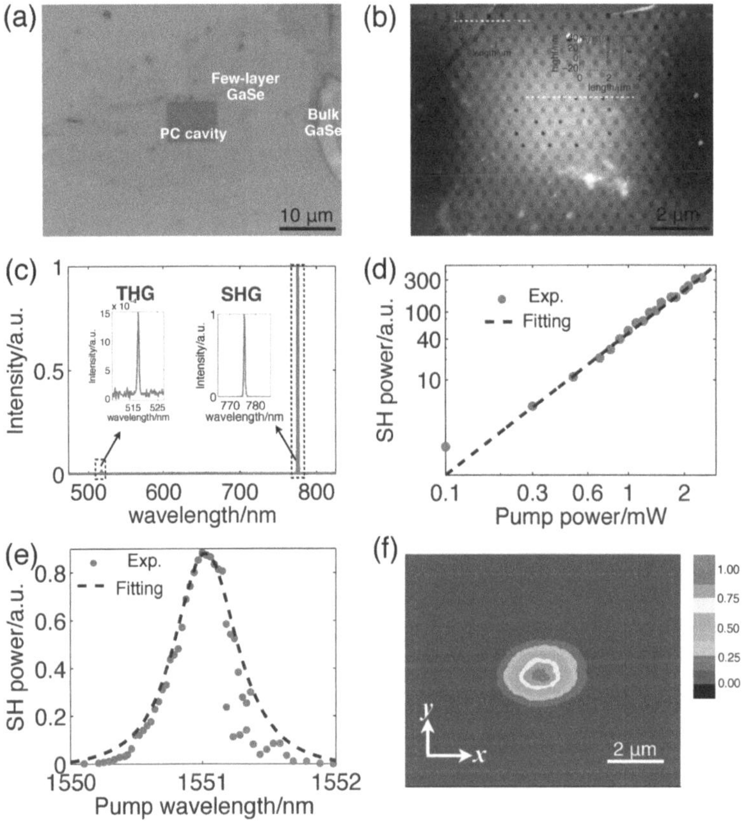

Figure 7.17 PPC cavity-enhanced SHG in few-layer GaSe. (a) Optical microscope image of a PPC cavity integrated with a few-layer GaSe flake. (b) AFM image of the device, indicating the height of GaSe's thickness. (c) Up-conversion spectrum acquired from the GaSe-PPC cavity, showing two peaks of the SHG and THG. (d) Pump power dependence of the measured SHG. (e) SHG spectra as the pump wavelength is tuned across the resonant mode. (f) Spatially resolved SHG image of the GaSe-PPC cavity, showing a strong signal around the defect region. Adapted with permission from Ref. [63]. Copyright © 2017, Springer Nature.

7.5 Summary

Two-dimensional materials' unique optical and electrical properties enable them to be promising active materials for various optoelectronic devices, including photo-detectors, modulators and lasers. However, if the devices operate in a normal-illumination configuration, the light–matter interaction only takes place over few atoms. Even the quantum yield in 2D materials is 100%, the device performance has no chance to be comparable with those fabricated in bulk semiconductors. Considering 2D materials' infinitely large planar geometry, an ideal way to improve the devices performance is to realise an architecture enabling in-plane light–matter interaction within 2D materials. Unfortunately, the atomic thickness of 2D materials cannot support the vertical confinements of optical modes to provide more sites for light–matter interaction.

Chip-integrated nanophotonic devices are fabricated in a sub-wavelength-thick slab, where light is vertically confined via the total internal reflection. Hence, optical fields in the on-chip devices are guided in the slab plane. Strong evanescent fields emerge around the slab surface, and it could be modulated by varying the slab thickness. If 2D materials are coated onto the planar photonic devices, the strong coupling between the evanescent field and 2D materials promises effective in-plane light–matter interactions within 2D materials, providing an ideal way to improve the performance of their optoelectronic devices.

In this chapter, the integrations of 2D materials with photonic wire waveguide, micro-ring resonators and PPC cavities were introduced. Relying on the extended coupling length, waveguides present a promising structure to enhance light-2D material interaction. Specifically, waveguides could operate over a broad spectral range and have functioned as a number of elements for photonic integrated circuits. The integration of 2D materials could expand these elements' active functionalities. In the resonators of micro-rings and PPC cavities, the integrated devices have a much more compact footprint. A PPC cavity as one of the resonators with the highest ratio of Q/V_{mode} has great potentials to study strong coupling between 2D materials and optical fields, for achieving ultralow power-consumed modulators, low-threshold lasers and micro-scale detections of Raman scattering. Compared to the PPC cavity, while micro-ring resonators have weaker localised optical fields, their fabrication precisions have low requirements. Also, optical fields in micro-rings could be effectively and simply coupled with a low-loss bus waveguide, facilitating device operations. Here, from the devices of chip-integrated 2D materials, strongly enhanced optical responses were addressed, such as light absorption, emission, Raman scattering and non-linear processes.

With advantages of a compact footprint, mass production, low cost and low-power requirements, chip-integrated nanophotonic devices have been widely recognised as a plentiful platform for telecommunication, data communication, sensing and quantum information processing. By integrating 2D materials, their application ranges could be expanded further by considering 2D materials as active media. For example, in Fig. 7.17c, there is another weak peak at a wavelength of 517 nm, corresponding to the wavelength of the THG of the pump laser. This THG peak was recognised as a frequency up-conversion signal from the third-order non-linearity of the silicon slab by comparing it with a control measurement on a bare silicon PPC cavity. Note that the SHG peak from GaSe is more than 650 times stronger than the THG peak from silicon. From the resonant mode simulation, the optical power inside the nine-layer GaSe flake is only 1.5% of that in the bulk silicon slab. The remarkably strong SHG signal can be explained by the ultra-high second-order non-linear susceptibility of GaSe. This example indicates that GaSe's high second-order non-linearity enables the chip-integrated non-linear effects to be much stronger than those in the silicon slab, though it only interacts with the evanescent field of the silicon slab. Therefore, integration of few-layer GaSe, with high non-linearity and negligible linear and non-linear losses in the telecom band, onto silicon nanophotonic devices may pave a new way towards developing high-performance non-linear devices for future on-chip interconnects.

References

1. Novoselov, K. S., Mishchenko, A., Carvalho, A. and Neto, A. H. C. (2016). 2D materials and van der Waals heterostructures, *Science*, **353**(6298), p. aac9439.
2. Geim, A. K. and Grigorieva, I. V. (2013). Van der Waals heterostructures, *Nature*, **499**(7459), pp. 419–425.
3. Liu, Y., Weiss, N. O., Duan, X., Cheng, H.-C., Huang, Y. and Duan, X. (2016b). Van der Waals heterostructures and devices, *Nat. Rev. Mater.*, **1**(9), p. 16042.

4. Xia, F., Wang, H., Xiao, D., Dubey, M. and Ramasubramaniam, A. (2014). Two-dimensional material nanophotonics, *Nat. Photonics*, **8**(12), pp. 899–907.
5. Mak, K. F., Sfeir, M. Y., Wu, Y., Lui, C. H., Misewich, J. A. and Heinz, T. F. (2008). Measurement of the optical conductivity of graphene, *Phys. Rev. Lett.*, **101**(19), p. 196405.
6. Eda, G. and Maier, S. A. (2013). Two-dimensional crystals: managing light for optoelectronics, *ACS Nano*, **7**(7), pp. 5660–5665.
7. Withers, F., Pozo-Zamudio, O. D., Mishchenko, A., Rooney, A. P., Gholinia, A., Watanabe, K., Taniguchi, T., Haigh, S. J., Geim, A. K., Tartakovskii, A. I. and Novoselov, K. S. (2015). Light-emitting diodes by band-structure engineering in van der Waals heterostructures, *Nat. Mater.*, **14**(3), pp. 301–306.
8. Hendry, E., Hale, P. J., Moger, J., Savchenko, A. K. and Mikhailov, S. A. (2010). Coherent nonlinear optical response of graphene, *Phys. Rev. Lett.*, **105**(9), p. 097401.
9. Rodrigues, M. J. L. F., de Matos, C. J. S., Ho, Y. W., Peixoto, H., de Oliveira, R. E. P., Wu, H.-Y., Neto, A. H. C. and Viana-Gomes, J. (2016). Resonantly increased optical frequency conversion in atomically thin black phosphorus, *Adv. Mater.*, **28**(48), pp. 10693–10700.
10. Malard, L. M., Alencar, T. V., Barboza, A. P. M., Mak, K. F. and de Paula, A. M. (2013). Observation of intense second harmonic generation from MoS_2 atomic crystals, *Phys. Rev. B*, **87**(20), p. 201401.
11. Li, Y., Rao, Y., Mak, K. F., You, Y., Wang, S., Dean, C. R. and Heinz, T. F. (2013). Probing symmetry properties of few-layer MoS_2 and h-BN by optical second-harmonic generation, *Nano Lett.*, **13**(7), pp. 3329–3333.
12. Janisch, C., Wang, Y., Ma, D., Mehta, N., Elías, A. L., Perea-López, N., Terrones, M., Crespi, V. and Liu, Z. (2014). Extraordinary second harmonic generation in tungsten disulfide monolayers, *Sci. Rep.*, **4**(1), p. 5530.
13. Seyler, K. L., Schaibley, J. R., Gong, P., Rivera, P., Jones, A. M., Wu, S., Yan, J., Mandrus, D. G., Yao, W. and Xu, X. (2015). Electrical control of second-harmonic generation in a WSe_2 monolayer transistor, *Nat. Nanotechnol.*, **10**(5), pp. 407–411.
14. Nair, R. R., Blake, P., Grigorenko, A. N., Novoselov, K. S., Booth, T. J., Stauber, T., Peres, N. M. R. and Geim, A. K. (2008). Fine structure constant defines visual transparency of graphene, *Science*, **320**(5881), pp. 1308–1308.

15. Furchi, M., Urich, A., Pospischil, A., Lilley, G., Unterrainer, K., Detz, H., Klang, P., Andrews, A. M., Schrenk, W., Strasser, G. and Mueller, T. (2012). Microcavity-integrated graphene photodetector, *Nano Lett.*, **12**(6), pp. 2773–2777.

16. Echtermeyer, T., Britnell, L., Jasnos, P., Lombardo, A., Gorbachev, R., Grigorenko, A., Geim, A., Ferrari, A. and Novoselov, K. (2011). Strong plasmonic enhancement of photovoltage in graphene, *Nat. Commun.*, **2**, p. 458.

17. Low, T., Chaves, A., Caldwell, J. D., Kumar, A., Fang, N. X., Avouris, P., Heinz, T. F., Guinea, F., Martin-Moreno, L. and Koppens, F. (2016). Polaritons in layered two-dimensional materials, *Nat. Mater.*, **16**(2), pp. 182–194.

18. Basov, D. N., Fogler, M. M. and de Abajo, F. J. G. (2016). Polaritons in van der Waals materials, *Science*, **354**(6309), p. aag1992.

19. Gao, L., Ni, G.-X., Liu, Y., Liu, B., Neto, A. H. C. and Loh, K. P. (2013). Face-to-face transfer of wafer-scale graphene films, *Nature*, **505**(7482), pp. 190–194.

20. Soref, R. and Lorenzo, J. (1985). Single-crystal silicon: a new material for 1.3 and 1.6 μm integrated-optical components, *Electron. Lett.*, **21**(21), p. 953.

21. Jalali, B. and Fathpour, S. (2006). Silicon photonics, *J. Lightwave Technol.*, **24**(12), pp. 4600–4615.

22. Hochberg, M. and Baehr-Jones, T. (2010). Towards fabless silicon photonics, *Nat. Photonics*, **4**(8), pp. 492–494.

23. Jones, A. M., DeRose, C. T., Lentine, A. L., Trotter, D. C., Starbuck, A. L. and Norwood, R. A. (2013). Ultra-low crosstalk, CMOS compatible waveguide crossings for densely integrated photonic interconnection networks, *Opt. Express*, **21**(10), pp. 12002–12013.

24. Xiao, S., Khan, M. H., Shen, H. and Qi, M. (2008). Silicon-on-insulator microring add-drop filters with free spectral ranges over 30 nm, *J. Lightwave Technol.*, **26**(2), pp. 228–236.

25. Xu, Q., Schmidt, B., Shakya, J. and Lipson, M. (2006). Cascaded silicon micro-ring modulators for WDM optical interconnection, *Opt. Express*, **14**(20), p. 9431.

26. Lipson, M. (2005). Guiding, modulating, and emitting light on silicon-challenges and opportunities, *J. Lightwave Technol.*, **23**(12), pp. 4222–4238.

27. Heck, M. J. R., Bauters, J. F., Davenport, M. L., Doylend, J. K., Jain, S., Kurczveil, G., Srinivasan, S., Tang, Y. and Bowers, J. E. (2013). Hybrid

silicon photonic integrated circuit technology, *J. Sel. Top. Quantum Electron.*, **19**(4), p. 6100117.

28. Keyvaninia, S., Verstuyft, S., Landschoot, L. V., Lelarge, F., Duan, G.-H., Messaoudene, S., Fedeli, J. M., Vries, T. D., Smalbrugge, B., Geluk, E. J., Bolk, J., Smit, M., Morthier, G., Thourhout, D. V. and Roelkens, G. (2013). Heterogeneously integrated III-V/silicon distributed feedback lasers, *Opt. Lett.*, **38**(24), pp. 5434–5437.
29. Chen, L. and Lipson, M. (2009). Ultra-low capacitance and high speed germanium photodetectors on silicon, *Opt. Express*, **17**(10), pp. 7901–7906.
30. Vivien, L., Osmond, J., Fédéli, J.-M., Marris-Morini, D., Crozat, P., Damlencourt, J.-F., Cassan, E., Lecunff, Y. and Laval, S. (2009). 42 GHz pin germanium photodetector integrated in a silicon-on-insulator waveguide, *Opt. Express*, **17**(8), pp. 6252–6257.
31. Luan, H.-C., Lim, D. R., Lee, K. K., Chen, K. M., Sandland, J. G., Wada, K. and Kimerling, L. C. (1999). High-quality Ge epilayers on Si with low threading-dislocation densities, *Appl. Phys. Lett.*, **75**(19), pp. 2909–2911.
32. Reed, G. T., Mashanovich, G., Gardes, F. Y. and Thomson, D. J. (2010). Silicon optical modulators, *Nat. Photonics*, **4**(8), pp. 518–526.
33. Kim, K., Choi, J.-Y., Kim, T., Cho, S.-H. and Chung, H.-J. (2011). A role for graphene in silicon-based semiconductor devices, *Nature*, **479**(7373), pp. 338–344.
34. Wang, J., Yao, Q., Huang, C.-W., Zou, X., Liao, L., Chen, S., Fan, Z., Zhang, K., Wu, W., Xiao, X., Jiang, C. and Wu, W.-W. (2016). High mobility MoS_2 transistor with low Schottky barrier contact by using atomic thick h-BN as a tunneling layer, *Adv. Mater.*, **28**(37), pp. 8302–8308.
35. Gan, X., Shiue, R.-J., Gao, Y., Meric, I., Heinz, T. F., Shepard, K., Hone, J., Assefa, S. and Englund, D. (2013c). Chip-integrated ultrafast graphene photodetector with high responsivity, *Nat. Photonics*, **7**(11), pp. 883–887.
36. Liu, M., Yin, X., Ulin-Avila, E., Geng, B., Zentgraf, T., Ju, L., Wang, F. and Zhang, X. (2011). A graphene-based broadband optical modulator, *Nature*, **474**(7349), pp. 64–67.
37. Li, H., Anugrah, Y., Koester, S. J. and Li, M. (2012). Optical absorption in graphene integrated on silicon waveguides, *Appl. Phys. Lett.*, **101**(11), p. 111110.
38. Midrio, M., Boscolo, S., Moresco, M., Romagnoli, M., Angelis, C. D., Locatelli, A. and Capobianco, A.-D. (2012). Graphene–assisted

critically-coupled optical ring modulator, *Opt. Express*, **20**(21), pp. 23144–23155.

39. Mikhailov, S. A. and Ziegler, K. (2007). New electromagnetic mode in graphene, *Phys. Rev. Lett.*, **99**(1), p. 016803.

40. Cai, H., Cheng, Y., Zhang, H., Huang, Q., Xia, J., Barille, R. and Wang, Y. (2016). Enhanced linear absorption coefficient of in-plane monolayer graphene on a silicon microring resonator, *Opt. Express*, **24**(21), pp. 24105–24116.

41. Ji, M., Cai, H., Deng, L., Huang, Y., Huang, Q., Xia, J., Li, Z., Yu, J. and Wang, Y. (2015). Enhanced parametric frequency conversion in a compact silicon-graphene microring resonator, *Opt. Express*, **23**(14), pp. 18679–18685.

42. Wei, G., Stanev, T. K., Czaplewski, D. A., Jung, I. W. and Stern, N. P. (2015). Silicon-nitride photonic circuits interfaced with monolayer MoS_2, *Appl. Phys. Lett.*, **107**(9), p. 091112.

43. Joannopoulos, J. D., Johnson, S. G., Winn, J. N. and Meade, R. D. (2011). *Photonic Crystals: Molding the Flow of Light* (Princeton University Press).

44. Gan, X., Mak, K. F., Gao, Y., You, Y., Hatami, F., Hone, J., Heinz, T. F. and Englund, D. (2012). Strong enhancement of light–matter interaction in graphene coupled to a photonic crystal nanocavity, *Nano Lett.*, **12**(11), pp. 5626–5631.

45. Gan, X., Gao, Y., Mak, K. F., Yao, X., Shiue, R.-J., van der Zande, A., Trusheim, M. E., Hatami, F., Heinz, T. F., Hone, J. and Englund, D. (2013a). Controlling the spontaneous emission rate of monolayer MoS_2 in a photonic crystal nanocavity, *Appl. Phys. Lett.*, **103**(18), p. 181119.

46. Gu, T., Petrone, N., McMillan, J. F., van der Zande, A., Yu, M., Lo, G. Q., Kwong, D. L., Hone, J. and Wong, C. W. (2012). Regenerative oscillation and four-wave mixing in graphene optoelectronics, *Nat. Photonics*, **6**(8), pp. 554–559.

47. Akahane, Y., Asano, T., Song, B.-S. and Noda, S. (2005). Fine-tuned high-q photonic-crystal nanocavity, *Opt. Express*, **13**(4), pp. 1202–1214.

48. Srinivasan, K., Barclay, P. E., Borselli, M. and Painter, O. (2004). Optical-fiber-based measurement of an ultrasmall volume high-Q photonic crystal microcavity, *Phys. Rev. B*, **70**(8), p. 081306.

49. Shiue, R.-J., Gan, X., Gao, Y., Li, L., Yao, X., Szep, A., Walker, D., Honc, J. and Englund, D. (2013). Enhanced photodetection in graphene-integrated photonic crystal cavity, *Appl. Phys. Lett.*, **103**(24), p. 241109.

50. Gan, X., Shiue, R.-J., Gao, Y., Mak, K. F., Yao, X., Li, L., Szep, A., Walker, D., Hone, J., Heinz, T. F. and Englund, D. (2013b). High-contrast electrooptic modulation of a photonic crystal nanocavity by electrical gating of graphene, *Nano Lett.*, **13**(2), pp. 691–696.

51. Gan, X., Shiue, R.-J., Gao, Y., Assefa, S., Hone, J. and Englund, D. (2014). Controlled light–matter interaction in graphene electrooptic devices using nanophotonic cavities and waveguides, *IEEE J. Sel. Top. Quantum Electron.*, **20**(1), pp. 95–105.

52. Mak, K. F., Lee, C., Hone, J., Shan, J. and Heinz, T. F. (2010). Atomically Thin MoS_2: a new direct-gap semiconductor, *Phys. Rev. Lett.*, **105**(13), p. 136805.

53. Han, S. W., Kwon, H., Kim, S. K., Ryu, S., Yun, W. S., Kim, D. H., Hwang, J. H., Kang, J.-S., Baik, J., Shin, H. J. and Hong, S. C. (2011). Band-gap transition induced by interlayer van der Waals interaction in MoS_2, *Phys. Rev. B*, **84**(4), p. 045409.

54. Kadantsev, E. S. and Hawrylak, P. (2012). Electronic structure of a single MoS_2 monolayer, *Solid State Commun.*, **152**(10), pp. 909–913.

55. Fujita, M., Takahashi, S., Tanaka, Y., Asano, T. and Noda, S. (2005). Simultaneous inhibition and redistribution of spontaneous light emission in photonic crystals, *Science*, **308**(5726), pp. 1296–1298.

56. Englund, D., Fattal, D., Waks, E., Solomon, G., Zhang, B., Nakaoka, T., Arakawa, Y., Yamamoto, Y. and Vučković, J. (2005). Controlling the spontaneous emission rate of single quantum dots in a two-dimensional photonic crystal, *Phys. Rev. Lett.*, **95**, p. 013904.

57. Wu, S., Buckley, S., Jones, A. M., Ross, J. S., Ghimire, N. J., Yan, J., Mandrus, D. G., Yao, W., Hatami, F., Vučković, J., Majumdar, A. and Xu, X. (2014). Control of two-dimensional excitonic light emission via photonic crystal, *2D Mater.*, **1**(1), p. 011001.

58. Wu, S., Buckley, S., Schaibley, J. R., Feng, L., Yan, J., Mandrus, D. G., Hatami, F., Yao, W., Vučković, J., Majumdar, A. and Xu, X. (2015). Monolayer semiconductor nanocavity lasers with ultralow thresholds, *Nature*, **520**(7545), pp. 69–72.

59. Li, Y., Zhang, J., Huang, D., Sun, H., Fan, F., Feng, J., Wang, Z. and Ning, C. Z. (2017). Room-temperature continuous-wave lasing from monolayer molybdenum ditelluride integrated with a silicon nanobeam cavity, *Nat. Nanotechnol.*, **12**(10), pp. 987–992.

60. Ota, Y., Moriya, R., Yabuki, N., Arai, M., Kakuda, M., Iwamoto, S., Machida, T. and Arakawa, Y. (2017). Optical coupling between atomically thin

black phosphorus and a two dimensional photonic crystal nanocavity, *Appl. Phys. Lett.*, **110**(22), p. 223105.

61. Cheng, R., Li, D., Zhou, H., Wang, C., Yin, A., Jiang, S., Liu, Y., Chen, Y., Huang, Y. and Duan, X. (2014). Electroluminescence and photocurrent generation from atomically sharp WSe_2/MoS_2 heterojunction p–n diodes, *Nano Lett.*, **14**(10), pp. 5590–5597.

62. Liu, C.-H., Clark, G., Fryett, T., Wu, S., Zheng, J., Hatami, F., Xu, X. and Majumdar, A. (2016a). Nanocavity integrated van der Waals heterostructure light-emitting tunneling diode, *Nano Lett.*, **17**(1), pp. 200–205.

63. Gan, X.-T., Zhao, C.-Y., Hu, S.-Q., Wang, T., Song, Y., Li, J., Zhao, Q.-H., Jie, W.-Q. and Zhao, J.-L. (2017). Microwatts continuous-wave pumped second harmonic generation in few- and mono-layer GaSe, *Light Sci. Appl.*, **7**(1), p. 17126.

Index